ALTERNATIVE TRANSPORTATION FUELS

Recent Titles from Quorum Books

The Coming Crisis in Accounting
Ahmed Belkaoui

Managing Human Service Organizations
Lynn E. Miller, editor

Behavioral Accounting: The Research and Practical Issues
Ahmed Belkaoui

Personnel Policies and Procedures for Health Care Facilities: A Manager's Manual and Guide
Eugene P. Buccini and Charles P. Mullaney

Designing the Cost-Effective Office: A Guide for Facilities Planners and Managers
Jack M. Fredrickson

The Employment Contract: Rights and Duties of Employers and Employees
Warren Freedman

Lobbying and Government Relations: A Guide for Executives
Charles S. Mack

U.S. Protectionism and the World Debt Crisis
Edward John Ray

The Social and Economic Consequences of Deregulation: The Transportation Industry in Transition
Paul Stephen Dempsey

The Accountant's Guide to Corporation, Partnership, and Agency Law
Sidney M. Wolf

Human Information Processing in Accounting
Ahmed Belkaoui

ALTERNATIVE TRANSPORTATION FUELS

An Environmental and Energy Solution

EDITED BY

Daniel Sperling

Quorum Books

New York • Westport, Connecticut • London

Library of Congress Cataloging-in-Publication Data

Alternative transportation fuels : an environmental and energy
   solution / edited by Daniel Sperling.
      p.   cm.
   Includes index.
   ISBN 0–89930–407–9 (lib. bdg. : alk. paper)
   1. Synthetic fuels.   I. Sperling, Daniel.
TP360.A47   1989
662'.66—dc19   89–3759

British Library Cataloguing in Publication Data is available.

Library of Congress Catalog Card Number: 89–3759
ISBN: 0–89930–407–9

First published in 1989 by Quorum Books

Greenwood Press, Inc.
88 Post Road West, Westport, Connecticut 06881

Printed in the United States of America

The paper used in this book complies with the
Permanent Paper Standard issued by the National
Information Standards Organization (Z39.48–1984).

10 9 8 7 6 5 4 3 2 1

Contents

ILLUSTRATIONS   vii

PREFACE   xi

ACKNOWLEDGMENTS   xv

1. Introduction   1
   Daniel Sperling

2. The Future of Oil: A Chevron View   11
   Thomas G. Burns

3. Natural Gas, Methanol, and CNG: Projected Supplies and Costs   21
   Michael F. Lawrence and Janis K. Kapler

4. Synthetic Fuel Costs in the Synfuels Era: A Chevron View   51
   Robert J. Motal

5. Distribution of Natural Gas and Methanol: Costs and
   Opportunities   65
   R. F. Webb, Carl B. Moyer, and M. D. Jackson

6. Hydrogen Vehicles   83
   Mark A. DeLuchi

7. Alternative Fuels as a Solution to the Air Quality Problem: An
   Environmentalist's Perspective   101
   Dick Russell

8. Motor Vehicle Emission Characteristics and Air Quality Impacts
   of Methanol and Compressed Natural Gas   109
   Jeffrey A. Alson, Jonathan M. Adler, and Thomas M. Baines

9. Compressed Natural Gas and Propane in the Canadian
   Transportation Energy Market   145
   Robert Sauvé

10. Transportation Fuels Policy Issues and Options: The Case of
    Ethanol Fuels in Brazil   163
    Sergio C. Trindade and Arnaldo Vieira de Carvalho, Jr.

11. Promoting Alternative Transportation Fuels: The Role of
    Government in New Zealand, Brazil, and Canada   187
    Jayant Sathaye, Barbara Atkinson, and Stephen Meyers

12. A Global Fuels Strategy: An Automotive Industry Perspective   205
    Albert J. Sobey

13. Regulated Utilities in the Vehicular Fuel Market: Toward a
    Level Playing Field   221
    Nelson E. Hay and Paul F. McArdle

14. Electric Vehicle Commercialization   235
    Gerald H. Mader and Oreste M. Bevilacqua

15. The Transition to Alternative Transportation Fuels in California   247
    Kenneth Koyama and Carl B. Moyer

16. Alternative Fuels Market Development: Elements of a Transition
    Strategy   263
    Barry McNutt

17. Is Methanol the Transportation Fuel of the Future?   273
    Daniel Sperling and Mark A. DeLuchi

18. The Role of Energy Efficiency in Making the Transition to
    Nonpetroleum Transportation Fuels   293
    Deborah L. Bleviss

19. Conclusion   309
    Daniel Sperling

    INDEX   317

    ABOUT THE CONTRIBUTORS   323

Illustrations

FIGURES

1.1	Forecasts of Petroleum Consumption, Production, and Imports	3
1.2	Conceptual Representation of Relationship Between Fuel Prices of Petroleum and Alternative Fuels	5
2.1	Chevron World Oil Consumption Forecasts	12
2.2	Crude Oil Production, 1960–2000	13
2.3	Crude Oil Price Outlook	14
2.4	Estimated World Resource Base	15
2.5	World Crude Oil Supply, High Demand	16
2.6	World Crude Oil Supply, Truncated Demand	16
2.7	Gasoline Cost Per Mile	18
3.1	Natural Gas Supply Curve for a Hypothetical Supply Region	22
3.2	World Natural Gas Resources, Proven and Additional Recoverable	25
3.3	Proven Natural Gas Reserves, Selected Regions: 1985	27
3.4	Natural Gas, Current and Estimated Production: 2000	28
4.1	Competitiveness of Alternative Transportation Fuels	60
5.1	Current CNG Prices in Canada	70
5.2	Shipping and Terminal Costs for Methanol to Los Angeles, 1988 (1995)	72
5.3	Estimated Shipping Costs for Methanol	76

8.1	Number of Persons Living in Counties with Air Quality Levels Above National Ambient Air Quality Standards in 1986	111
9.1	Whither Transportation Fuels in Canada?	148
9.2	Car Costs in Canada, 1987	150
9.3	Kilometers Traveled and Fuel Used by Light Vehicles in Canada	153
9.4	Gaseous Fuel Vehicle Conversion Costs	154
9.5	Gaseous Fuel Vehicle Conversion Costs and Paybacks	154
10.1	Sales of Neat Ethanol Vehicles as Percentage of Total Vehicle Sales	175
11.1	New Zealand Program CNG and LPG Conversions	190
11.2	Brazil: Fuel Use in Road Transport	193
11.3	Brazil: Vehicle Sales by Fuel Type	196
11.4	Canada: Monthly LPG Tank and Vehicle Sales	199
14.1	Introduction of Prototype Electric Vans	237
14.2	EV Market Survey Results	243
15.1	Retail Prices for Methanol and Gasoline, 1987 Dollars	255
15.2	Projected Prices, Regular Unleaded and Premium Unleaded, Dollars per Gallon	255
15.3	Ranges of Methanol Competitiveness Based on Landed Methanol Costs and Crude Oil Prices	256
18.1	Non–Eastern Bloc Light-Vehicle Fuel Use for Various Levels of Fuel Economy	307

TABLES

3.1	Projected Natural Gas Supply Regions	26
3.2	Average and Marginal Supply Price, Q_1	36
3.3	Average and Marginal Supply Price, Q_2	38
3.4	Methanol and LNG Production Costs	42
3.5	Methanol and LNG Shipping Costs	43
3.6	Methanol and CNG: Delivered Supplies and Prices	44
4.1	Remote Gas Development Costs	54
4.2	Parameters for Cost Comparisons	56
4.3	Basis for Cost of Methanol from New Remote Gas Plants	59
5.1	Long-Term CNG Pump Prices in Los Angeles	66

5.2   Shipping and Terminal Costs for Methanol to Los Angeles
       (1988 Cents per U.S. Gallon)   73

5.3   Ocean Transport and Terminal Costs for Methanol (1988
       Cents per U.S. Gallon)   73

5.4   Methanol Transport and Terminal Costs by Rail and Pipe
       (1988 Cents per U.S. Gallon)   74

5.5   Costs of Shipping and Terminaling Methanol in 1988,
       8-Million Gallon Shipment from Alberta to Houston, Texas,
       via Panama Canal   74

5.6   Methanol Shipping Cost Estimates for California   77

5.7   Estimated Methanol Fuel Prices for Los Angeles   80

6.1   Characteristics of Hydrogen Storage Systems   86

6.2   Emissions of Hydrogen Vehicles, Percent Change Relative to
       Gasoline Vehicles   90

6.3   Emissions of Greenhouse Gases from Alternative Vehicular
       Fuels   93

6.4   Comparison of Important Components of the Total Life-Cycle
       Cost, Gasoline and Hydrogen Vehicles   95

6.5   Gasoline Prices at Which Hydrogen and Gasoline Vehicle
       Life-Cycle Costs Are Equal   96

8.1   Average Zero-Mile Emissions from Gasoline Cars   113

8.2   Average 50,000-Mile Emissions from Gasoline Cars   113

8.3   Reaction Rates with Hydroxyl Radicals Relative to Butane   115

8.4   Summary of Exhaust Emissions from Current-Technology
       Methanol Vehicle Data Base   117

8.5   Projected Total Organic Emissions from Current-Technology
       Methanol Vehicles   118

8.6   Emissions from Low-Mileage Methanol Toyota Carinas   120

8.7   Exhaust Emissions from CNG/Gasoline Dual-Fuel Cars   124

8.8   Exhaust Emissions from Low-Mileage CNG and Gasoline
       Ford Rangers   126

8.9   Diesel Transit Bus Engine vs. Chassis Emissions   130

8.10 Low-Mileage Exhaust Emissions from Diesel Bus Engines   131

8.11 Low-Mileage Exhaust Emissions from Methanol Bus Engines   133

8.12 Low-Mileage Exhaust Emissions from Brooklyn Union Gas
       CNG Bus Engine   135

8.13 Projected Ambient Urban Ozone and Carbon Monoxide
       Reductions   141

9.1   Canadian Energy Demand Structure   147

9.2   Gaseous Fuel Conversion Grants and Subsidies, 1988   151

9.3   Retail Price Structure: Toronto, April 1988   152

9.4   CNG Experience   158

10.1 Transportation Fuels Consumption in Brazil, 1973–1987   164

10.2 Brazilian Petroleum Imports, 1973–1987   169

10.3 Brazilian Trade Balance and Foreign Debt, 1973–1987   170

10.4 Production and Consumption of Oil Derivatives, 1986–1991   173

10.5 Production and Demand Annual Growth Rates for Oil
      Products Through 1995   173

10.6 Ethanol and Gasoline Exports, 1980–87   174

10.7 Brazilian Vehicle Sales in the Domestic Market, 1973–87   176

10.8 Brazilian Vehicle Sales in the Domestic Market, 1973–87   176

10.9 Anhydrous Alcohol Production Costs   180

11.1 New Zealand: Government CNG Incentives   189

14.1 Griffon, G-Van, and TEVan   241

15.1 FFV Emission Data   252

15.2 Components of Retail Methanol Price   254

15.3 Equivalent Retail Prices for Various Landed Methanol Prices   257

17.1 Natural Gas Reserves in 1985 in TCF   276

17.2 Start-up Barriers for Methanol and CNG Relative to Gasoline   281

18.1 Estimated Dates of Production Readiness for Advanced Light-
      Vehicle Fuel-Efficient Technologies   298

18.2 High Fuel Economy Prototype Vehicles   300

Preface

Public policies often seem alarmingly irrational and misbegotten, especially in hindsight. In some cases, undue influence by self-interested parties distorts the policy-making process. More often, though, policy-makers do not adequately understand the problems they are addressing nor appreciate the implications of their actions.

Legislators and government administrators are not necessarily at fault; all too often the responsibility lies with the research community. Simply stated, the knowledge does not always exist to make informed decisions.

That has been the case with energy policy. Governments around the world have struggled in the 1970s and 1980s to respond to a rapidly changing energy market—changes that continue to restructure economies and societies and alter the physical environment. Plagued by a poor understanding of the energy market and the implications of selecting a particular option, legislators and other policy-makers made many mistakes. They continue to do so.

One particular energy activity, possibly the most problematic facing this planet, is of concern here: the use of petroleum for powering motor vehicles. The question is how much longer we can and should rely on petroleum as a transportation fuel. While considerable research and numerous conferences have addressed the subject of alternative fuels, most of those efforts were intensive and narrow, generally focusing on motor-vehicle technology, sources of energy, or energy conversion processes. It was the narrowness of these investigations and explorations that motivated this broader examination of the subject of alternative fuels.

This book is a product of a meeting held on July 17–19, 1988, in Monterey, California, at the Asilomar Conference Center. The symposium was motivated by a perception that if broad judgments were to be made about what fuels to pursue, and when and where, then a broad investigation of alternative fuels was critical. The goal, therefore, of this meeting, "Alternative Transportation Fuels in the 1990s and Beyond," was to move beyond narrow technical discussions and to begin creating a policy agenda for introducing alternative fuels.

An explicit effort was made to resolve conflicts over facts and to untangle ideologies, economic interests, and technical judgments. The meeting was unique and proved highly productive largely because of the mix of participants: influential executives and administrators from all interested parties—the motor vehicle and energy industries; federal, state, and local governments; environmental groups; and leading researchers in the fields of air quality analysis, motor vehicle technology, and energy policy. Prominent participants included Donna Fitzpatrick, undersecretary of the U.S. Department of Energy, James Lentz, executive director of the South Coast Air Quality Management District, Betsy Ancker-Johnson, vice-president of General Motors, and Jerard Myer, president of Chevron Research Company. Approximately 100 invited individuals participated in the meeting.

The meeting was organized by the Alternative Transportation Fuels Subcommittee of the Transportation Research Board, an arm of the National Research Council and National Academy of Sciences, and the University of California, Davis. Funding for the meeting was provided by the University of California Energy Research Group, U.S. Department of Energy, Canada Ministry of Energy, Mines, and Resources, and the South Coast (Los Angeles area) Air Quality Management District.

The symposium was only the third major conference in the world that focused specifically on alternative transportation fuels policies and strategies. The first one, sponsored by the Society of Automotive Engineers (SAE) in 1978 in Santa Clara, California, focused mostly on gasohol. The second conference on alternative transportation fuels was held in 1985 in Washington, D.C. It was organized by Gene Ecklund and others, again under the auspices of SAE. A sense of pessimism surrounded that meeting; only about 45 people attended, and discussions were diffuse with most attention being given to synthetic fuels made from coal, oil shale, and tar sands. This meeting, the third, represented a dramatic turnaround; there was a palpable enthusiasm and optimism that was missing at the meeting three years earlier.

We believe the symposium was a great success. It brought together people with widely varying knowledge and responsibilities and allowed them to interact in a collegial, pressure-free environment (lubricated by ethanol refreshments). It was educational for them to learn the broader implications of what they were investigating and making decisions about. Most importantly, they expanded their appreciation of the many self-interests at stake and began exploring what strategies and policies might be more effective and desirable in accomplishing goals of energy security, economic growth, and environmental quality.

The chapters in this book are a subset of papers commissioned for the California symposium, with the exception of the chapter by Dick Russell that presents a citizen's and environmentalist's perspective on alternative fuels and the chapter coauthored by Mark DeLuchi and myself. The remaining chapters underwent extensive review for accuracy and content, first at the symposium and then through peer review. Hopefully, this volume will contribute to a broader understanding of transportation fuel choices and to intelligent and sound energy and transportation policies.

Acknowledgments

The intellectual and organizational genesis of this book was within the Alternative Fuels Subcommittee of the Transportation Research Board (National Research Council). I thank the committee members, several of whom are authors of chapters in this book, for their help in identifying and soliciting authors and in organizing the symposium, "Alternative Transportation Fuels in the 1990s and Beyond" (Monterey, California, July 17–19, 1988).

Funding for the symposium and this book was provided by the University of California Energy Research Group, U.S. Department of Energy, Canada Ministry of Energy, Mines and Resources, and South Coast Air Quality Management District (AQMD). The people who provided leadership and enthusiasm from those organizations and deserve special recognition are Mike Lederer, Carl Blumstein, and Richard Gilbert from the Energy Research Group; Barry McNutt and Donna Fitzpatrick from DOE; Roy Sage from the Canada Ministry of Energy, Mines, and Resources; and Allan Lloyd and James Lents from the South Coast AQMD. We greatly appreciate the willingness of these organizations to fund a wide-ranging exploration of options and strategies even when they disagreed with the positions of some of the authors.

Special thanks go to all those who reviewed the chapters, including the anonymous reviewers and those who served as discussants at the symposium. Thanks also go to the staff of the Transportation Research Group at the University of California, Davis, especially Pam Pickering and Gina Gomez, and to David Hungerford and Kevin Nesbitt, graduate students at UC Davis, for helping with the organization and management of the symposium.

The authors would especially like to thank Linda Dayce and Carol Kozlowski of the UC Energy Research Group for taking primary administrative responsibility in organizing the symposium, and Delores Dumont of UC Davis for preparing, and in many cases improving, the final tables and text. We thank all three for doing a superb job with efficiency and enthusiasm. And lastly, we want to thank Patricia M. Davis for an outstanding index; it is comprehensive and reflects a deep understanding of our book.

ALTERNATIVE TRANSPORTATION FUELS

1   DANIEL SPERLING

Introduction

The world is not running out of petroleum. Large amounts of petroleum are available but most of the easily accessible and inexpensive supplies are located around the Persian Gulf. Those countries without supplies of inexpensive petroleum, including the United States, are becoming increasingly dependent on imported petroleum. Whether this import dependency is acceptable or not is a difficult question to answer because of uncertainties about the future availability and price of petroleum.

Reliance on petroleum has other costs and risks as well. Because of the erratic and unpredictable nature of petroleum prices, petroleum dependency creates uncertainty and therefore large indirect costs to the economy. And, of increasing concern in the United States, petroleum use in transportation is the leading contributor to urban air pollution and a major contributor to global warming.

The transportation sector, mostly motor vehicles, is by far the largest and fastest-growing consumer of petroleum. Unlike industrial, commercial, residential, and electricity-generating users, transportation users have remained almost totally dependent on petroleum fuel. In the United States, transportation relies on oil for 97 percent of its energy. In contrast, none of the other energy-consuming sectors relies on petroleum for more than 30 percent of its fuel needs. Only in Brazil, New Zealand, and South Africa have nonpetroleum fuels gained a significant share of the transportation fuels market. Reliance on petroleum has resulted in transportation increasing its share of the petroleum market—in the United States from about 53 percent in the mid–1970s to 63 percent in 1987. The principal burden of a transition away from petroleum fuels therefore falls on the transportation sector.

The subject of this book is the introduction of clean-burning alternative transportation fuels. The focus is on North America, although the book draws upon lessons and experiences in other parts of the world. The focus is roughly the late twentieth and early twenty-first centuries. The goal is to answer the following questions: Which alternative fuels are most attractive? How and when could the transition occur? What are the roles to be played by various public and private organizations and institutions? What are the implications of pursuing different energy options, including increased fuel efficiency in vehicles?

THE NON-PROBLEM

As Thomas G. Burns demonstrates (Chapter 2), the energy problem is not that petroleum supplies will soon be used up. Proven reserves of world oil have been increasing steadily, with new discoveries keeping pace with increasing consumption. If one were willing to rely on Persian Gulf countries for their oil supply and if the Persian Gulf countries could be relied upon to supply oil at their cost of production, there would be no need to worry about oil for many decades. Even if future oil discoveries begin to lag significantly behind consumption, there are many other energy resources that could be used to manufacture transportation fuels.

Indeed, because of the availability of these other resources, it will be a very long time before future prices of transportation energy exceed 1981 oil prices on a sustained basis. Michael F. Lawrence and Janis K. Kapler (Chapter 3) show that natural gas can be economically used as compressed or liquefied gas or converted into methanol when oil prices are considerably less than $43, the price (adjusted to 1988 dollars) that existed in 1981. At about that 1981 price, coal and biomass could be economically converted into methanol, substitute natural gas, and possibly petroleum-like liquids, and oil shale could be processed into gasoline and diesel fuel.[1] Since natural gas, coal, and oil shale are all available in larger quantities than petroleum, worldwide as well as in the United States, sufficient energy resources are available at or near 1981 prices for at least another century.

After that time a permanent transition could be made to renewable resources: hydrogen made from water using photovoltaic solar energy and liquid and gaseous fuels made from biomass. Although biomass fuels would cost about the same as coal-based fuels and be environmentally superior, their production should probably be limited so as not to exacerbate soil erosion. Mark A. DeLuchi (Chapter 6) shows that while the cost of producing hydrogen would be somewhat higher than for other fuel options, it does provides external benefits of much lower pollution. The point is that the world is not in imminent danger of running out of energy, and that with a well-functioning market system energy prices will not increase dramatically in the foreseeable future.

IMPERATIVES FOR ALTERNATIVE TRANSPORTATION FUELS

This does not mean there is no petroleum energy problem. Indeed a problem does exist; it exists because the international petroleum market does not allocate

Figure 1.1
Forecasts of Petroleum Consumption, Production, and Imports

HISTORY AND OUTLOOK FOR THE U.S.

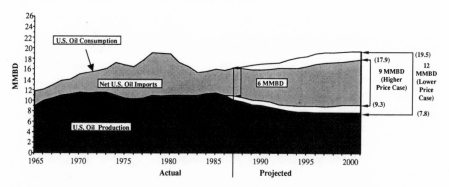

Source: U.S. Energy Information Administration, *Annual Energy Outlook 1986* (Washington, D.C.: 1987), 37.

Note: Post–1986 numbers are base-case forecasts.

oil resources efficiently nor in a socially optimal manner. The petroleum market does not account for large environmental impacts and is erratic and politicized, distorting energy decisions through inappropriate price signals and uncertainty. More specifically, the petroleum problem is due to dependence on insecure petroleum suppliers, high indirect economic costs, increased global warming, and urban air pollution. These are four reasons to initiate a transition to alternative transportation fuels before the market says it is the economical thing to do.

First, consider the oil import situation. The United States is becoming increasingly dependent on oil imports. The trend is unmistakable: domestic oil production is on a downward trajectory, and domestic oil consumption is increasing. These trends are graphically presented in Figure 1.1. In the scenario of high oil prices, net oil imports increase from 6.3 million barrels per day in 1988 to 8.5 million barrels per day in 2000. In the scenario of low oil prices, oil imports increase to 11.8 million barrels per day, representing 60 percent of total consumption.

It is unclear how important this import dependency problem is. The severity of the problem depends on one's view of the future: Will OPEC be able to regain market control and escalate oil prices? Will Saudi Arabia succumb to revolution? Will radicalized oil producers decide to use oil as a political weapon? The cost of oil dependency is difficult to measure; it depends not only on determinations of the probabilities of the foregoing types of events occurring but also on how the cost of military expenditures in the Middle East and other important supply regions are allocated, the cost of maintaining the Strategic Petroleum Reserve (now containing over 500 million barrels), the risk of supply disruptions, and

losses in national income from contraction of demand for U.S. goods and services. These costs have been estimated at 21–125 billion dollars per year.[2]

Import dependency will probably not be a principal motivation for initiating
a transition to alternative fuels in the near future unless an unexpected disruption
or price escalation occurs. Nevertheless, oil-import dependency will continue to
increase and therefore gain greater political attention, resulting in the dependency
issue becoming an even more important force in motivating an energy transition.

Dependency on oil imports is not just a problem of security, however. It also
creates the second problem of large indirect economic costs caused by increasing
world oil prices resulting in increased revenues for exporters and increased costs
to importers, and by increasing price volatility. The availability of a credible
alternative (and/or reduced petroleum consumption) would dampen oil price
volatility and restrain oil price increases. Price volatility is due in part to the
uncertain cost and availability of still undiscovered oil, but more so to the
concentration of easily accessible (and therefore low-cost) oil in a few sparsely
populated countries. The finite nature of the resource and, for a few fortunate
countries in the Middle East, huge supplies of cheap oil, tempt those countries
to manipulate oil prices and supplies.

Price volatility creates uncertainty and distorts investment decisions, resulting
in a preference for short-term investments. Erratic petroleum prices also result
in faulty forecasts and wasted investments such as delays in introducing energy-
efficient equipment in the 1960s and early 1970s, and billions of dollars of losses
on overenthusiastic investments in synthetic fuel plants in the late 1970s and
early 1980s.

The absence of a credible alternative to petroleum transportation fuels also
results in oil prices being higher than they would otherwise be. This effect,
illustrated in Figure 1.2, holds for the long term as well as in response to rapid
price escalations. Initial efforts at modeling the effect of alternative fuels on
world petroleum prices indicate that substituting an alternative fuel for 2 million
barrels per day of gasoline fuel would lower the world oil price by about $2 per
barrel from $34 per barrel.[3] Thus the price suppression benefit in reduced import
payments to the United States of those 2 million gasoline-equivalent barrels
would be about $20 million per day in 2000 (assuming 10.2 million barrels per
day of imports) or $7.5 billion per year.

The effect is even more dramatic for short-term price spikes. If, for instance,
petroleum prices were to increase quickly to 1981 levels, which is plausible once
excess world capacity is used up in the 1990s or later (one respected industry
chief executive expects prices to rise to $75 per barrel in the 1990s before
collapsing back to around $35 per barrel) then oil importers would be faced with
steeper spikes that dropped off slower than otherwise. If oil importers wait for
the higher prices, they will not be able to react with substituted fuels for many
years. High prices could be maintained for 20 years or more as the United States
and other oil importers struggle to expedite the transition to nonpetroleum fuels
and to replace vehicles that consume only gasoline and diesel fuel.

Figure 1.2
Conceptual Representation of Relationship Between Fuel Prices of Petroleum and Alternative Fuels

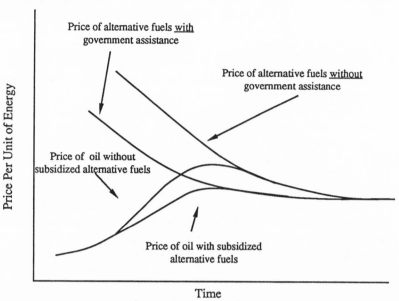

Indirect economic costs are a powerful motivation for introducing alternative fuels, but because the costs cannot be accurately quantified and because they are so diffuse, they are not likely to play a principal role in motivating the introduction of new fuels.

The third problem, global warming, is caused by emissions of carbon dioxide and other ''greenhouse'' gases; it attracts much more attention than energy security or indirect economic impacts, in part because the potential costs are much greater though more speculative. At this time a strong commitment does not exist to mitigate the greenhouse effect, in the United States or elsewhere, in large part because of uncertainty over the severity, location, and timing of the impacts. The scientific community is in general agreement that both the globe's temperature and the concentration of carbon dioxide and other greenhouse gases in the atmosphere are increasing. Still uncertain is how fast this is happening and how climatic patterns will change. It is expected that the warming will be disproportionately greater near the poles, causing melting of ice masses and increases in ocean levels of about 1 meter in the next 60 years or so. Gradual but ultimately dramatic changes will occur in local and regional climates. Rainfall will increase in some areas, decrease in others, and atmospheric temperatures will change, increasing in most but not all locations. Unfortunately, these climatic changes can not be predicted accurately with existing meteorological models. In any case the potential for devastating environmental and economic damage is huge.

The principal source of carbon dioxide and other greenhouse-gas emissions are carbon-bearing fossil fuels: oil, coal, and natural gas. Transportation accounts for about 25 percent of greenhouse gases emitted in the United States. If scientific evidence confirming the greenhouse effect becomes more certain, the possibility exists that a strong commitment will be made to reduce the use of carbon fuels. It is unlikely that carbon dioxide emissions could be reduced economically using control technologies on vehicles or refineries. The only feasible strategy for reducing carbon dioxide emissions from transportation is less consumption of petroleum, either through fuel efficiency or by the use of nonfossil fuels, including biomass, hydrogen made from water with nonfossil electricity, or electricity made from nonfossil fuels.

The fourth imperative for introducing alternative transportation fuels is, in the United States, politically the most potent: air pollution reduction. The use of petroleum for transportation results in large quantities of pollutant emissions from vehicles, refineries, and fuel stations. These external costs amount to 11–187 billion dollars per year, the large range due mostly to uncertainty about the number of deaths and illnesses due to pollution and the monetary value assigned to deaths and illnesses.[4] But what makes the air pollution imperative most salient is not the estimated costs but the existence of a set of institutions and rules for improving air quality. In the near term two specific requirements will be the driving force for introducing clean-burning transportation fuels: stringent new emission standards for diesel vehicles and enforcement of existing ambient air quality standards for ozone and carbon monoxide. These are described in Chapter 8.

The problem associated with continued reliance on petroleum fuels, therefore, is not necessarily long-run supply but rather ignored social costs (especially air pollution and global warming) and economic losses resulting from unpredictable oil prices, inflexible responses to oil price changes, and absence of substitute fuels. Because the price of petroleum does not take into account these social costs and economic losses and because of the disjointed and conservative nature of transportation energy systems, alternative fuels and increased vehicular efficiency are uneconomically delayed.

In summary, if market mechanisms were operating efficiently, then optimal consumption and production of oil would follow. But that is not the case. Efficiency improvements and alternative fuels are delayed by uncertain and low oil prices beyond the time when they would otherwise be economically attractive.

Moreover, as shown in later chapters, there are also large start-up barriers to alternative fuels, Because of the start-up barriers and flawed market, new fuels will only be introduced if they receive strong support from government. Significant government intervention will come about only if there is a consensus that action must be taken to protect the public interest. The problems likely to elicit such intervention are the public-good concerns listed above: the greenhouse effect, dependency on foreign oil supplies, economic benefits of lower energy prices, and urban air pollution.

THE SOURCE OF CONTROVERSY

The decision of how, when, and where to initiate a transition to alternative transportation fuels is wrapped in controversy and conflict. There is no consensus and no obvious answer of what to do. The difficulty is that future petroleum prices and availability are uncertain, and many of the costs and benefits associated with introducing alternative fuels cannot be easily quantified. Given this uncertainty and lack of knowledge, it is not surprising that there is considerable disagreement among individuals and organizations regarding the initiation of a transition.

To offer a better understanding of the debate over alternative fuels, I suggest that disagreements can occur on three levels: disagreements about facts, disagreements about values, and disagreements about beliefs. For instance, what are the facts about air quality impacts of methanol? We should be able to resolve disagreements over facts. Disagreements over values and beliefs are more intractable. But we need to appreciate that differences in values underlie and help explain opinions. Values such as self-sufficiency vs. international integration, equity vs. efficiency, growth vs. stability, conservation, free enterprise, individual initiative, and public health all play an important role in forming opinions about alternative fuel policy.

And what about the different beliefs we hold? Surely individual and organizational beliefs about the future of oil prices, revolutions in Iran, and other world events influence opinions of what to do about alternative fuels. Likewise beliefs held by many people that technological innovations will allow us to adapt to increased global warming and to lower the cost of manufacturing synthetic fuels, hydrogen, and batteries play an important role.

Values and beliefs are often held very strongly and are not always responsive to rational analysis. This book will not change people's values and beliefs. What it can do, in addition to resolving conflicts over facts, is untangle ideologies, economic interests, and technical judgments. That was one of the objectives in putting this book together. The means of accomplishing that objective was to solicit chapters from experts in key organizations. While an explicit point-by-point debate between these individuals was not attempted, the reader will easily observe very different perspectives in this book. We did not rewrite to impose total agreement for the simple reason that agreement did not always exist.

ORGANIZATION OF BOOK

Chapter 2 by Thomas G. Burns, manager of economics for Chevron Corporation, argues that huge petroleum resources will continue to be economically available worldwide. In Chapter 3, Michael F. Lawrence and Janis K. Kapler explore the supply of another important energy resource, natural gas, and analyze the cost of using it to supply methanol and compressed natural gas fuels to the

U.S. transportation market. This analysis, which is part of a continuing study for the U.S. Department of Energy (DOE), has played an important role in the decision by the U.S. Department of Energy and the California Energy Commission to promote methanol as the preferred alternative transportation fuel.[5]

Chapter 4 by Robert Motal explains why cost estimates of alternative fuels cover a large range. In particular, he discusses why he and Chevron believe that the method of analysis used by Lawrence and Kapler is inaccurate and underestimates the cost of remote natural gas.

In Chapter 5, Reginald Webb, Carl Moyer, and Michael Jackson, consultants to the Canadian and California governments, analyze the distribution costs of methanol and compressed natural gas (CNG), and the cost of delivering the fuel to the vehicle operator. As they demonstrate, distribution costs will be a major part of total fuel costs and will strongly influence where and how these fuels are used.

While the focus of this book is compressed natural gas and methanol because they are the most attractive alternative fuels in the next 40 years or so, other options do exist, as suggested earlier. Perhaps the most important of these is hydrogen, not because it will be economically competitive with compressed natural gas or methanol in the foreseeable future, because it probably will not be, but because, as Mark DeLuchi shows in Chapter 6, it has important advantages: it is potentially renewable and can be virtually pollution free. Thus it is important to keep hydrogen in mind as we explore more near-term and perhaps interim options.

Chapter 7 is an unusual contribution for a book like this; it is by a journalist, Dick Russell, who articulates the frustrations in reducing air pollution from the perspective of an informed and environmentally concerned citizen. Russell gives a sense of the pressures being imposed on the political system to do something about urban air pollution and why alternative fuels are receiving so much attention.

In Chapter 8, Jeffrey A. Alson and his colleagues at the U.S. Environmental Protection Agency provide a thorough analysis of the emissions and air-quality impacts of methanol and CNG. It is the most up-to-date analysis of what is known and what is not known. It presents a basis for assessing the technical merits of the arguments put forth by Russell in Chapter 7 for using alternative fuels as an air-quality control strategy.

The next three chapters analyze the experiences with alternative fuels in Canada, Brazil, and New Zealand. They focus on what can be learned from those experiences. The chapter on gaseous fuel use in Canada was prepared by Robert Sauvé of that country's Ministry of Energy, Mines and Resources, the chapter on ethanol in Brazil by Sergio Trindade and Arnaldo Vieira de Carvalho, Jr., and a chapter that synthesizes the experiences in Canada, Brazil, and New Zealand was prepared by Jayant Sathaye and his colleagues from Lawrence Berkeley Laboratory.

Chapter 12 is an exploration by Albert J. Sobey, former director of energy economics and advanced product development at General Motors, of strategies for introducing alternative fuels in less-developed countries. He argues that it would be in the interest of the U.S. automotive industry to work with other public and private-sector organizations to promote CNG and methanol vehicle sales in less affluent oil-importing countries.

Chapter 13 by Nelson E. Hay, chief economist and policy analysis chief at the American Gas Association, and a colleague, Paul F. McArdle, explains why compressed natural gas has not received much attention in the past and why that is changing.

In Chapter 14, Gerald Mader, president of a corporation formed by the electric utility industry to promote and market electric vehicles, and Oreste M. Bevilacqua, consultant, describe the strategies and goals this organization is pursuing to establish the support infrastructure and to develop the market for electric vehicles.

Chapters 15 and 16 are policy analyses of the transition to alternative fuels prepared by Kenneth Koyama of the California Energy Commission and Barry McNutt of the U.S. Department of Energy. While they do not necessarily represent the official positions of their respective organizations, they do provide a clear sense of the beliefs and values with which those organizations identify. It is instructive to note differences in methanol cost estimates used by these two authors and by other authors in this book (see chapters 2, 3, 4, and 6).

In Chapter 17, Mark DeLuchi and I address the question of whether methanol truly is a superior alternative transportation fuel, and in Chapter 18, Deborah L. Bleviss explores the relationships between energy conservation and alternative fuels. She argues that fuel efficiency improvements in vehicles should be an integral, even first step in the transition to alternative fuels.

Finally, in Chapter 19, I examine the diversity of views and issues presented in this book and explore their implications for the coming decades.

NOTES

1. Daniel Sperling, *New Transportation Fuels: A Strategic Approach to Technological Change* (Berkeley, Calif. and London: University of California Press, 1988).

2. Mark A. DeLuchi, Daniel Sperling, and Robert A. Johnston, *A Comparative Analysis of Alternative Transportation Fuels,* Institute of Transportation Studies, University of California, Berkeley, California, UCB-ITS-RR–87–13, 1987.

3. Carmen Difiglio, initial results from U.S. Department of Energy Alternative Fuels Trade Model, report in preparation.

4. DeLuchi et al., *A Comparative Analysis of Alternative Transportation Fuels.*

5. See U.S. Department of Energy, *Assessments of Costs and Benefits of Flexible and Alternative Fuel Use in the U.S. Transportation Sector,* Progress Report One, Washington, D.C., 1988; California Energy Commission, *Fuels Report,* Sacramento, Calif., 1987.

2

THOMAS G. BURNS

The Future of Oil: A Chevron View

One of the most difficult aspects of trying to think about the future is shaking off the old ideas that derive from our experiences of the past. We are all aware of the adage that the general staff is perfectly prepared to fight the last war—when it is really the next war that they should be concerned about! There is a danger that energy policies may be influenced too much by conditions that, while well-remembered, no longer exist. The 1970s were characterized by rising prices and supply disruptions. The 1980s have been a time of oversupply and falling prices. What will the 1990s bring? It is instructive to look at some of our old forecasts in order to see how our image of the future has changed in only 15 years (see Figure 2.1). And it should help us to understand whether our current thinking is more in tune with the outmoded ideas of the 1970s or with the actual experiences of the 1980s. For over 20 years prior to the first oil shock in 1973, energy and GNP had grown hand-in-hand at 4 to 5 percent per year, with oil growing even faster. During that entire period, oil consumption forecasts were consistently low—what forecaster could believe that the 7 to 8 percent annual growth rates of the 1950s and 1960s could be sustained year after year?

At the time, the generally accepted model was that energy and oil were essential components of economic growth. Oil supplies were cheap, seemingly limitless, and made the "good life" possible.

Just as forecasters became comfortable with this model, the first oil shock hit. Energy markets haven't been the same since. Sharply higher prices led to lower demand and a renewed search for new sources of oil to replace supplies that were vulnerable to political actions.

The times were characterized by great differences of opinion about the future of oil supplies. A typical forecast of this period made in 1978 reflected the transition in thinking that was going on. At the time it was generally believed

Figure 2.1
Chevron World Oil Consumption Forecasts (Excluding Centrally Planned Economies)

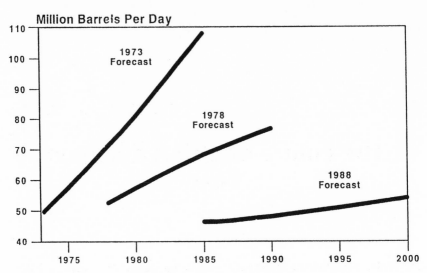

that GNP, energy, and oil growth would slow to a uniform rate of about 4 percent per year.

The Iranian revolution in 1979, followed by the Iran-Iraq War which began in 1980, confirmed everyone's worst fears about oil—by then everyone was convinced that oil was both expensive and risky.

These events and the related price increases gradually led to the conventional view of the 1980s that oil, energy, and GNP growth are no longer locked in their historical one-to-one relationship. Now most observers believe that, even if world GNP grows at 3 percent per year, energy consumption will grow 2 percent per year and oil consumption less than 1 percent per year. Contributing to this change was the fact that decontrol of oil and gas prices in the United States removed some of the distortions that led to the shortages and uneven price responses in the 1970s.

In addition, acceptance of the idea that oil was expensive and risky led to widespread changes in the way people used and thought about oil. Every factory appointed an energy conservation "czar." New automobile and appliance efficiency standards were introduced. Programs designed to develop alternative sources of energy supply were launched with great fanfare and enormous subsidies.

Anything that helped to reduce dependence on imported oil—and particularly on OPEC oil—was thought of as good almost regardless of its cost. It was widely believed that inevitable increases in the price of oil would eventually turn such programs into moneymakers and that the initial investment subsidies by government were appropriate ways to launch new and vital industries.

At the time, the oil industry didn't respond much differently. Many of our

Figure 2.2
Crude Oil Production, 1960–2000

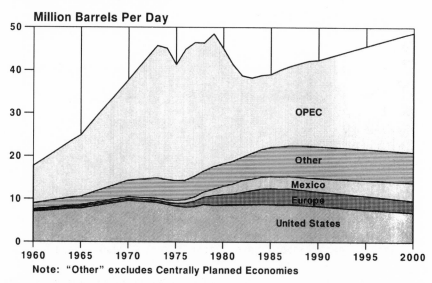

Note: "Other" excludes Centrally Planned Economies

projects were predicated on the basis of high and rising future oil prices. At the same time that oil consumption was falling sharply (see Figure 2.1), expanded oil exploration and development programs led to a significant increase in oil supplies from a widely diversified base of countries. As shown in Figure 2.2, non-OPEC crude oil production has risen by more than 55 percent, or over 8 million barrels per day since 1973. In comparison, the total contribution by all forms of alternative sources of synthetic liquid fuels amounts to only about 0.5 million barrels per day. Only conservation has made a contribution larger than that of conventional oil to the improved oil supply-and-demand balance.

Unfortunately, the model developed in the 1970s—which says that oil is both costly and risky—still influences much of our thinking on energy issues today. It is time to reexamine the premises of this model in order to be sure that this approach is still appropriate as the 1980s end and the 1990s begin. Just as it was painful to recognize and deal with the changes that occurred in the 1970s, it is also uncomfortable to reconsider some of the precepts that form the basis of our so-called conventional wisdom today. I think we are facing nothing less than a second major paradigm shift on energy matters. Oil has once again returned to its historical status as a commodity. Although the 1970s demonstrated that, in the short run, politics can play a role, in the long run, economics are more important.

The long-term history of crude prices shown in Figure 2.3 is interesting to contemplate. After 90 years of volatile but relatively level crude oil prices (averaging about $10 per barrel in 1988 dollars) the volatility suddenly disappeared about 1950. A major factor in smoothing oil prices from 1950 to 1973

Figure 2.3
Crude Oil Price Outlook

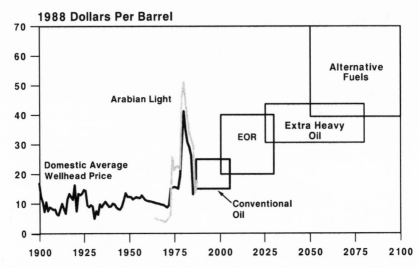

was the Texas Railroad Commission, which regulated oil production rates in Texas—although it was unable to prevent a steady deterioration in real prices.

It is understandable that shocks of the magnitude experienced during this period could lead the general public, the government, and the oil industry to conclude that the sky was falling. And I have to confess that the oil industry helped to spread this news in the 1970s. It was obvious—to us at least—that something was dreadfully wrong. Price and allocation controls in the United States were distorting the system so badly that ordinary textbook economic principles no longer seemed to apply. Price controls sent a message to consumers that oil was cheap and desirable; producers interpreted these same signals as a disincentive to produce and treated them as a subsidy for imports. While no government can suppress the laws of supply and demand forever, it often can do so long enough to get its economy into serious trouble.

In order to draw attention to this situation, the oil industry pointed out that prices would have to rise substantially if growing world demand was to be met. This message was accepted by a wide variety of interest groups, each with its own prescription to cure the problem. Unfortunately for the oil industry, most of these prescriptions ran counter to our original goals. They included conservation, rationing, and subsidies for energy alternatives such as wind, solar, ethanol, methanol, and oil from shale and coal. Who can forget the Synfuels Corporation or the Windfall Profit Tax? Fortunately, however, decontrol of the highly regulated oil industry occurred before the other initiatives had much influence. Decontrol unleashed conventional market forces, increasing supply and reducing demand. The result has been falling oil prices through the 1980s.

Even though circumstances have changed dramatically for the better in the

Figure 2.4
Estimated World Resource Base

Trillion Barrels of Recoverable Oil

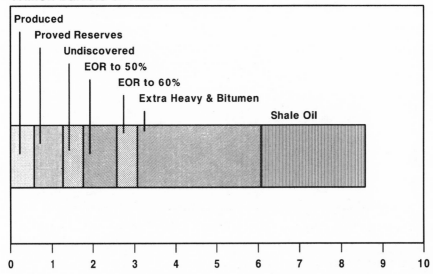

1980s, it would appear that the outdated "sky is falling" message of the 1970s still has far too much influence on our thinking about energy issues.

It is important to recognize that although oil is a depletable resource, the resource base is huge. In the 130-year period since Colonel Drake spudded his first well, the world has produced and consumed 600 billion barrels of oil. As shown in Figure 2.4, proven reserves today are 920 billion barrels, and the industry expects to discover and develop another 500 billion by conventional oil recovery processes.

This volume of conventionally produced oil, a total of almost 2 trillion barrels, will only bring us to the point of having produced about one-third of the original oil in place. Enhanced oil recovery (EOR) methods will eventually permit extraction of much higher percentages through the use of steam, chemicals, surfactants, polymers, miscible gases, and microbes.

If EOR can increase the yield of these reservoirs to 50 percent, it will add about 800 billion barrels to reserves; if 60 percent recovery can be attained, it will mean an additional 500 billion barrels.

Beyond this oil resource base lies an even larger resource of heavier hydrocarbons. Extra heavy oil and bitumens—of the types found in California as well as in Alberta and Venezuela—could add 3 trillion barrels of recoverable oil to the resource base. Shale oil, much of it in the United States, might add another 2.5 trillion barrels.

Figure 2.4 arranges the segments of the resource base in order of increasing

Figure 2.5
World Crude Oil Supply, High Demand

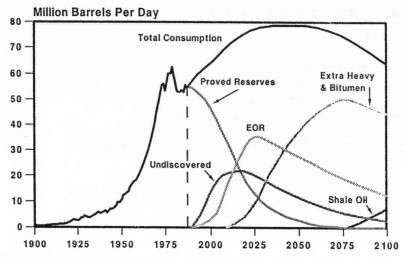

Figure 2.6
World Crude Oil Supply, Truncated Demand

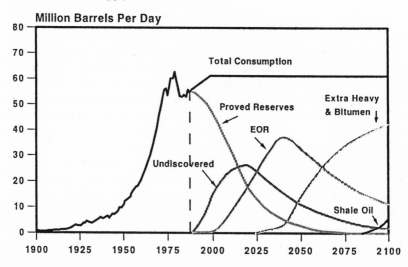

specific gravity which is roughly proportional to the increasing cost of extracting the resource from the ground and refining it into finished products.

In order to put these volumes into context, we have examined two alternative oil demand cases: rising oil demand until the middle of the twenty-first century, (see Figure 2.5), and rising oil demand until 2000 followed by flat demand thereafter (see Figure 2.6). Using these two scenarios, we were able to examine the impact of widely differing demand assumptions on the energy production profiles.

We discovered that although demand is important, it does not lead to a terribly great difference in these supply curves, particularly in the next 20 to 30 years. Of course, if demand is even lower—which could occur for a variety of reasons— it would only serve to prolong the life of the least costly parts of the oil resource base and to make oil even more price competitive compared to other energy sources.

What does this suggest for oil prices? We think that the 1970s were important. The dislocations were real. Although it is certainly possible that oil prices could return to the historical $10 per barrel level, that does not seem likely based upon what we know today about the nature of the resource base. Although we in the oil business lose a lot of sleep worrying about very low oil prices, we have concluded that, at $10 per barrel, demand was simply growing too fast for supply to keep up for many more years.

So, although we may see brief periods during which oil prices could fall back into this range, it is not likely that they will stabilize there, partly because of the real costs of finding and developing new sources of oil. On the other hand, prices much above $25 per barrel clearly attract too much supply and suppress demand to the point where they, too, become unstable. At the moment, the oil balance is still burdened by the legacy of the signals to increase supply and reduce demand that were sent out by the high prices prevailing from 1975 to 1985.

On the time scale used in Figure 2.3, even a range of $15 to $25 per barrel seems like a fairly sharp break with the past. It would appear that while the first price shock of 1973 may have been a necessary correction to bring the long-term picture into better balance, the second one in 1979 and 1980 was a substantial overshoot. In any event, we do not think that real prices will reach the high levels of the 1970s on a sustained basis for many years to come.

How does a mid-price trend through the conventional oil box indicated in Figure 2.3 translate into gasoline prices at the pump? The current gasoline price of about $1 per gallon in 1988 is about as low in real terms as it has ever been. And it does not seem likely to increase much in the 1990s. This is the competitive target against which all alternative transportation fuels must be measured. Alternatives will not be a commercial success unless they can be economic at prices like these.

But there is an even more dramatic way to show just how big a bargain gasoline is today. Remember that the average fleet miles-per-gallon has been increasing steadily at the same time that gasoline prices have been falling.

As shown in Figure 2.7, this translates into an extremely low cost for the gasoline required to move a passenger car one mile. The current cost of about $0.05 per mile driven is far lower in real terms than it has been in decades. We don't see this changing much in the future. Even at these relatively low gasoline prices, we expect to see vehicle fuel efficiency continue to improve by about 1 percent a year through the end of the century.

Now that we have examined the issues of price, what about security of supply? Weren't the dislocations caused by the oil-supply disruptions of the 1970s worse

Figure 2.7
Gasoline Cost Per Mile

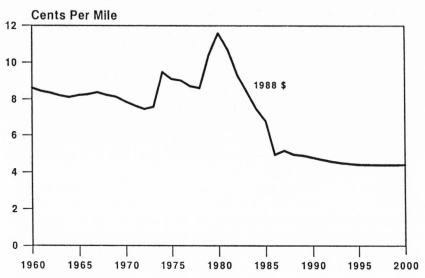

than the price effects? Can this happen again due to the concentration of oil resources in the hands of OPEC?

I would answer possibly—maybe even probably. But we have to keep this in perspective. And we certainly need to take out the appropriate amount of insurance to protect us against such an eventuality. At the same time, as a nation, we have to be extremely cautious that we do not bankrupt ourselves with so much insurance coverage that we cannot both pay the premiums and generate the necessary investment capital to provide for a growing standard of living for a growing population.

We now have the Strategic Petroleum Reserve, which is designed to provide insurance against short-term disruptions. And we have even more insurance in the form of the already discovered two-thirds of the original oil in place around the world which is just waiting for the price signal that will make it economic to produce. It is this oil that remains in place after primary and secondary recovery that is susceptible to enhanced oil recovery (EOR) techniques. This is the part of the resource base for which technology is developing rapidly. Many projects that were started in a higher price environment remain economic today.

Furthermore, the producers of extra heavy oil, tar sands, and bitumens are also making great strides to reduce the costs of producing, transporting, and using their resources. For example, the Venezuelan product, Orimulsion, is currently undergoing successful blending, shipping, and combustion tests.

Although about three-fourths of conventional oil reserves are located in the OPEC nations, these countries have learned a few lessons about markets over the past decade and a half. Their numerous recent moves into downstream refining and marketing operations in the consuming countries come as a result of their recognition that, if they expect to stay in the oil business in the future—and what else do most of them have?—they have to have markets for their oil.

This means that they have to price competitively to ensure that they do not encourage premature investments in alternatives that would then take away their markets. They also need to keep prices low in order to encourage consumption growth. This involvement in the entire market—with at-risk investments of their own scarce capital—will tend to make them better suppliers because their interests will be linked more closely to those of their customers.

When we put the entire picture together, we see a future of slowly growing oil prices—just over 1 percent per year in real terms over the next century. The resource economics appear overwhelming. There is simply too much oil (and similar hydrocarbon resources) available at relatively low costs to permit prices to rise any faster. If anything, technology will probably tend to reduce those anticipated costs in the future. Furthermore, the overlaps in the expected costs of the various segments are generally great enough to keep prices in a relatively narrow range during any individual period.

It is also important to note that future oil supplies, perhaps to an even greater degree than in the past, will be dominated by resources that have large initial capital costs and relatively low incremental operating costs. This means that a miscalculation on the part of a would-be cartel, as OPEC learned in the 1970s, can lead to a new increment of low-cost production that can come back to haunt the cartel for years to come.

This analysis suggests that oil is here to stay. The fundamental economics of oil supply and demand and the self-interest of the parties involved should coincide to make oil and the related parts of the resource base the lowest cost source of most transportation fuels for a long time to come.

Whatever transition to other sources of liquid fuels is ultimately required will probably take a long time and is likely at least in a historical sense to be relatively smooth. Although the transition zones may, indeed, be punctuated by brief periods of dislocation like the 1970s, these can probably be smoothed out by letting the natural forces of the market work freely.

My suggestions for the 1990s and beyond include:

- Let market forces be the dominant mechanism for selecting among fuel choice alternatives.
- Be cautious about subsidies of all types beyond the research and development level.
- Work to improve the international political climate to encourage economic interdependence.
- Build a reasonably sized Strategic Petroleum Reserve that can be used to smooth out short-term disruptions in oil supplies.
- Recognize that, because oil is where you find it, artificial restrictions on the search for and development of oil resources may carry a significant cost for the economy as a whole.

3   MICHAEL F. LAWRENCE
and JANIS K. KAPLER

Natural Gas, Methanol, and CNG: Projected Supplies and Costs

This chapter is concerned with two potential alternative fuels, methanol and compressed natural gas (CNG). Both of these fuels have proven vehicle technologies, offer environmental benefits, and have strong constituencies. If public authorities or entrepreneurs are to make decisions that will lead to the widespread use of either of these fuels, several questions must be addressed, among them the following fuel supply issues:

- How much natural gas feedstock is available, and where is it located?
- What is the minimum price that owners of these resources will require to make them available?
- How much will it cost to convert natural gas into methanol or liquefied natural gas?
- How much will it cost to ship these products to U.S. ports and make them available to the fuel distribution network?

While only an active future market for these products can provide definitive answers, there is sufficient information available to develop reasonable answers today. In this chapter we present data on proven reserves and additional recoverable sources of natural gas around the globe, and select potential supply regions where domestic opportunities for use of the resources are limited. For each potential supply region we utilize data on natural gas production costs to develop marginal cost (supply) curves. By adding methanol conversion or gas liquefaction costs to fuel-specific transportation costs for delivery from each supply region to U.S. ports, we are able to estimate the minimum long-run delivered price required by these resource owners for them to supply natural gas-based fuels to the U.S. transportation market.

Figure 3.1
Natural Gas Supply Curve for a Hypothetical Supply Region

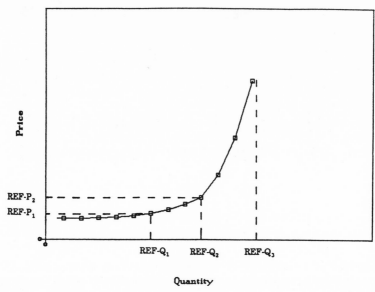

Quantity

Natural gas supply curves of the form shown in Figure 3.1 are developed for each region. Reference quantity one (REF-Q_1) for each region is the actual 1985 natural-gas production level. This output corresponds to the minimum price (REF-P_1) required to induce the 1985 production level. Reference quantity three (REF-Q_3) is the estimated limit to annual production in the year 2000 caused by physical constraint or political/economic factors. The other identified point on the supply curve (REF-P_2, REF-Q_2) represents a moderate expansion in production. Curves of the following form are developed for each supply region:

$$P = a + b/c - q \qquad (1)$$

Where:

  P = price
  q = quantity supplied
  a,b,c = parameters
  c = the asymptote

It will be shown that there is an abundance of low opportunity-cost natural gas in countries likely to make these supplies available to the U.S. transportation market. Furthermore, the costs of converting gas supplies to methanol or LNG and of shipping these products to U.S. ports appear to be reasonable enough to permit alternative fuels to compete with a gasoline (wholesale) supply price of $0.51 to $0.85 per gallon. Finally, it is estimated that enough moderately-priced alternative fuels can be made available to the United States to displace 2.2 to

2.6 million barrels of gasoline per day by the year 2000. These results assume that the infrastructure necessary to permit such extensive alternative fuel consumption is provided in a timely fashion. The estimated alternative-fuel costs provided in this chapter do not include infrastructure costs.

WORLDWIDE NATURAL GAS RESOURCES

The total natural gas resource base is composed of both "conventional" and "unconventional" sources of gas. Conventional gas is extracted from hydrocarbon reservoirs made up of sedimentary rock characterized by moderate to high degrees of porosity and permeability. Unconventional gas is found in environments for which the extraction technology is less mature and more costly than that employable at conventional sites. Unconventional sources of natural gas include tight or low-permeability sands, coal deposits, organic shale, high-pressured subsurface brines, and others. Because the recovery technology is relatively underdeveloped and costly, production from unconventional sources constitutes a minuscule portion of global production, and estimates of the gas contained in these sources are highly speculative. For these reasons, the natural gas supply to be considered as a feedstock available for alternative fuels is confined to conventional gas in this chapter. Nevertheless, it should be noted that the 1985 World Gas Conference has estimated that unconventional gas supply could reach 6 to 9 percent of world conventional supply in the year 2000, and 9 to 11 percent of conventional supply in the year 2020.[1] The feasibility of unconventional gas development will depend greatly upon the advent of improved recovery technologies.

Estimates of the magnitude of the conventional gas resource base in each geographic region vary in degree of reliability depending upon the extent of hydrocarbon development already undertaken and the productive history of the area. Resource estimates are categorized, in descending order of reliability, as "proven," "probable," "possible," and "speculative."[2] Proven reserves are estimated with the highest degree of certainty because capital has already been invested in drilling, geological and geophysical evaluation, and development of these reserves. For the same reason, however, proven reserves provide a very conservative estimate of future supply potential.

Proven reserves are considered to be a conservative measure for several reasons. Most importantly, the magnitude of proven reserves in a given hydrocarbon field typically expands for several years after discovery, as drilling and evaluation continue. "Probable resources" is the category of future resources to be provided both by the extension of discovered pools through further development drilling and by the discovery of new pools within existing fields. For example, it has been estimated that U.S. and Canadian gas discoveries made before the 1980s have appreciated on average by a factor of 3 or 4.[3] Additional gas expected to be recovered from new field discoveries in known productive provinces is categorized as "possible resources." Gas expected to be recovered from new pool or new field discoveries in formations or provinces not yet known to be productive is referred to as "speculative resources."

Proven reserves are an especially conservative estimate of exploitable gas in many developing countries where capital investment for exploration, or for drilling and evaluation following the discovery of a gas field, has been relatively light. The dearth of investment for "proving" reserves in developing countries is not explained by an expected paucity of resources. It has been estimated that 47 percent of the prospective hydrocarbon areas of the world is located in developing countries.[4] In the non-OPEC part of the developing world, more than 40 percent of the prospective hydrocarbon area has been subject to minimal exploration. In addition, there is evidence that success rates (productive wells as a percentage of total wells drilled) and finding rates (quantity of resource discovered per well drilled) in the developing world greatly exceed those of the developed world, and are increasing at faster rates.[5] Nevertheless, costly infrastructure must be built before many developing countries can utilize their discovered gas resources domestically. In addition, international oil companies have been reluctant to invest in gas exploration and development except in the few cases where large-scale export potential has been identified.[6] The World Bank has embarked upon a program of assisting petroleum exploration and development in developing countries, with a particular emphasis on gas production for the purpose of achieving greater energy self-sufficiency in oil-importing countries.[7] Such activity could lead to dramatic percentage increases in the estimates of proven reserves in the developing world, although it is likely that most newly developed gas resources would be absorbed by the needs of domestic growth.

Proven reserves include only the portion of the discovered resource that is recoverable under contemporary technological constraints. Although only about 32 to 33 percent of in-place oil is recovered worldwide on the average, for conventional gas the recovery rate is generally 80 to 90 percent.[8] Consequently, technological improvement will not significantly expand proven reserves of non-associated natural gas. However, approximately 25 percent of the world's proven gas reserves is in the form of associated gas, i.e., gas found in oil reservoirs.[9] It has been estimated that as enhanced oil recovery technology becomes available and economically feasible, the quantity of recoverable associated natural gas could increase by one-fourth.[10]

Proven reserves are by definition economically recoverable gas resources. Falling energy prices of the last few years had the effect of rendering portions of discovered gas subeconomic. If that price trend proves to be temporary, rising prices will tend to move more discovered gas resources into the economically recoverable pool and lead to an upward revision in proven reserves estimates, provided of course that price increases exceed cost increases.

Figure 3.2 demonstrates the quantities of worldwide proven gas reserves as of January 1987 as well as estimates of "additional recoverable resources." [11] This latter category includes probable, possible, and speculative resources. The

Figure 3.2
World Natural Gas Resources, Proven and Additional Recoverable

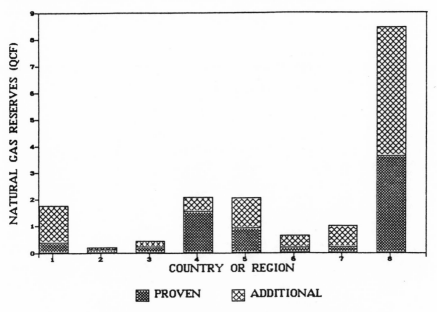

Legend	Proven	Additional
1 NORTH AMERICA	0.4	1.4
2 CENTRAL & SOUTH AMERICA	0.1	0.1
3 WESTERN EUROPE	0.2	0.2
4 EASTERN EUROPE & USSR	1.6	0.5
5 MIDDLE EAST	0.9	1.2
6 AFRICA	0.2	0.5
7 FAR EAST AND OCEANIA	0.2	0.8
8 WORLD TOTAL	3.6	4.8

SOURCES: EIA, *International Energy Annual 1986*
          IGT, *World Reserves Survey 1984*

exhibit shows that development to date indicates a highly certain worldwide supply of 3.6 quadrillion cubic feet (QCF) of gas, with at least an equal amount potentially available. This represents a 25-percent increase in the world's proven reserves over the last five years, despite a recent history of depressed hydrocarbon prices.[12] The fastest-growing sources of reserves in the last five years were Western Europe and the Far East. Rapid growth in the Far East was largely the result of an 80-percent rate of additions to reserves over five years in Indonesia and a 160-percent rate of additions in Malaysia. The European reserves growth was due primarily to Norway's 109-percent addition to reserves. The exhibit indicates that the lion's share of the world's proven reserves is to be found in the USSR (43 percent) and the Middle East (26 percent).

Naturally, much of this gas supply is unavailable as a feedstock for alternative

Table 3.1
Projected Natural Gas Supply Regions

OECD:

    Alaska
    Canadian Arctic
    Norwegian Arctic
    Australia/New Zealand

PERSIAN GULF OPEC:

    Iran
    Kuwait, Qatar, Saudi Arabia, and U.A.E.

OTHER OPEC:

    Algeria
    Libya
    Nigeria/Gabon
    Indonesia
    Venezuela/Ecuador

OTHER:

    Malaysia
    Thailand
    Mexico
    Trinidad and Tobago
    Tierra Del Fuego
    China

fuels in the United States due either to domestic demands for gas or to political considerations. Consequently, the feedstock resource base is defined here as the subset of global reserves found outside the USSR and Eastern Europe, in countries or remote sites with domestic supplies substantially in excess of expected future domestic demand. The world regions chosen as potential future suppliers of natural gas are listed in Table 3.1. OPEC members are included because of their demonstrated eagerness for export earnings. Naturally, developing political considerations must be taken into account in assessing the likelihood and degree of supply forthcoming from these countries. The quantities of gas reserves proven in these areas as of 1985 are indicated in Figure 3.3. Total reserves in these potential supply sites amounted to 1,375 trillion cubic feet (TCF), or more than one-third of 1987 global reserves.

A caveat concerning the estimates of gas reserves is in order. While proven reserves may constitute a conservative appraisal of exploitable gas resources, even this conservative figure should not be interpreted as an indication of immediately available supply. Proven reserves of natural gas do not constitute a "supply." Reserves is a stock concept; the production that flows from this stock constitutes supply. The distinction is important since technical and economic constraints permit only a small fraction of total reserves to be produced in a

Figure 3.3
Proven Natural Gas Reserves, Selected Regions: 1985

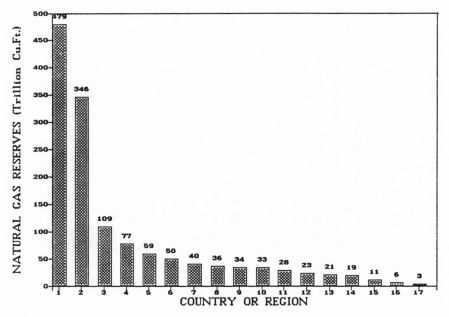

Legend
------
1 IRAN	10 CHINA
2 OTHER PERSIAN GULF	11 TIERRA DEL FUEGO
3 ALGERIA	12 AUSTRALIA/NEW ZEALAND
4 MEXICO	13 LIBYA
5 VENEZUELA/ECUADOR	14 NORWEGIAN ARCTIC
6 MALAYSIA	15 TRINIDAD & TOBAGO
7 INDONESIA	16 THAILAND
8 NIGERIA/GABON	17 CANADIAN ARCTIC
9 ALASKA	

SOURCES: U.S. Crude Oil, Natural Gas,
         and Natural Gas Liquids Reserves
         1985; COGLA; Roland; *International
         Energy Annual 1984*; IGT.

single year. Morris A. Adelman and Michael C. Lynch of MIT's Energy Laboratory cite an industry rule of thumb setting annual production at one-fifteenth (or 6.7 percent) of reserves.[13]

Figure 3.4 shows the 1985 level of gross natural gas production in the selected supply regions along with an estimate of potential maximum production levels in the year 2000. In many cases this maximum quantity represents a quadrupling

Figure 3.4
Natural Gas, Current and Estimated Production: 2000

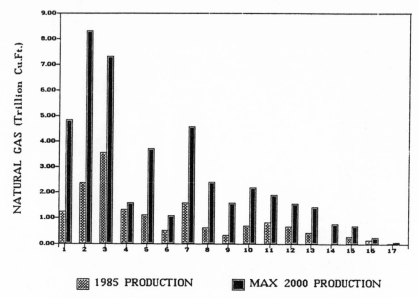

Legend
------
1 IRAN	10 CHINA
2 OTHER PERSIAN GULF	11 TIERRA DEL FUEGO
3 ALGERIA	12 AUSTRALIA/NEW ZEALAND
4 MEXICO	13 LIBYA
5 VENEZUELA/ECUADOR	14 NORWEGIAN ARCTIC
6 MALAYSIA	15 TRINIDAD & TOBAGO
7 INDONESIA	16 THAILAND
8 NIGERIA/GABON	17 CANADIAN ARCTIC
9 ALASKA	

SOURCES: U.S. Crude Oil, Natural Gas,
and Natural Gas Liquids Reserves
1985; COGLA; Roland; *International
Energy Annual 1984*; IGT.

of 1985 capacity based on a World Bank projection except where this would
generate a rate of production in excess of 6.7 percent of 1985 reserves.[14]

A comparison of Figures 3.3 and 3.4 is instructive. While Iran and the countries
aggregated as "Other Persian Gulf" contain 60 percent of the 1985 proven
reserves found in the potential supply regions, they are projected to contribute
only 32 percent of maximum production in the year 2000. This is because future
production is estimated as a function of current depletion rates, and these rates
are comparatively low in the Persian Gulf.

These estimates represent gross production, and as such overstate the quantity
of gas available to the U.S. transportation fuels market. To obtain a measure of
potential feedstock supplies, gross production should be reduced to account for

competing demands for gas, the constraints imposed on associated gas production by oil production requirements, and political considerations. Nevertheless, even if the gross estimates are halved, these figures indicate a large potential supply of feedstock gas to the United States. Half of the estimated maximum output in the year 2000 would yield 246 billion gallons of methanol per year, i.e., enough methanol to displace almost 9 million barrels of gasoline per day. The most optimistic scenarios of U.S. methanol demand in the year 2000 do not suggest that demand could exceed 61 billion gallons per year.[15] Alternatively, if the gas were converted to LNG for use as CNG in the United States, half of the maximum 2000 supply would provide enough LNG to replace over 10 million barrels of gasoline per day.[16]

The future feasibility of alternative vehicle fuels such as methanol or CNG depends upon the likely market prices of these products as well as upon potential supplies. In turn, future market prices will be a function of production costs, market demand, and the opportunities for feedstock or fuel producers and their national governments to extract economic rents. In the following section, the distinction between supply price (or cost-based price) and market price is discussed, and estimates of the future supply price of natural gas feedstock in the designated surplus regions are presented.

THE FUTURE SUPPLY PRICE OF NATURAL GAS

This section presents estimates of the marginal supply price of natural gas from the designated supply regions in the year 2000. Supply price is distinct from, and provides a floor for, market price. Failure to appreciate the distinction between the two will lead to a misunderstanding of the results reported here. Consequently, a brief discussion of price theory precedes the presentation of price estimates.

Supply Price, Market Price, and User Cost

Supply price is essentially a "cost price." It is the remuneration just sufficient to reimburse a producing agent for all of its per-unit costs of production, including an acceptable rate of return on its capital investments. In economic theory, it is generally assumed that any unit increment to output costs more than the last, an assumption that is treated in more detail below. This incremental cost is known as "marginal cost," and when marginal cost exceeds average total cost, the producing agent must receive a unit price at least equal to marginal cost in order to induce that agent to continue producing at an undiminished intensity. If the product market is perfectly competitive, no producing agent has any control over market price, which is set by the interaction of many suppliers and many purchasers. Consequently, the only decision variable facing the producer is the appropriate level of production. The profit-maximizing solution is to produce that quantity at which marginal cost is just equal to the prevailing market price.

If circumstances cause market price to rise, the producer can and should increase production until, once again, its higher marginal cost is just equal to the higher market price. Thus, a marginal cost curve or supply curve maps out a rising level of output that will be forthcoming, at higher cost, as a rising market price permits those higher costs to be recouped. In this perfectly competitive model, supply price is synonymous with market price.

In the case of imperfect competition, a few producers acting individually or collectively control such a large proportion of total supply that their output decisions affect market price. Given a state of demand that is inelastic, or relatively unresponsive to price changes, these producers can push market price above marginal cost by restricting supply. Market price then includes a component over and above the cost of production called "economic rent."

Typically, firms also experience noneconomic expenditures in the course of production. Two prime examples are taxes and resource royalties. These are economic "transfers," not costs, as they represent a profit-sharing arrangement with government or with owners of mineral rights. In other words, expenditures on taxes and royalties, unlike expenditures on labor and equipment, do not elicit supply and are not considered economic costs of production. They are merely a means by which the producers share the profits of resource production with others. Furthermore, transfers tend to be more substantial in cases where the industry is able to capture some economic rents. Adverse market conditions can work to decrease or eliminate rents and transfers. Adelman and Lynch cite a recent trend for gas-producing governments to reduce their shares of gas revenues in the face of weakened world demand and downward pressure on prices.[17]

It is likely that natural-gas market prices at the turn of the century will include some rents and transfers and will thus exceed supply prices. The existence and magnitude of these premiums will depend upon the state of world demand in the energy market and the degree of authority achievable by OPEC or any other producer cartel. The core oil producers of the OPEC oil cartel ("Other Persian Gulf") control only 15 percent of total gas production of the designated supply regions and only 22 percent of OPEC gas production. Therefore, it does not appear that OPEC is situated to exert as much dominance in the gas market of the future as it was able to exercise in the oil market of the 1970s. This chapter does not estimate the likely supplements to the supply price of natural gas. However, the U.S. Department of Energy is preparing a document that will address this component of the future market price of gas.

Supply prices are derived by apportioning the costs of land, labor, and capital to the quantity of resource thereby produced. However, there is an additional component to the supply price of natural gas because of the fact that gas is a nonrenewable resource. User cost is the opportunity cost of using a unit of a finite resource today, thereby relinquishing the value that might have been obtained by producing the same unit at a later date. The economic theory of supply is therefore amended for the case of exhaustible resources. The profit-maximizing rule is to produce that quantity at which market price (p) equals marginal production cost (mc) plus user cost (λ):

$$p = mc + \lambda \tag{2}$$

The classic Hotelling Principle, derived by Harold Hotelling in 1931, was one of the first theoretical treatments of the path of user cost.[18] The Hotelling Principle simply states that the undiscounted user cost must rise at the rate of interest (r).

$$d \lambda_t/dt = r \lambda_t \tag{3}$$

The logic of this rule follows from the fact that an unextracted resource such as natural gas is a capital asset and must yield a return comparable to that enjoyed by owners of other assets. In the simple Hotelling model, the only way that the unextracted resource can yield a return is by appreciating in value, since it does not earn a dividend.

The implication of the Hotelling Principle is that rising user cost must continuously pull up prices over time unless the other components of marginal cost continuously drop at a rate that is faster than the increase in user cost. Unfortunately, this basic rule has received broader application than is warranted.[19] The Hotelling Principle is derived from a basic model incorporating several simplifying assumptions. The model treats only proven reserves and assumes that the quantity of these is fixed forever and is commonly known. The decision variable is the rate of extraction, which is set such that price covers the marginal cost of extraction plus the appropriate user cost. Knowledge concerning future prices is assumed, and the necessary investment to produce the desired rate of extraction is based on this knowledge. The model ignores the economics of exploration, the effect of additions to reserves, and uncertainty.

The theoretical work following Hotelling has been devoted to removing many of these assumptions.[20] The result of this work is a theoretical indeterminacy regarding the value and path of user cost. Future events such as technological improvements and additions to reserves can decrease future extraction costs, thus yielding a type of dividend on unproduced resources held in the ground. These possibilities destroy the notion that there is any simple relationship between user cost and the rate of interest. Extending the model to encompass the effects of uncertainty introduces a wild card, since the value and path of user cost then become partial functions of expectations about future market prices. For example, expectations of a future price decline should significantly reduce user cost, and could produce, in effect, a negative user cost. In other words, it is possible that user cost could cause future marginal supply price to rise more slowly than actual extraction costs or to fall absolutely. Adelman and Houghton point out, based on a 1962 Resources for the Future study, that prices of nonrenewable resources have exhibited a long-term tendency to fall.[21] The EIA has observed that depletion decisions governed by the Hotelling Principle are rare in a competitive market, and risky in any event since future prices and demand are so uncertain.[22] Withholding a unit of resource from the market until $p = mc + \lambda$ requires future real-price increases large enough to offset the reduction in the present value of total revenues resulting from the delayed production schedule.

These theoretical and practical difficulties with the user cost concept pose problems for the estimation of that component of supply price. In the price estimation procedure discussed below, a simple model that ignores uncertainty and improved technology is adopted to approximate user cost, and the future path of user cost is left unresolved.

The Marginal Supply Price Model and Results

The marginal supply price model is adapted largely from models derived by Adelman and colleagues to assess future oil supply costs in the Persian Gulf and current natural gas supply costs in Western Europe and in the East Asia/Pacific region.[23] Since supply price is quantity specific, the model is employed to estimate year 2000 supply prices corresponding to Q_1 (1985 production) and Q_2 (moderate expansion). All prices and costs are expressed in 1987 U.S. dollars.

In brief, the structure of the model is as follows. Average wellhead supply price is composed of average capital (development), operating, and user cost, except that nonassociated gas is assigned no capital cost because its production is subordinate to that of oil, the primary economic product. Development and operating costs associated with Q_1 are estimated on the basis of region-specific drilling effort, success rates, and average well productivity in the 1980s, and upon 1985 actual U.S. costs per well. The supply price associated with Q_1 is estimated as the price that would elicit 1985 production levels if all of the investment necessary to produce that quantity were to take place in 1987. For supplies beyond 1985 levels, a price elasticity of .76 is assumed, as explained later. The assumed escalation in supply price is not disaggregated into development cost and user cost components. Marginal supply price is derived as a function of average price, estimated rate of depletion of reserves, and assumed rate of return.

Unit Capital Costs. Estimated yearly capital expenditure on hydrocarbon development in each supply region is based upon the average annual number of wells drilled in the years 1983–85, average well depth, and the location of the gas onshore or offshore or in arctic areas.[24] Capital expenditures include drilling and equipping costs, nondrilling costs (lease equipment, overhead, etc.), and gathering costs. Since country-specific expenditures outside the United States are largely unavailable, U.S. average expenditures per well for each well-depth category are used, with an adjustment for arctic regions.

Total annual hydrocarbon investment is allocated between gas and oil according to the percentage of total successful wells accounted for by gas wells in 1983–85. This method of allocation means that all development expenditures

on oil wells that contain associated gas are allocated to oil. No capital costs are
assigned to associated gas. If an international gas market develops as a result
of the transportation use of gas, some of this investment should be allocated to
gas production. Procedures that utilize the inverse of demand elasticities for joint
products (Ramsey pricing) could be applied.[25]

The new nonassociated gas production resulting from this capital investment
is calculated by multiplying each country's average annual number of new successful gas wells drilled in 1983–85 by its average annual gas well productivity.
Projecting additions to annual capacity in this manner employs certain simplifying
assumptions, following Adelman and Lynch.[26] Specifically, it is assumed that
all of the capital investment necessary to elicit new supply takes place in one
year and that the new supply is produced at maximum potential in the following
year. Production then falls off at a constant rate of decline in succeeding years.
In reality, capital expenditures are usually spread over several years, and the
rate of production of new supply tends to rise in the initial years, level off, and
then decline. Nevertheless, Adelman and Paddock have shown that this simplified
approach yielded an excellent prediction of the actual unit capital costs of gas
experienced in the North Sea.[27]

Average capital cost per unit of new nonassociated gas is then expressed as
follows:[28]

$$(K/Q) (a + r) \tag{4}$$

K is capital investment, Q is the resulting initial annual production, r is the rate
of return, and a is the annual rate of decline, which is approximated by the
depletion rate, i.e., the ratio of Q to the volume of proven reserves.[29] The equation
is derived as follows. With an annual decline rate of a, output in any year Q_t
is Qe^{-at}. The minimum acceptable average supply price P, abstracting from
nondevelopment costs, is that which equates investment outlays with the net
present value (NPV) of the multiyear stream of revenues resulting from the
investment:

$$NPV = PQ \int_0^\infty e^{-(a+r)t} dt - K = 0 \tag{5}$$

$$PQ/(a+r) = K \tag{6}$$

$$P = (K/Q)(a+r). \tag{7}$$

Assumed real rates of return combine a risk-free return with a risk premium.
According to Adelman and Lynch, the necessary rate of return for hydrocarbon
investment in the United States and Canada is about 10 percent.[30] We have
assigned this value to Alaska and the Canadian Arctic. Adelman argues that
nonindustrial oil-exporting nations should discount future oil revenues at real
rates of at least 20 percent, and more likely in excess of 25 percent.[31] He calculates

this unusually large risk premium assuming insignificant diversification, considerable leveraging, and political instability. Upon this premise, we have assigned a 24-percent rate of return to all OPEC members and to Mexico. The remaining countries have been assigned 12 percent, the rate that Adelman and Lynch suggest for Western Europe and for the East Asia/Pacific region due to the higher degree of geological risk in these areas compared to that experienced in the United States.[32]

Since capital cost is assigned to nonassociated gas only, capital cost per MCF of total supply is derived by treating the typical unit of produced gas as part nonassociated and part associated, according to each region's ratio of nonassociated gas reserves to total reserves. Thus, investment cost per "average" MCF is each region's unit investment cost for nonassociated gas, equation (6) above, multiplied by its nonassociated gas factor. Estimated average capital costs associated with Q_1 and Q_2 are shown on Tables 3.2 and 3.3.

Unit Average and Marginal Supply Prices: Total Supply (Nonassociated and Associated Gas). Average supply price is the sum of average capital, user, and operating costs per unit. The difficulties involved in estimating user cost were discussed above. If uncertainty and the possibility of technological improvements are ignored, Adelman and Shahi have demonstrated that the upper limit to user cost can be approximated by average capital cost.[33] That is the approach adopted in this chapter. Therefore, the user costs displayed in Tables 3.2 and 3.3 are identical to average investment costs. This procedure results in very low supply prices for regions with an abundance of associated gas, since no capital costs are assigned to associated gas. Should the allocation of some investment to associated gas become warranted, then estimated user cost and supply price would increase for Alaska and the Central and South American supply regions, as well as for Mexico, Libya, and parts of the Persian Gulf.

Operating costs per MCF of gas are assigned to each region on the basis of average 1985 U.S. operating costs per well by well-depth category.[34]

The estimated average supply prices associated with Q_1 and Q_2 are demonstrated on Tables 3.2 and 3.3. The average price of Q_1 is capital plus user plus operating cost. The average price of Q_2 must be adjusted to reflect the assumption that expanded levels of hydrocarbon production entail higher costs per unit. This assumption is problematic.

As a general proposition, it is believed that the long-run costs of expanding the production of extractive resources must rise for the following reasons:[35]

1. Rapid as opposed to moderate depletion of a given reserve is more costly because it requires more capital investment for the extraction of a fixed total amount.
2. As a given reserve is depleted, extraction of the remainder becomes more costly as gas pressure dwindles.
3. It is believed that the better deposits are found and produced before the inferior ones.[35]

However, these considerations do not close the matter. New discoveries offset the rising costs experienced in existing reserves due to reasons 1 and 2. Also, technological improvements over the course of time tend to reduce production costs. Technology also tends to lower the cost of discovering the less obvious deposits and sometimes reverses the sequence assumed in reason 3. Therefore, the long-run path of extraction cost, as well as that of user cost discussed above, is theoretically indeterminate.

The marginal supply price model incorporates a middle-of-the-road approach to the path of natural gas costs over the next two decades. The price associated with Q_1 is estimated as that which would elicit 1985 production levels if all of the investment necessary to produce that quantity were to take place in 1987. For supplies beyond 1985 levels, a price elasticity of .76 is assumed. That is, a 1.0 percent price increase is necessary to elicit a 0.76 percent increase in supply. This figure is drawn from the EIA, which has estimated that the price elasticity of supply of gas wells in the United States between 1950 and 1984 was 0.76.[36] The assumed escalation in supply price adopted here is not disaggregated into development-cost and user-cost components.

If the costs of hydrocarbon supply are rising more rapidly in the OECD countries than the rate of U.S. inflation, then average price as estimated for them is probably too low. On the other hand, if costs are rising slowly outside the OECD, then the estimated price of Q_1 may be a reasonable approximation for the rest of the world.[37] However, a price elasticity of supply of .76 probably overstates the cost of increased supply (Q_2) for these countries. Column D of Table 3.3 reflects the average price of supplying Q_2 as adjusted by the assumed price elasticity of supply.

Average prices are converted to marginal prices in the final columns of Tables 3.2 and 3.3. These results indicate that there is an abundance of gas that can be produced at wellhead costs that compete very favorably with the price of oil. Approximately 87 percent of Q_2, or 26 TCF per year, can be supplied for less than $1.00 per MCF. The exhibit also indicates a very great disparity in supply price among regions.

The highest-cost regions identified are Argentina and Chile (Tierra del Fuego) and the arctic regions, followed by Australia, Thailand, and Mexico. Argentina and Chile are identified as high cost because of excessive unit operating costs, more than four times those shown for Alaska, due to an extremely low average well productivity. Operating costs are also unusually high in the arctic regions, due to harsh conditions. Large per unit investment demands in the Canadian Arctic and Norwegian Arctic are accentuated by the estimated significant concentrations of nonassociated gas in those areas. P_2 for Alaska, a function of high operating costs and relatively high investment expenditures, is also pushed up by the estimate of a sizeable capacity expansion from Q_1 to Q_2.[38] High supply prices for Australia, Thailand, and Mexico are a function of deep wells and estimated low well productivity.

Turning to the low-cost countries, most of the OPEC countries and Malaysia exhibit minuscule costs because of very high estimates of well productivity. High productivity significantly reduces per-unit investment and operating costs. Ven-

Table 3.2
Average and Marginal Supply Price, Q_1

	A Per Average MCF (Q1) (1987 $/MCF)	B User Cost Per MCF (1987 $/MCF)	C Operating Cost Per MCF (1987 $/MCF)	D Price Per MCF (Q1) (1987 $/MCF)	E Price Per MCF (Q1) (1987 $/MCF)
OECD:					
1. ALASKA	0.11	0.11	0.085	0.31	0.34
2. CANADIAN ARCTIC	0.18	0.18	0.065	0.42	0.45
3. NORWEGIAN ARCTIC	0.18	0.18	0.006	0.37	0.37
4. AUSTRALIA/NEW ZEALAND	-	-	-	0.29	0.36
Australia	0.14	0.14	0.016	0.29	0.36
New Zealand	0.07	0.07	0.002	0.14	0.17
PERSIAN GULF OPEC:					
5. IRAN	0.11	0.11	0.004	0.22	0.22
6. OTHER PERSIAN GULF	-	-	-	0.05	0.06
Kuwait	0.00	0.00	0.003	0.00	0.00
Qatar	0.02	0.02	0.001	0.04	0.04
Saudi Arabia	0.01	0.01	0.003	0.01	0.01
United Arab Emirates	0.02	0.02	0.003	0.05	0.06

36

OTHER OPEC:

7. ALGERIA	0.06	0.06	0.001	0.12	0.13
8. LIBYA	0.01	0.01	0.001	0.02	0.02
9. NIGERIA/GABON	-	-	-	0.03	0.03
Gabon	0.01	0.01	0.001	0.01	0.01
Nigeria	0.02	0.02	0.001	0.03	0.03
10. INDONESIA	0.03	0.03	0.004	0.07	0.08
11. VENEZUELA/ECUADOR	-	-	-	0.04	0.04
Venezuela	0.00	0.00	0.011	0.02	0.02
Ecuador	0.01	0.01	0.025	0.04	0.04

OTHER:

12. MALAYSIA	0.01	0.01	0.001	0.01	0.01
13. THAILAND	0.14	0.14	0.019	0.30	0.36
14. MEXICO	0.15	0.15	0.025	0.33	0.35
15. TRINIDAD & TOBAGO	0.00	0.00	0.012	0.01	0.02
16. TIERRA DEL FUEGO	-	-	-	0.52	0.70
Argentina	0.05	0.05	0.347	0.45	0.55
Chile	0.08	0.08	0.366	0.52	0.70
17. CHINA	0.11	0.11	0.015	0.23	0.27

Table 3.3
Average and Marginal Supply Price, Q_2

	A Capital Cost Per Average MCF ($Q2$) (1987 $/MCF)	B User Cost Per MCF (1987 $/MCF)	C Operating Cost Per MCF (1987 $/MCF)	D Adjusted Average Supply Price Per MCF ($Q2$) (1987 $/MCF)	E Marginal Supply Price Per MCF ($Q2$) (1987 $/MCF)
OECD:					
1. ALASKA	0.13	0.13	0.085	1.37	1.76
2. CANADIAN ARCTIC	0.19	0.19	0.065	1.46	1.65
3. NORWEGIAN ARCTIC	0.21	0.21	0.006	1.41	1.65
4. AUSTRALIA/NEW ZEALAND	-	-	-	0.73	1.01
Australia	0.16	0.16	0.016	0.73	1.01
New Zealand	0.08	0.08	0.002	0.32	0.45
PERSIAN GULF OPEC:					
5. IRAN	0.11	0.11	0.004	0.72	0.73
6. OTHER PERSIAN GULF	-	-	-	0.13	0.16
Kuwait	0.00	0.00	0.003	0.01	0.01
Qatar	0.02	0.02	0.001	0.12	0.12
Saudi Arabia	0.01	0.01	0.003	0.04	0.05
United Arab Emirates	0.03	0.03	0.003	0.13	0.16

OTHER OPEC:

7. ALGERIA	0.06	0.06	0.001	0.25	0.30
8. LIBYA	0.01	0.01	0.001	0.06	0.07
9. NIGERIA/GABON	-	-	-	0.11	0.13
Gabon	0.01	0.01	0.001	0.05	0.05
Nigeria	0.02	0.02	0.001	0.11	0.13
10. INDONESIA	0.04	0.04	0.004	0.19	0.25
11. VENEZUELA/ECUADOR	-	-	-	0.13	0.13
Venezuela	0.00	0.00	0.011	0.05	0.06
Ecuador	0.01	0.01	0.025	0.13	0.13

OTHER:

12. MALAYSIA	0.01	0.01	0.001	0.03	0.03
13. THAILAND	0.15	0.15	0.019	0.61	0.77
14. MEXICO	0.15	0.15	0.025	0.48	0.51
15. TRINIDAD & TOBAGO	0.00	0.00	0.012	0.03	0.05
16. TIERRA DEL FUEGO	-	-	-	1.03	1.44
Argentina	0.06	0.06	0.347	1.03	1.44
Chile	0.08	0.08	0.366	0.88	1.28
17. CHINA	0.12	0.12	0.015	0.70	0.96

ezuela and Ecuador, also very low-cost countries, are estimated to have conspicuously lower productivity than the rest of OPEC with correspondingly higher unit investment and operating costs. However, this is offset by the fact that negligible investment and user costs are assigned to these countries because of an estimated tiny percentage of nonassociated gas. The same pattern applies to Trinidad and Tobago.

Even if it may be assumed that the estimated prices are sufficiently accurate, the conclusion that cheap gas is available can be misleading. While gas that is processed in its country of origin may be exchanged at a rate close to supply price, the advent of any significant international market in gas would tend to establish a world market price that balances supply and demand. Assuming away rents and transfers for a moment, the world price would lie somewhere on the aggregate world supply curve, thus yielding a producers' surplus, or a return above supply price, to the participating suppliers who enjoy lower costs. The stronger the demand for traded, unprocessed natural gas, the higher the market-clearing world price would be.

Market conditions prevailing at the turn of the century will determine the magnitude of rents (in addition to supply price) that may be extracted by governments or oligopolistic producers. These noncost components of market price will be estimated in a forthcoming U.S. Department of Energy report (1989) using a Stackelberg model of likely variations of oligopolistic behavior over the next two decades. Furthermore, most likely end-uses of this gas will require considerable additional capital investment in order to transport or transform the gas. This is where the analysis of competing alternative uses such as pipeline delivery or LNG or methanol transformation must pick up. The following section treats the likely costs of converting natural gas to LNG or methanol and of transporting these products to the United States.

CONVERSION AND TRANSPORT COSTS: METHANOL AND LNG

Chem Systems, Inc. has estimated the costs of converting natural gas to methanol or LNG in the designated natural-gas supply regions and of transporting the end products to the United States.[39] Their results are summarized here.

Although the current world-scale capacity of methanol plants is in the range of 2,000 to 2,500 metric tons per day (MTPD), much larger and more technologically advanced plants are likely to be built in gas-rich world regions if methanol becomes a preferred automotive fuel or fuel component. Chem Systems has divided the designated natural-gas supply regions into four categories on the basis of the magnitude of investment required to establish a large, advanced-technology methanol plant in each. Category I requires the lowest investment as these sites are already developed and are situated within an established industry environment, such as the U.S. Gulf Coast. The remaining categories require increasing amounts of investment due either to high labor rates or low productivities, unusual terrain, and/or lack of infrastructure. Category IV requires the most investment due either to the offshore location of natural gas and high gathering costs, or to harsh climatic conditions.

The unit cost of producing fuel-grade methanol in a new advanced-technology, 10,000 MTPD plant is expressed as a function of site category and the cost of feedstock natural gas. The energy loss in the conversion process is 7 percent.

Site Category	Methanol Production Cost, $/gal
I	$0.09\ G^* + 0.17$
II	$0.09\ G + 0.19$
III	$0.09\ G + 0.24$
IV	$0.09\ G + 0.33$

*G is the cost of natural gas in $/MMBtu.

Table 3.4 displays Chem Systems's site category designations and the estimated cost of producing methanol, exclusive of feedstock, in each supply region.

The final column of Table 3.4 shows estimated LNG production costs, exclusive of feedstock, for each supply region. These costs assume new construction of a 1-billion standard cubic feet per day (SCFD) plant incorporating some improvements over current state-of-the-art technology. The energy loss in the conversion process is 9 percent. Regasification costs are not included here; they are treated separately in the next section of this chapter. Unit production costs are based on the following Chem Systems equations.

Site Category	LNG Production Cost, $/MMBtu
I, II, or III	$1.10\ G + 1.70$
IV	$1.10\ G + 2.30$

The cost of shipping methanol or LNG from supply regions to the United States, as estimated by Chem Systems, is a function primarily of tanker size, shipping distance, and tanker fuel consumption. It is assumed that all vessels must be newly constructed, since any significant U.S. automotive use of methanol or LNG will require shipments that exceed current shipping capacity. Furthermore, because the vessels must be designed and built to reflect the specific nature of the methanol and LNG cargoes, estimated shipping costs must reflect the assumption of empty tanks on the return trip.

Table 3.5 displays the estimated costs of shipping methanol and LNG from various supply regions to the United States (Los Angeles). Although today methanol is generally transported in tankers that do not exceed 40,000 deadweight tons (DWT), the costs in Table 3.5 are based on a 250,000 DWT tanker to accommodate the dramatic increase in methanol shipped that must accompany

Table 3.4
Methanol and LNG Production Costs

	A	B	C
		Methanol Production Cost (excluding feedstock) (1987 $/gal)	LNG Production Cost (excluding feedstock) (1987 $/MCF)
	SITE CATEGORY*		
OECD			
1. ALASKA	IV	0.33	2.30
2. CANADIAN ARCTIC	IV	0.33	2.30
3. NORWEGIAN ARCTIC	IV	0.33	2.30
4. AUSTRALIA/NEW ZEALAND		-	-
Australia	III/IV	0.33	2.30
New Zealand	II/III	0.24	1.70
PERSIAN GULF OPEC			
5. IRAN	II/III	0.24	1.70
6. OTHER PERSIAN GULF		-	-
Kuwait		0.24	1.70
Qatar		0.24	1.70
Saudi Arabia	II/III	0.24	1.70
United Arab Emirates		0.24	1.70
OTHER OPEC			
7. ALGERIA	II/III	0.24	1.70
8. LIBYA	III	0.24	1.70
9. NIGERIA/GABON		-	-
Gabon		0.24	1.70
Nigeria	III	0.24	1.70
10. INDONESIA	III/IV	0.33	1.70
11. VENEZUELA/ECUADOR		-	-
Venezuela	II/III	0.24	1.70
Ecuador	II/III	0.24	1.70
OTHER			
12. MALAYSIA	III/IV	0.33	2.30
13. THAILAND	III	0.24	1.70
14. MEXICO	II/III	0.24	1.70
15. TRINIDAD & TOBAGO	I	0.17	1.70
16. TIERRA DEL FUEGO		-	-
Argentina	III	0.24	1.70
Chile	III	0.24	1.70
17. CHINA	III	0.24	1.70

*Where a region is assigned to more than one site category, the higher-cost category
    is used.

Table 3.5
Methanol and LNG Shipping Costs

	A	B
	Methanol Shipping Cost (1987 $/gal)	LNG Shipping Cost (1987 $/MCF delivered)
OECD		
1. ALASKA	0.008	0.36
2. CANADIAN ARCTIC	0.008	0.36
3. NORWEGIAN ARCTIC	0.027	1.17
4. AUSTRALIA/NEW ZEALAND	0.027	1.15
Australia		
New Zealand		
PERSIAN GULF OPEC		
5. IRAN	0.022	0.95
6. OTHER PERSIAN GULF	0.022	0.95
Kuwait		
Qatar		
Saudi Arabia		
United Arab Emirates		
OTHER OPEC		
7. ALGERIA	0.025	1.10
8. LIBYA	0.025	1.10
9. NIGERIA/GABON	0.027	1.18
Gabon		
Nigeria		
10. INDONESIA	0.023	1.00
11. VENEZUELA/ECUADOR	0.013	0.56
Venezuela		
Ecuador		
OTHER		
12. MALAYSIA	0.025	1.10
13. THAILAND	0.026	1.11
14. MEXICO	0.015	0.65
15. TRINIDAD & TOBAGO	0.014	0.62
16. TIERRA DEL FUEGO	0.019	0.83
17. CHINA	0.021	0.92

Assumptions:

Product shipped to Los Angeles

Heavy fuel oil @ $13.50 per barrel

Methanol transported in 250,000 DWT tankers

LNG transported in 125,000 cubic meter tankers

Table 3.6
Methanol and CNG: Delivered Supplies and Prices

	Gas Netback ($/MCF)	Gas Netback ($/gal meth)	Conversion Cost ($/gal meth)	Shipping Cost ($/gal meth)	Delivered Price ($/gal meth)	Gas Supplied (TCF)	Methanol Supplied (bill gals/yr)
METHANOL HIGH DEMAND:							
Other Persian Gulf	0.20	0.02	0.24	0.022	0.28	2.7	30
Nigeria/Gabon	0.14	0.01	0.24	0.027	0.28	0.8	9
Venezuela/Ecuador	0.30	0.03	0.24	0.013	0.28	1.7	19
Trinidad & Tobago	1.07	0.10	0.17	0.014	0.28	0.3	3
Total	-	-	-	-	-	5.5	61
METHANOL LOW DEMAND:							
Other Persian Gulf	0.09	0.01	0.24	0.022	0.27	1.0	11
Nigeria/Gabon	0.03	0.00	0.24	0.027	0.27	0.2	1
Venezuela/Ecuador	0.19	0.02	0.24	0.013	0.27	2.4	16
Trinidad & Tobago	0.96	0.09	0.17	0.014	0.27	0.3	3
Total	-	-	-	-	-	2.8	31

	Gas Netback ($/MCF)	Conversion Cost ($/MCF)	Shipping Cost ($/MCF)	Delivered Price ($/MCF)	Gas Supplied (TCF)	CNG Supplied (TCF)
CNG HIGH DEMAND:						
Other Persian Gulf	0.32	1.70	0.95	4.27	3.0	2.61
Nigeria/Gabon	0.09	1.70	1.18	4.27	0.3	0.25
Venezuela/Ecuador	0.71	1.70	0.56	4.27	1.9	1.71
Trinidad & Tobago	0.65	1.70	0.62	4.27	0.2	0.21
Total	-	-	-	-	5.5	4.78
CNG LOW DEMAND:						
Other Persian Gulf	0.11	1.70	0.95	4.06	0.7	0.64
Nigeria/Gabon	.00	.00	.00	.00	.0	.00
Venezuela/Ecuador	0.50	1.70	0.56	4.06	1.8	1.61
Trinidad & Tobago	0.44	1.70	0.62	4.06	0.2	0.20
Total	-	-	-	-	2.8	2.45

a significant automotive-fuel end use. LNG is assumed to be carried in 125,000 cubic meter vessels since the high cost of larger ships appears to preclude their use in the near future. The heavy fuel oil to power both types of vessel is assumed to cost $13.50 per barrel. Finally, LNG cost is adjusted to reflect the boil off of gas that occurs during transport as a function of the length of the trip. Estimated boil-off rates from the selected supply regions to Los Angeles range between 1.6 percent for Alaska and Canada to slightly over 6 percent for Australia and Nigeria. The estimated boil-off rate from the Persian Gulf is 5 percent.

DELIVERED COSTS OF METHANOL AND CNG

Combining the feedstock, conversion, and shipping cost estimates, it is possible to construct aggregate supply curves (comprising the designated supply regions) for delivered methanol and CNG. The delivered cost of CNG includes $0.50 per MCF for terminaling and regasification of LNG, and $0.80 per MCF for compression of the gas. CNG compression costs are included in order to obtain a delivered cost for a finished automotive fuel comparable to methanol. Overland domestic transport costs associated with delivery of imported gas to regional compression plants are not included.

Points on the supply curves can then be selected to demonstrate the marginal supply prices of methanol and CNG that would correspond to various alternative-fuel demand scenarios for the United States at the turn of the century. Two such scenarios are depicted in Table 3.6.

The low-demand scenario assumes an annual U.S. demand for 31 billion gallons of automotive methanol or for 2.4 billion MCF of CNG. The methanol and CNG cases are presented in the alternative. This means that in the methanol case, for example, it is assumed that there is no competing demand for natural-gas feedstock for conversion to LNG. However, there is the potential for the development of a substantial LNG demand to supply electric power plants, particularly if and when countries overcome the aversion to domestic LNG facilities. In such a case, methanol plants would have to compete for the gas feedstock, which would then command a higher price.

In the low-demand scenario, either the methanol or the LNG delivered to the United States would consume 2.79 TCF of natural gas feedstock, or 6 percent of the estimated maximum gas production from the supply regions. Thirty-one billion gallons of methanol would replace 1.1 million barrels of gasoline per day. Although the thermal value of this quantity of methanol is only 93 percent of that of the gasoline replaced, the replacement factor reflects the automotive-motor efficiency gain achieved by substituting methanol for gasoline. In the alternative, the CNG supply in the low-demand scenario would replace 1.3 million barrels of gasoline per day. The CNG/gasoline replacement factor used here is based on thermal equivalency and does not reflect any anticipated motor efficiency gain from substituting CNG for gasoline.

It is assumed that the feedstock cost is the marginal supply price of the 2.79 TCF of natural gas produced above and beyond those quantities necessary to satisfy all other projected demands for gas. Future demand is estimated as actual 1985 production levels augmented by DOE's estimated demand expansion for OECD, OPEC, and Rest of World to the year 2000.[40] As is shown in Table 3.6, the estimated supply price of methanol (exclusive of rents and transfers) in the low-demand scenario is $0.27 per gallon. Since 1.8 gallons of methanol replace 1.0 gallon of gasoline, the methanol price is the equivalent of a supply price of $0.49 per gallon of gasoline. Alternatively, CNG could be delivered to the United States (including compression costs) at a supply price of $4.06 per MCF,[41] equivalent to a gasoline supply price of $0.51 per gallon.

In the methanol case, the bulk of the product is provided by Venezuela/Ecuador (51 percent) and the Persian Gulf group, excluding Iran (37 percent). Trinidad supplies less than 10 percent, and West Africa supplies a negligible amount. In the alternative CNG case, West Africa drops out as a supplier, and Persian Gulf supplies are reduced because of the higher unit shipping costs associated with LNG. Venezuela/Ecuador supplies 65 percent of the LNG delivered to Los Angeles, the Persian Gulf supplies 27 percent, and Trinidad supplies less than 10 percent.

The high-demand scenario assumes an annual U.S. demand of 61 billion gallons of methanol or 4.7 billion MCF of CNG. Either of these would require 5.5 TCF of natural gas feedstock, or 12 percent of the supply regions' estimated maximum production in the year 2000. Sixty-one billion gallons of methanol would replace 2.2 million barrels of gasoline per day; the CNG equivalent would replace almost 2.5 million barrels per day.

Table 3.6 shows that, in the high-demand scenario, methanol could be delivered to the United States at a supply price of $0.28 per gallon, the equivalent of a gasoline supply price of $0.50. The increase in the methanol delivered price is minuscule because feedstock cost rises by only $0.01 per gallon. In the alternative, CNG could be delivered at $4.27 per MCF, the equivalent of gasoline at $0.53 per gallon. The delivered price of CNG is 5 percent higher than in the low-demand case. This is because feedstock cost is a tiny component of the delivered price.

The source-of-supply pattern in the high-demand scenario is similar to that observed in the low-demand case. All of the suppliers increase their shipments to meet the higher U.S. demand. The Persian Gulf and Venezuela/Ecuador are the sources of most of the methanol and CNG delivered to the United States. The Persian Gulf supplies 50 percent of the methanol delivered, while Venezuela/Ecuador supplies 31 percent. Trinidad provides less than 10 percent, while West Africa provides about 15 percent.

In the CNG case, the Persian Gulf supplies 55 percent and Venezuela/Ecuador supplies 35 percent of delivered LNG. West Africa and Trinidad combined supply about 10 percent.

CONCLUSIONS

This chapter has shown that there is an abundance of low-cost natural gas feedstock available for conversion into methanol or CNG for the U.S. automotive fuel market. In addition, reasonable conversion and shipping costs associated with advanced technologies make these alternative fuels obtainable at supply prices competitive with the cost of gasoline. It is possible for the United States to displace up to 2.5 million barrels of gasoline consumption per day by the year 2000 by turning to these fuels. Of course, this outcome would require additional investment in vehicle conversion and in fuel-delivery infrastructure. Those costs are beyond the scope of this chapter.

The prices estimated in this paper are supply prices. It is likely that they will be augmented by rents and transfers, the multitude of which will be determined by market conditions. Market prices for alternative fuels will probably be a function of gasoline market prices because the experience in the latter industry will set the standard for rents paid to resource owners. For example, high gasoline prices resulting from robust demand relative to supply will presumably enable natural gas and fuel suppliers to extract higher rents than would be possible in a depressed energy market.

NOTES

The authors wish to thank Carmen Difiglio, Robert Motal, Michael Rothkopf, and Daniel Sperling for helpful comments on drafts of this work. We are particularly indebted to Marshall Frank and Chem Systems, Inc. for their estimates of conversion and transportation costs.

1. *Report of Task Force, World Gas Supply and Demand, 1983–2020,* Report to the 16th World Gas Conference, Munich, 1985 (Munich: International Gas Union, 1985).

2. Potential Gas Committee, *Potential Supply of Natural Gas in the United States (December 31, 1986)* (Golden, Colorado: Potential Gas Agency, Colorado School of Mines, 1987).

3. See David H. Root, "Historical Growth of Estimates of Oil- and Gas-Field Sizes," in *Oil and Gas Supply Modelling,* ed. Saul I. Gass (Washington, D.C.: U.S. Department of Commerce, 1982); Russell S. Uhler, "Costs and Supply in Petroleum Exploration: The Case of Alberta," *Canadian Journal of Economics,* 9 (February, 1976): 72–90. Cf. Morris A. Adelman and Michael C. Lynch, "Supply Aspects of North American Gas Trade," in *Final Report on Canadian–U.S. Natural Gas Trade,* 85–013 (Cambridge: Massachusetts Institute of Technology, 1985).

4. United Nations, "Development of the Energy Resources of Developing Countries, Report of the Secretary-General," A/40/511, New York, August 13, 1985.

5. C. R. Blitzer, P. E. Cavoulacos, D. R. Lessard, and J. L. Paddock, "Oil Exploration in the Developing Countries: Poor Geology or Poor Contracts?" *Natural Resources Forum,* 9 (1985): 293–302.

6. The World Bank, *The Energy Transition in Developing Countries* (Washington, D.C.: The World Bank, 1983), 32.

7. Ibid., 26–42.

8. Institute of Gas Technology, *IGT World Reserves Survey As of December 31, 1984* (Chicago: Institute of Gas Technology, 1986), 52–53; G. Dolton et al., *Estimates of Undiscovered Recoverable Conventional Resources of Oil and Gas in the United States, U.S. Geological Survey Circular 860* (Washington, D.C.: U.S. Department of the Interior, 1981), 7.

9. Don Hedley, *World Energy: The Facts and the Future* (2d ed.; New York: Facts on File, 1986), 198.

10. International Energy Agency, *Natural Gas Prospects* (Paris: OECD/IEA, 1986), 67.

11. Energy Information Administration (EIA), *International Energy Annual 1986* (Washington, D.C.: EIA, 1987); IGT, *IGT World Reserves Survey*. Where IGT estimates a range of potential resources, Figure 3.2 shows the upper limit to the range.

12. EIA, *International Energy Annual 1981* (Washington, D.C.: EIA, 1982).

13. Morris A. Adelman and Michael C. Lynch, "Natural Gas Trade in Western Europe: The Permanent Surplus," in *Western Europe Natural Gas Trade, Final Report,* MIT EL 86–010 (Cambridge: Massachusetts Institute of Technology, 1986).

14. World Bank, *Energy Transition.*

15. Carmen Difiglio, "Timing of Methanol Supply and Demand: Implications for Alternative Fuel Policies" (paper presented to the Conference on Transportation Fuels in the 1990s and Beyond, Monterey, California, July 17–19, 1988).

16. A small portion of LNG loaded would be lost or "boiled off" during shipment to the United States. The actual magnitude of the boil off is a function of the distance between the supplier and Los Angeles. See section on conversion and transport costs in this chapter.

17. Adelman and Lynch, "Natural Gas Trade in Western Europe."

18. Harold Hotelling, "The Economics of Exhaustible Resources," *Journal of Political Economy,* 39 (April 1931): 137–75.

19. M. A. Adelman, John C. Houghton, "Introduction to Estimation of Reserves and Resources," in M. A. Adelman, John C. Gordon Kaufman, and Martin B. Zimmerman, *Energy Resources in an Uncertain Future* (Cambridge: Ballinger, 1983).

20. See Anthony C. Fisher, *Resource and Environmental Economics* (Cambridge: Cambridge University Press, 1981); P. S. Dasgupta and G. M. Heal, *Economic Theory and Exhaustible Resources* (Welwyn, Great Britain: Nisbet and Cambridge University Press, 1979); Adelman and Houghton, "Introduction to Estimation of Reserves."

21. Adelman and Houghton, "Introduction to Estimation of Reserves."

22. Information Administration (EIA), *An Economic Analysis of Natural Gas Resources and Supply,* DOE/EIA–0481 (Washington, D.C.: EIA, 1986).

23. M. A. Adelman, "The Competitive Floor to World Oil Prices," *The Energy Journal,* 7 (1986), 9–31; M. A. Adelman and Michael C. Lynch, "Natural Gas Supply in the Asia-Pacific Region," in *East Asia/Pacific Natural Gas Trade, Final Report,* 86–005 (Cambridge: Massachusetts Institute of Technology, 1986); Adelman and Lynch, "Natural Gas Trade in Western Europe."

24. No information is available upon which to apportion hydrocarbon activity in Argentina and Chile between mainland sites and Tierra del Fuego, the 16th designated supply source. Consequently, data for both countries are simply aggregated to yield estimated values for Tierra del Fuego.

25. F. Ramsey, "A Contribution to the Theory of Taxation," *Economic Journal,* 37

(March 1927): 47–61; William J. Baumol and David F. Bradford, "Optimal Departures from Marginal Cost Pricing," *The American Economic Review,* 60 (1970): 265–83.

26. Adelman and Lynch, "Natural Gas Trade in Western Europe"; Adelman, "The Competitive Floor."

27. See Adelman and Lynch, "Natural Gas Trade in Western Europe."

28. This treatment of capital cost is taken from Adelman, "The Competitive Floor."

29. The depletion rates in the model are pseudo-depletion rates in that the divisor in each case is 1985 proven reserves. In other words, the model incorporates a somewhat conservative assumption that reserves will increase at a rate just sufficient to replace projected production. The only evidence available to date on this matter is aggregate worldwide data. From 1980 through 1986, including those years in which falling oil prices were devaluing natural gas assets, absolute reserve levels increased by 35 percent while production increased 19 percent. See EIA, *International Energy Annual,* 1979–86. This suggests that reserves may grow faster than has been assumed here. If so, then regional depletion rates will be lower than those adopted in this chapter, and average capital costs are correspondingly overestimated. (See Equation 7.)

30. Adelman and Lynch, "Natural Gas Trade in Western Europe."

31. M. A. Adelman, "Oil Producing Countries' Discount Rates," *Resources and Energy,* 8 (1986): 309–29.

32. Adelman and Lynch, "Natural Gas Supply in the Asia-Pacific Region"; Adelman and Lynch, "Natural Gas Trade in Western Europe."

33. M. A. Adelman and Manoj Shahi, *Oil Development-Operating Cost Estimates 1955–1985,* MIT-EL 88–008 WP (Cambridge: Massachusetts Institute of Technology, 1988).

34. Energy Information Administration (EIA), *Costs and Indices for Domestic Oil and Gas Field Equipment and Production Operations 1985,* DOE/EIA–0185 (85) (Washington, D.C.: EIA, 1986).

35. Adelman and Houghton, "Introduction to Estimation of Reserves."

36. EIA, *An Economic Analysis.*

37. The World Bank has stated that the long-run marginal cost of gas is unlikely to rise in most developing countries before the turn of the century. See World Bank, *The Energy Transition;* Afsaneh Mashayekhi, *Marginal Cost of Natural Gas in Developing Countries: Concepts and Applications,* Energy Department Paper no. 10 (Washington, D.C.: The World Bank, 1983).

38. Q_2 is a function of Q_3. The estimated Q_3 for Alaska is taken from Policy Evaluation and Analysis Group, American Gas Association (AGA), *The Gas Energy Supply Outlook Through 2010* (Arlington, Virginia: AGA, 1985).

39. Chem Systems, Inc., "Methanol and LNG Production and Transportation Costs, Draft Report" (report presented to Alternative Fuels Supply Project, U.S. Department of Energy, July 1988).

40. Office of Technology Policy, U.S. Department of Energy (DOE), *Long-Range Energy Projections* (Washington, D.C.: DOE, August 1987).

41. The CNG supply price is slightly underestimated because the feedstock and LNG conversion costs are not adjusted to reflect the subsequent boil-off of LNG during shipment. In this model, the bulk of the LNG purchased comes from the Persian Gulf and Venezuela. The corresponding estimated boil-off rates are 5 percent and 2.8 percent.

4

ROBERT J. MOTAL

Synthetic Fuel Costs in the Synfuels Era: A Chevron View

There are numerous alternatives to oil for powering a vehicle. Liquid fuels can be made from oil shale, coal, gas, and biomass in addition to oil. The fuel can be made to look like conventional gasoline or diesel fuel—or it could be an alcohol. Some of these fuels, such as methanol, offer the potential for reduced vehicle emissions of some pollutants. Electricity is an option, and solid fuels like powdered coal can also be used, although there are even more imposing technical, economic, and environmental problems associated with using solid fuels.

New alternatives are proposed regularly. At Chevron we focus our evaluations on a subset of these possibilities—those that are most likely to be developed first.

EVALUATION BASIS

Most economists use a crystal ball. Analysts, like myself, use mirrors; we are constantly looking at the same project in different ways.

We have numerous mirrors at our disposal. We can change the return on our investment or ignore a return on our investment. We can finance internally or assume that some of the plant is financed by others. We can change the interest rate charged on the loans. We can alter our view on how the product will be valued in the marketplace. We can change the cost of our feedstock. We can change the size of the proposed project.

There is no shortage of ways to estimate costs, and much of the continuing confusion in synthetic fuels costing is due to different assumptions used in project evaluation.

SYNFUELS ERA COSTING

In our screening evaluations at Chevron, we assume that synthetic fuel must compete in the marketplace without special government incentives. Some people call this a level playing field.

We also assume that feedstock prices must be consistent with the resulting energy prices. This means that if a fuel can only compete with gasoline when oil costs $100 per barrel, then all other assumptions must be consistent with this $100-per-barrel economy. This has a strong effect on fuels such as ethanol that use a large amount of energy in the production process. Ethanol production uses large amounts of energy in planting, fertilizing, cultivating, and harvesting corn and making ethanol from it. This energy will be increasingly expensive, and ethanol will be more expensive than it is now. We are consequently pessimistic on ethanol.

We also assume feedstock prices are consistent with each other. This is particularly important for domestic gas. What does one assume for the price of domestic gas when oil costs much more than it does now? We believe that it will command a premium and will be priced considerably higher than coal. The premium is likely to be determined by competition with coal in the electricity-generation industry.

It is also important to use the same ultimate consumer in comparing alternatives. Our recent studies have used a fuels terminal in Chicago as the common basis for comparisons. Consequently, we equate the value of the synthetic fuel with an oil-based product in Chicago. This includes adjusting for the different engine efficiencies one might expect. This works to the advantage of fuels like methanol and compressed natural gas that have high octane values.

It helps to walk through an example. Consider shale oil costs. First we determine the cost of syncrude derived from oil shale in Colorado. This cost represents the mining, retorting, and upgrading of the raw shale oil. This syncrude is piped to an existing refinery on the Gulf Coast, and the products are piped to Chicago. Alternatively, if new pipelines are built, the syncrude could be transported directly to a Chicago-area refinery. Typical pipeline tariffs are added. Upon refining, the shale oil syncrude is converted to salable gasoline, jet fuel, and diesel. It is a relatively straightforward process to bracket the cost to convert syncrude into this basket of products; allocating refining costs to the various products is not straightforward. For this study we set the value of these oil look-alike fuels equal on a Btu basis.

We used the same product values for the basket of liquid products that would be produced by refining West Texas Intermediate crude oil at a Midwest refinery. The value of oil at the refinery gate is then back-calculated based on the product basket value. We used West Texas Intermediate as the reference oil because it is widely quoted. Then by subtracting the pipeline tariff we arrive at the value of West Texas Intermediate at Cushing, Oklahoma, which is the reference location for quoting the price of this oil.

We did the same for the various alternatives, which allows us to compare the options directly. The use of a Midwest clearing location was arbitrary. Use of a different clearing market such as a coastal location in California or Japan would favor resources closer to these markets. This is particularly true for Pacific Rim- or Middle East–based gaseous fuels derived from remote gas because the transportation of liquefied natural gas is high compared to the cost of moving liquids.

A DIFFERENT APPROACH TO FEEDSTOCK COSTS

An example of how different assumptions and parameters can be used to arrive at different cost estimates is provided in Chapter 3. For the following reasons we at Chevron believe those estimates of remote natural gas costs are low:

Their investment costs are much lower than the costs reported for gas field developments around the world. The majority of their investment costs cluster in the $0.20–1.00 per thousand cubic feet per year (MCFY) range. In contrast, Table 4.1 shows that costs in the literature reported are generally $4.00–6.00/ MCFY. Chevron's actual experience tends to substantiate the literature numbers. (The actual costs of several Chevron-affiliated projects and internal engineering studies have also been shown in the table.) If one assumes that these gas field developments represent the most attractive projects, then the disparity is even greater.

- They assume that remote gas drilling costs are the same as U.S. gas drilling costs (except for the Arctic). This is generally not the case. One needs to set up construction camps, bring in drilling and production equipment from further distances, provide for greater self-sufficiency, and deal with inhospitable terrain. Local labor may also need to be employed and trained.

- They assume that new gas developments can capitalize on the existing infrastructure (ostensibly previously paid for by oil production). This varies. It certainly is not true in areas that are gas prone like the Northwest Shelf of Australia where very little oil infrastructure is in place; it is only partly true in mature oil-producing regions like West Africa. For example, new offshore platforms may need to be built for treating and compressing the gas since the existing oil production platforms are already loaded with oil production equipment.

- It does not appear that sufficient allowances have been made for offshore production platforms, supply bases, or pipelines.

- The analysis implies that the gas is already found, delineated, and waiting to be sold. It does not appear to include the costs to process the gas, and it is not clear if all exploration costs are included.

- We are also concerned that the assumed well productivities are high in a number of locations.

Their assumed real rate of return (RROR) appears low. The geological, technical, marketing, and political risks are simply too great for companies routinely to fund exploration and production projects based on a 10 percent RROR.

Table 4.1
Remote Gas Development Costs

Country/Area	Reserves (TCF)	Onshore/Offshore (Miles to Shore)	Production (MMCFD)	Capital Costs (MM 1988 $)			Reference Source
				Platforms, Wells, Production Facilities	Pipelines	Cost/MCFY 1988$	
Australia-North West Shelf	8.3	Offshore (80)	1,600	2,000	500	4.30	North West Shelf Brochure
Australia-Barrow Island	1.0	Onshore	100	114	60	4.70	Chevron-Internal
Australia Offshore (Near Barrow Is.)	10+	Offshore	600	840	210	4.80	Chevron-Internal
Thailand - "B" Block	1.8-6.0	Offshore (110)	150	500	150	11.90	Int'l Gas Rpt. (4/83)
Qatar-North Field (Phase One)	150+	Offshore (40)	800	520	520	3.60	World Gas Rpt. (4/83) Arab Oil & Gas (1988)
Norway-Troll	47	Offshore (820)	3,180	7,800		6.70	Int'l Pet. Encyclopedia (1987)

U.S.-Gulf of Mexico (Garden Banks)	?	Offshore (140)	80	130		20	5.10	Int'l Pet. Encyclopedia (1980)
Trinidad-NMCA	2.8	Offshore (25)	500?	500		300	4.40	Dept. of Transportat. Study (1987)
Nigeria-Assoc. Gas	?	Onshore	370		600		4.40	Chevron-Internal
Nigeria-Assoc. Gas	?	Offshore (10)	170		350		5.60	Chevron-Internal
Indonesia-Typical	1	Offshore	170	464		320	12.60	Chevron-Internal
Indonesia-Typical	0.5	Onshore	70		100		3.90	Chevron-Internal
Tunisia-Miskar	1.1	Offshore	400		600		4.10	Int'l Oil Ltr. (August 15, 1988)
Netherland-North Sea	1.8	Offshore (320+)	1,700	1,100		1,000	3.40	Oil & Gas Journal (Aug. 15, 1988)

Notes: TCF = Trillion cubic feet;
MMCFD = Million Cubic fee/day;
MCFY = Thousand cubic feet/year;
1 cubic foot = 1,000 Btu.

Table 4.2
Parameters for Cost Comparisons

Case	Special Situation	Standard Economics	
		Low	High
After Tax Constant $ Rate-of Return, %	10	15	15
Feedstock Cost:			
Remote Gas, $/MCF	0.50	0.50	2.00
Coal, $/MMBtu	1.00	1.50	2.00
Corn, $/bushel[3]	1.50	2.00	2.00 (WTI[1] = $20/bbl)
			3.00 (WTI = 40/bbl)
Byproduct Value:			
Dried Distillers Grain $/ton[3]	120	120	100 (WTI = $20/bbl)
			150 (WTI = $40/bbl)

Vehicle Characteristics:

	Optimized for each fuel Fleet (30,000 MPY)	Optimized for each fuel Normal (12,000 MPY)	Optimized for each fuel Normal (12,000 MPY)
Engine			
Usage			
Fuel Storage Added Cost (CNG only), $/vehicle	600	600	600
Service Station Added Investment (CNG only), $/gal/month	2.6	3.5	3.5

Notes:
1. West Texas Intermediate FOB Cushing, Oklahoma.
2. Vehicle costs are included based on a capital charge factor = 0.40.
3. Corn and dried distillers' grain prices are sensitive to oil prices. The values shown were derived when oil prices were less than $40/bbl; it is unclear if they apply when oil reaches "synfuels age" prices.

Assumptions: Costs are wholesale costs delivered to Chicago-area distributor. Costs do not include excise tax, local fuel distribution, or service station markup. They include cost for vehicle gas cylinders and service station compressive/ dispensing equipment for compressed natural gas.

Crude and syncrudes are refined in existing facilities.

35% combined tax rate based on post–1988 rules under 1986 Tax Reform Act.

This is particularly important since the period of time between lease acquisition and initial production can be very long. (The initial leases for the Northwest Shelf were acquired in 1963, the initial discoveries were in 1971, and large-scale sales won't start until 1989.) We would also caution that alternative uses are being developed for remote gas. Saudi Arabia is a case in point. They have spent billions of dollars to build a master gas system. This system supplies gas for their desalination plants, petrochemical plants, refineries, and other local industries. At current oil-production rates, this gas system is unable to supply existing gas uses from associated gas production. They are currently drilling for more expensive nonassociated gas. One can argue that the price of gas should be set at the higher alternative value (fuel-oil displacement) and is not determined by a marginal cost model that Lawrence and Kopler have used.

THREE VIEWS OF SYNTHETIC FUELS

We calculate the cost of synthetic transportation fuels under three sets of economic assumptions to determine the sensitivity of our cost estimates to our economic and resource assumptions. The three scenarios are special situations, low end of standard economics, and high end of standard economics; this representing the range of likely costs for a new facility.

The assumptions used for each of these three cases are shown in Table 4.2. Because of the high current interest in methanol, I show in Table 4.3 the specific assumptions used for estimating the cost of methanol production.

We believe that natural-gas feedstock cost ranges from $0.50 to $2.00 per thousand cubic feet at remote sites. Construction conditions range from a favorable developed site such as in Jubail, Saudi Arabia, to an undeveloped remote tropical site such as in Indonesia. This still doesn't define the high end. For example we believe that construction in particularly hostile environments like the North Slope of Alaska could easily double these costs. The special-situation case also assumes low construction costs due to depressed conditions in the construction and equipment-fabrication industries. We have not, however, included economics for the other side of the coin, representing unusually high costs resulting from boom conditions in these industries.

We would characterize our high standard economics case as defining projects that would likely be developed when the synfuels age matures. This price may represent a plateau for energy prices since this price could support a large number of projects. I might add that this price however is not sufficient to develop all resources. For example the gas reserves on the North Slope of Alaska would not be developed at these prices.

The constant-dollar rate of return ranges from 10–15 percent after taxes. The 15 percent return was selected to represent a rate likely to attract private capital on a large scale, assuming a reasonable commercial risk. The 10 percent return recognizes that lower effective capital costs could result in special situations. For example, some foreign governments with unmarketable energy resources

Table 4.3
Basis for Cost of Methanol from New Remote Gas Plants (1800 MTPD Plant Size)

	Special Situation	Standard Economics	
		Low End	High End
After-tax constant $ rate of return	10	15	15
Feedstock cost, $/million BTU	0.50	0.50	2.00
Site conditions	Developed Jubail, Saudi Arabia	Developed Jubail, Saudi Arabia	Undeveloped Indonesia
Construction & fabrication industry conditions	Depressed	Normal	Normal
% of normal costs	85	100	100
Investment, $ million	320	380	550

Figure 4.1
Competitiveness of Alternative Transportation Fuels

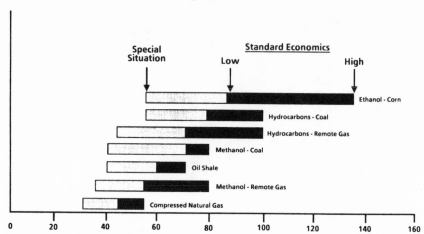

Competitive when West Texas Intermediate Crude Oil Costs, 1987 $/barrel
* Includes ⌂ Vehicle and Service Station Costs

may be motivated by public-sector economics or politics to provide a low-cost loan to assist in the financing of a synthetic fuels plant.

We have assumed that methanol will be produced with commercially-proven technology.[1]

A similar methodology was used for the other alternative transportation fuels. In some cases, technology has not been satisfactorily demonstrated. This is true for both oil shale and direct coal liquefaction. In these cases we have taken the most advanced technology that we feel has a good chance of working at the commercial scale.

In Figure 4.1, the alternative transportation fuel options are ranked by cost. The prices shown on the lower axis are the crude oil prices required before the synthetic fuel can compete with oil-derived products. Product prices are higher than oil prices because of refining and transportation costs.

Each alternative transportation fuel is shown by a bar. The dark-shaded portion of the bar indicates the range between the low standard economics scenario and the high standard economics scenario. In other words, this shows the range in estimated costs one might expect due to resource differences. We have varied the resource characteristics while keeping the economic parameters the same. The left portion of the bar reflects the differences between standard economics and the special-situation economic parameters.

We expect that limited quantities of synthetic fuels may be introduced at special-situation prices. Large-scale development of synthetic fuels will require that crude oil prices are at least equal to low standard economics prices.

COMPRESSED NATURAL GAS AND METHANOL

At this time compressed natural gas and methanol appear to be the most attractive synthetic fuels from an economic aspect. We believe that they could compete when oil prices reach $40–60 per barrel. As Burns suggests in Chapter 2, these prices are unlikely to be reached until oil resources are largely used up.

There are two caveats. Both compressed natural gas and methanol are new fuels. The existing transportation-fuel distribution systems and the vehicles themselves will need to be extensively modified or built specially for these fuels. Vehicle manufacturers would like to see the fuel available before they commit to large-scale production of these vehicles. Likewise fuel suppliers would like to see a developed fuel market before they build expensive synthetic fuels plants. This is commonly referred to as the ''chicken-and-egg'' problem. Therefore even higher oil prices may be required to induce industry to develop this new market.

Chevron does not feel that this will prevent the market from developing. We believe that if economic fundamentals support a new fuel, then industry will take the necessary risks to evolve without government intervention. For example private industry undertook development and marketing of video cassette recorders, compact disc players, and personal computers without government assistance.

The second caveat is that large quantities of remote gas are available at $0.50 per thousand cubic feet. The value is a best case based on the expected cost to gather a large centralized volume of flared gas, treat it, and make it available to the synfuels plant. This gas is associated with oil production, and the entire cost of the exploration, drilling, and production activities are borne by oil sales. This gas will be less and less available as producing countries take steps to use it. Local gas use in Saudi Arabia, for example, exceeds the associated gas production; they are currently drilling for gas to meet existing local gas demand.

Producing countries read the *Wall Street Journal*; without a great degree of expertise they can calculate the approximate netback value of their resources. And producing countries can and do act when the netback value greatly exceeds the cost of production. An example in the oil industry is Saudi Arabia where oil can be produced for about $2 per barrel. Do they sell it for that price? Of course not. They sell it at the market clearing price, currently around $15 per barrel. Similarly one might expect that producing countries would extract any significant ''economic rents'' for remote gas that may occur.

I have assumed that internal combustion engines will be designed to run specifically for both compressed natural gas and methanol. Consequently I have included a credit for increased fuel efficiency relative to gasoline. These efficiency credits may need to be reduced during the introductory period since automakers may opt to supply engines that are optimized for gasoline and then retrofitted to run on these alternative fuels.

OTHER SYNTHETIC FUELS

The most economic domestic synthetic fuel is shale oil. It has the added advantage that products produced from oil shale look, feel, and act like gasoline, jet fuel, and diesel produced from conventional oil. In industry jargon, they are "fungible." They can be used by existing vehicles and supplied from existing service stations and terminals.

This advantage should not be underestimated. Fuels suppliers are faced with a variety of risks—geological, political, market, financial, among others. All of these risks are factored in the analysis used to assess a given project. To the extent that one or more of these risks can be eliminated, a project can be justified at a lower product price. Since shale oil–derived products can be sold just like oil-derived products, the project developer will minimize the contingency for market uncertainty.

Methanol from coal is only modestly more expensive than products from shale oil. We would expect both to evolve as acceptable fuels in the future. Shale oil products will dominate fuel markets where high energy-density or superior compression-ignition characteristics are important, as with jet fuel and diesel. Methanol could move into markets where superior spark-ignition characteristics are important. We have not shown estimated costs for methanol made from domestic gas because of the uncertainty over the premium that gas will command over coal. Methanol from domestic gas will cost somewhat less to make than methanol from coal if gas premiums are set by competition with coal for base-load power generation. Methanol from gas would cost more if the price of gas is set by competition with fuel oil or distillate.

This choice of gas or coal is also true for direct coal liquefaction. We feel that the hydrogen required will be produced from the plant off-gas and probably supplemented by coal instead of producing hydrogen from purchased natural gas.

Ethanol is quite expensive. The costs we show in Figure 4.1 are consistent with the capital and operating costs derived from our experience with the Kentucky Agricultural Energy Corporation plant in Franklin, Kentucky. We have not included investment tax credits, gasohol blending credits, or other government incentives that led us to build the plant. The big uncertainty is the cost of corn and value of the dried distillers' grain. We have attempted to extrapolate these costs to an energy price environment that could support ethanol. While the extrapolation is crude, the conclusion is the same—ethanol is very expensive.

PRODUCT COSTS ARE SENSITIVE TO CAPITAL COSTS

Capital costs account for 50 percent to 75 percent of the total costs shown for these synthetic fuels. These costs include a return on our investment. This sensitivity to capital costs leads to an interesting conclusion. Synthetic fuels plants will not be turned on and off to moderate energy price manipulation. The sunk costs are simply too expensive. These plants will run as long as they can recover their out-of-pocket operating costs. Their only impact on oil prices will be as a reduction to base-load oil demand—not as a short-term replacement source of substitute oil products.

CONCLUSION

Synthetic fuels now cost a great deal more than fuels from oil. At this time we estimate that synfuels will compete only when oil doubles or triples in price. This could be reduced if significant technological breakthroughs occur. We would expect them to be able to compete in the next century.

Many of the most attractive resources for synthetic fuels are located outside of North America. We expect that the United States will continue to rely on foreign countries for a significant portion of its fuel needs.

Synthetic fuels are unlikely to reduce oil price volatility. Most synfuels cost less than $20 per barrel to make if one excludes the initial investment in plant and resource development. Once built, the plants will run at capacity. They will not be turned off and on in response to oil price fluctuations.

The economic ranking among alternatives is not significantly affected by the economic basis chosen. This is not surprising since the major costs are capital charges on these very capital-intensive projects.

NOTES

I would like to gratefully acknowledge the ongoing assistance and counsel that Bruce Beyaert in Chevron U.S.A.'s Strategic Planning and Business Evaluation group has provided in the development of this chapter.

1. A methanol plant cost study is underway involving Chevron, government agencies, methanol producers, energy companies, and a research institute to improve our collective understanding of methanol plant investment costs. The investment and operating costs for methanol plants used in this chapter reflect our understanding prior to completion of the study. The per-gallon capital costs estimated in that study tend to be somewhat lower because it assumes multiple plants built on one site, creating economies of scale for certain infrastructure and equipment.

5

R. F. WEBB, CARL B. MOYER
and M. D. JACKSON

Distribution of Natural Gas and Methanol: Costs and Opportunities

The number of vehicles operating on methanol and natural gas fuels in North America total less than 100,000 and most of these vehicles are natural gas conversions operating in the United States. The number of refueling stations for these fuels is very limited in comparison to petroleum fuels. Canada, for example, has 150 compressed natural gas refueling stations for 18,000 vehicles and California has 20 refueling stations for their 500–plus methanol vehicles. The infrastructure to move large quantities of these fuels is also limited. In the case of methanol, tanker ships, harbor facilities, and possibly pipelines are required. A large network of pipelines is already in place to transport natural gas, however, but even this network does not reach many North American locations. Because of the sparse fuel distribution infrastructure, the cost of distributing methanol and natural gas fuels can more than double the delivered price of those fuels.

This chapter compares the infrastructure and distribution costs for compressed natural gas (CNG) and methanol. Of particular interest is the volume of fuels that have to be introduced to be economic and competitive with gasoline and diesel fuels. This chapter focuses on the delivery of fuels to California, likely to be the first large market for methanol and CNG.

COMPRESSED NATURAL GAS (CNG)

Natural Gas

The principal and lowest-cost source of natural gas in the United States will continue to be domestic gas (current wellhead price about $1.70 per mmBtu), but this will be augmented for expanded CNG use by imports from Canada (current border price about $1.83 per mmBtu), a resumption of pipeline imports

Table 5.1
Long-Term CNG Pump Prices in Los Angeles

	Price U.S. $/MMBtu	Gasoline Equivalent Price ($/Gallon)
Uncompressed Gas Prices		
Gas Price at Canadian Border	2.00	
City Gate Price	2.70	
(1,400 Mile Pipeline to Los Angeles)		
Gas Price to Fueling Station for Station Serving:		
10 Automobiles	4.40	
(Dual Fuel 90% CNG)		
45 Automobiles (Dual Fuel)	3.55	
40 City Buses, Monofuel	3.20	
(or 1,000 Autos Dual Fuel)		
CNG Pump Price (Without Taxes)		
Public Service Station [a]	6.05	72
Private Fleet Station [b]	5.10	61

Notes: [a]Public service station serving 350 to 400 CNG vehicles.

[b]Private refueling station serving 1,000 dual fuel autos or 40 urban transit buses.

from Mexico, and, at a much later date, the possible import of high-cost liquid natural gas (LNG) (for example, from Algeria at prices above $3.75 mmBtu, the current price representing a $1.00 per mmBtu minimum shipping charge and a low netback to the producer of $2.75 per mmBtu).

Other high cost gas, comparable in price to LNG, would be Canadian Arctic gas and gas imported from Alaska by pipeline through Canada, as well as small quantities of biogas from waste-treatment facilities. The cost of this gas could be rolled into the total gas resource base, especially for CNG use, but would not be economic for the production of methanol.

Distribution Costs for Natural Gas

Table 5.1 summarizes our estimate of the long-term CNG pump prices in the Los Angeles area. As indicated, we have broken the estimate into the following categories: uncompressed gas prices at the Canadian border, at Los Angeles city limits, and to the service station (these distribution prices are dependent on the gas volume). Compression prices were then estimated and added to station prices for two scenarios: a public service station serving 350 to 400 CNG vehicles and a private fueling station serving either 1,000 dual-fueled vehicles or 40 urban transit buses.

The most likely source of added volumes of gas for the California market is Canadian gas imported through expansion of the existing pipeline systems; these systems currently deliver 620 billion cubic feet of gas annually to California— equivalent in energy terms to 430 million gallons of gasoline, 37 percent of total California gasoline demand. This quantity of gas could be doubled and still be available at an estimated future border price of $2.00 per mmBtu (1988 dollars).[1]

Incremental expansion of the existing 1,400-mile pipeline system between the border and Los Angeles to accommodate an eventual doubling of gas shipments would result in a city gate price of $2.70 per mmBtu, as indicated in Table 5.1. The distribution tariff set by the local utilities varies with the location, type, and annual demand of each customer. In Canada, where the majority of CNG vehicles are privately owned automobiles and light trucks refueled at public service stations, the gas utilities and producers have established a market-building incentive price for gas used as CNG. Similar incentives for CNG have been established by at least 30 gas utilities in the United States.

Under these incentive conditions, a refueling station with a gas demand equivalent to annual use of 20,000 gallons of gasoline (sufficient for 45 to 50 private automobiles fueled 90 percent by CNG and 10 percent by gasoline) would pay about $3.55 per mmBtu or about $0.85 per mmBtu for distribution costs and return on investment for uncompressed gas. At a large refueling station with an annual gas demand equivalent to 440,000 gallons of gasoline (sufficient for about 1,000 dual-fuel automobiles) or 368,000 gallons of diesel (equivalent to the requirements of about 40 transit buses), the price of uncompressed natural gas would be in the vicinity of $3.20 per mmBtu or about $0.50 per mmBtu for the gas distributor.

Refueling Station Capital Cost

A CNG refueling station requires a compressor to increase gas pressure to 3,000 psi and a dispenser to transfer the gas at this pressure to the CNG vehicle. While slow overnight refueling is possible in limited applications, a 5- to 10-minute fast-fill period is essential at public and most fleet service stations. The fast-fill market, therefore, requires the addition of a cascade CNG storage system, a larger compressor than is used for slow fill, and preferably a backup compressor. Fast-fill service stations are capital intensive (typically requiring about $700 per automobile for a public station serving 350 to 400 vehicles); capital service charges typically amount to 27 to 30 percent of the total delivered cost. Total compression costs are about 50 percent of the total CNG pump price. The opportunities for capital savings through economies of scale are limited by the requirement for multiple compressors to provide capacity and redundancy in case of equipment downtime.[2]

Strategies for the introduction of CNG should therefore emphasize rapid market development with minimum facilities, sales to the public from existing service stations, as well as new facilities built to service bus or taxi fleets. In Italy, start-up costs at new stations have frequently been reduced by the use of daughter stations that retail CNG from trailer-mounted storage tanks filled from a mother compression station that is also the main CNG retail outlet in the area.

The mother-daughter station concept and advanced low-pressure storage concept currently under development requires further investigation as a means to reduce the capital cost in an aggressive CNG-based clean fuel program. If adequate fuel storage can be obtained at pressures around 600 psi using carbon or other absorbents, the capital costs for compression can be significantly reduced. These low-pressure storage systems are currently being developed.

Other means of reducing the cost of these stations is to provide cost sharing either by government or gas utility incentives. In Canada, for example, the high cost of refueling facilities has been partially overcome by government grants to companies that construct and operate a public CNG refueling station. The $40,000 (Can. $50,000) grants typically provide about 16 to 18 percent of the capital cost of a CNG facility and are equivalent to the nonrecoverable portion of the investment.[3] Gas utilities in the United States have also proposed to rate base the compression costs, thus substantially lowering the costs to the vehicle user.

Long-Range CNG Costs in the Southern California Market

As shown in Table 5.1, the price of CNG at a full-service station in the Los Angeles area is estimated at about $6 per mmBtu. This cost assumes that utilities will provide incentive-priced gas over the long term. As previously indicated, $6 per mmBtu is equivalent to a gasoline price of $0.72 per gallon without taxes, but including a return on investment throughout the distribution network. This price compares favorably with the mid–1988 untaxed price of unleaded gasoline in California of about $0.75 per gallon. The high capital cost of the CNG facility makes the fuel price extremely sensitive to throughput; if annual sales of the self-service station were about one-fifth that assumed above (the energy equivalent of 90,000 gasoline gallons per month rather than 440,000 gallons), then the untaxed CNG cost would rise from $0.72 to $0.86 per gallon gasoline equivalent.

As shown in Table 5.1, the cost of gas compression and vehicle refueling at large fleet locations will be about 16 percent lower than at public service stations. Because of predictable fuel demand at private fleet stations, capacity can be added incrementally, further reducing capital exposure.

CNG has less of a cost advantage relative to diesel fuel than to gasoline. While the gasoline equivalent price at a private station is $0.61 per gallon, the diesel equivalent is $0.73 per gallon. (The prevailing California diesel rack price (untaxed) is $0.35 per gallon.) The diesel and gasoline equivalent prices cited above were calculated to reflect fuel economy of natural gas and gasoline use in spark-ignition engines, and of diesel fuel in the more energy-efficient compression ignition engines. For instance, in a typical city bus, a gallon of diesel fuel can be replace by 150 cubic feet (150 mBtu) of natural gas, while in a typical dual-fueled automobile, 126 cubic feet of gas (126 mBtu) replaces a gallon of gasoline.[4]

In summary, the high cost of CNG service stations and the difference in value of CNG as a replacement for gasoline and diesel fuel indicate that optimum market development strategy will involve the replacement of both gasoline and diesel fuel.

CNG Cost Experience in Canada

CNG has a much larger presence in Canada than in the United States. The predominant source of natural gas in Canada is Alberta; a major pipeline system transports the gas 2,500 miles or more to be distributed by large gas utilities. As illustrated in Figure 5.1, the gross margins available to CNG retailers are excellent, and CNG is cost competitive at the pump with untaxed gasoline at all major centers. Since CNG is not taxed and gasoline is heavily taxed, an attractive pump price spread is available to CNG users. Untaxed CNG is available at between 65 and 50 percent of the taxed price of unleaded gasoline providing cost savings of up to $0.87 per gallon (U.S.) of gasoline equivalent fuel. A survey of the payback experience of personal CNG vehicle users in Canada showed payback periods averaging 17 months.[5] Commercial vehicle users (who purchase gasoline at lower prices) have an average 20-month payback on the conversion investment. Fuel cost savings amounting to $2,800 to $3,000 (U.S.) per year are being achieved by the Hamilton, Ontario, transit bus company, even including the cost of converting diesel buses to CNG.[6] Still-higher savings will be available after 1991 when the cost of diesel fuel is expected to rise in response to the need to lower the sulfur and aromatic levels of diesel fuel so that buses newly equipped with particulate traps can meet the tightened 1991 heavy duty, emissions standards. The added capital and operating costs of the post–1991 diesel bus make CNG an attractive option for buses in Canada despite the cost of the fueling facility and the conversion cost.

METHANOL

Status of Methanol Bulk Transport

The availability of low-cost natural gas at many locations and the economies of refinery-scale methanol production leave transportation and distribution as the key unresolved cost impediment to worldwide trade in fuel methanol. The dis-

Figure 5.1
Current CNG Prices in Canada

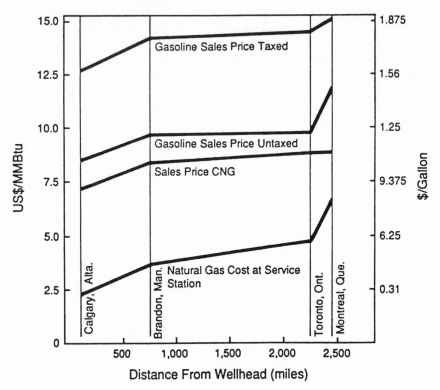

tribution system developed for chemical-grade methanol is aimed at delivering dry product free of contaminants since they poison the sensitive catalysts used in manufacturing formaldehyde. As methanol gains wider use as a fuel, new production techniques will be developed to manufacture methanol that is allowed to contain contaminants and water; production will be less costly, as will distribution. Eventually, transportation costs should be comparable to those now encountered with LPG and, in the distant future, with gasoline.

Ocean movements of methanol in lots of 2,000 to 6,000 tons in stainless steel tankers have already been superseded by 20,000 to 40,000 ton cargoes moved in dedicated tankers, in epoxy and urethane-coated tankers, and in LPG tankers. A large-scale trial of methanol shipment in the high-pressure-rated Cochin ethane-LPG pipeline has shown that methanol can be batched in LPG pipelines, many of which have spare capacity, particularly in the summer when the demand for heating is low.[7] Methanol has also been successfully batched between gasoline buffers in a demonstration using the trans-mountain crude oil pipeline system between Edmonton and Vancouver.[8] Studies are underway to define the cost and feasibility of a dedicated pipeline from a future cluster of methanol plants located at the Canada–United States border using the right-of-way of the 1,400 miles of existing gas pipeline that leads from Kingsgate, Alberta, to the Los Angeles market.

Meanwhile, before the fuel market can support these new tankers and pipelines, intercontinental methanol movements will continue using unit trains and a combination of rail and coastal tankers. These will compete with ocean tanker deliveries from plants already constructed or under construction in other countries.

Methanol for the U.S. market will be manufactured outside the lower 48 states due to natural gas feedstock costs and availability. Natural gas in the United States already has a variety of markets ranging from heating to electric generation. These markets provide competition for gas and also set the price. It is currently too expensive to produce methanol from this gas, and it is expected that this will remain true in the future.

Methanol can be manufactured cheaper in regions of the world where the natural gas feedstock is less expensive. In these regions there are no current uses of the gas and, therefore, its value is lower than gas sold in the lower 48 states. There is a tradeoff, however, between the lower feedstock costs in these remote regions and the higher costs to ship methanol to the United States. Even so, there are many gas-rich regions where methanol shipping costs plus terminaling costs plus feedstock costs can be substantially less than the cost of producing methanol in the United States.[9]

For Los Angeles, the high shipping cost areas and modes are identified in Figure 5.2 and Table 5.2. Tables 5.3 and 5.4 provide some specific examples for other destinations. These costs include terminaling charges of about $0.05 per gallon per month; it is assumed the fuel is held in storage for one month.

The lowest-cost transport mode is ocean freight, but charges for loading and unloading at the terminals, product storage, and insurance must be added. In addition, freight charges from the plant to the dockside are incurred that can be substantial if the plant is not located nearby. Canal tolls are also substantial: the Suez toll amounts to $99,250 per ship plus $3.37 per ton (a 2.5 cent per gallon charge for a 20,000 ton cargo); Panama tolls are $4.30 per ton (1.3 cents per gallon of methanol). Table 5.5 illustrates these components for an 8-million gallon cargo of chemical grade methanol shipped from Alberta to Houston—by rail to the port of Kitimat in British Columbia and by ship to Houston via the Panama Canal. The 760-mile rail movement, including tank car rental, amounts to 48 percent of the total cost; ocean shipping amounts to 25 percent. Loading, unloading, and storage at the two terminals are also a major cost component—19 percent of the total cost in this example.

The examples of shipping costs consider only a fairly large tank ship (say 30,000 DWT, a fairly small oil tanker by today's standards). Unfortunately, even a ship of this size, delivering perhaps 28,000 tons (about 8 million gallons) of methanol per voyage, requires a big market to support it. A dedicated 30,000-ton tanker running between Kitimat and Los Angeles could deliver almost 200 million gallons per year, enough fuel for 200,000 cars regularly using methanol.

Figure 5.2
Shipping and Terminal Costs for Methanol to Los Angeles, 1988 (1995)

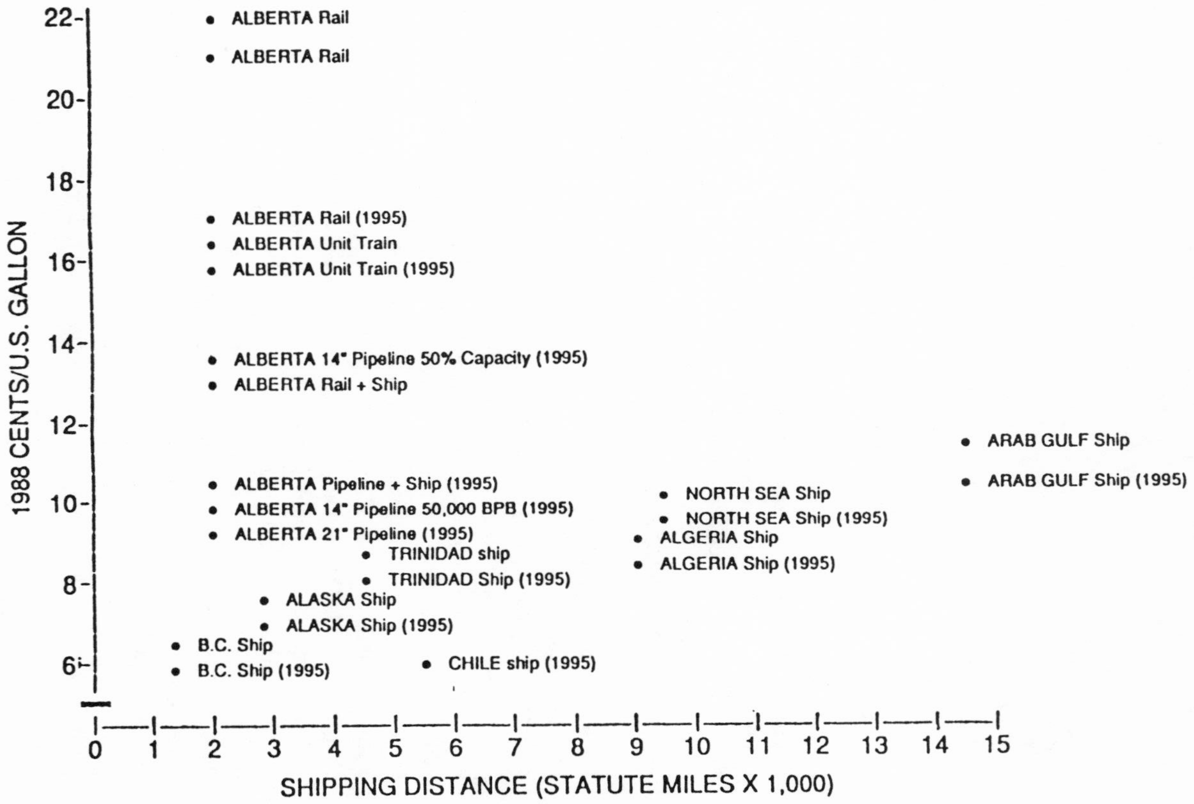

- ALBERTA Rail
- ALBERTA Rail

- ALBERTA Rail (1995)
- ALBERTA Unit Train
- ALBERTA Unit Train (1995)

- ALBERTA 14" Pipeline 50% Capacity (1995)
- ALBERTA Rail + Ship

- ALBERTA Pipeline + Ship (1995)
- ALBERTA 14" Pipeline 50,000 BPB (1995)
- ALBERTA 21" Pipeline (1995)
  - TRINIDAD ship
  - TRINIDAD Ship (1995)
  - ALASKA Ship
  - ALASKA Ship (1995)
- B.C. Ship
- B.C. Ship (1995)   • CHILE ship (1995)

- NORTH SEA Ship
- NORTH SEA Ship (1995)
- ALGERIA Ship
- ALGERIA Ship (1995)

- ARAB GULF Ship
- ARAB GULF Ship (1995)

1988 CENTS/U.S. GALLON

22 20 18 16 14 12 10 8 6

SHIPPING DISTANCE (STATUTE MILES X 1,000)

0 1 2 3 4 5 6 7 8 9 10 11 12 13 14 15

Table 5.2
Shipping and Terminal Costs for Methanol to Los Angeles (1988 Cents per U.S. Gallon)

Origin	Mode	1988	1995	Note
Low Cost				
British Columbia	Ship	6.3	6.0	26,000 Ton
Chile	Ship	--	6.0	30,000 Ton
Alaska	Ship	7.5	7.3	40,000 Ton
Medium Cost				
Trinidad	Ship	8.5	8.0	Via Panama
Algeria	Ship	9.3	8.5	Via Panama
Alberta	14" Pipeline	--	9.7	50,000 BPD
Alberta	21" Pipeline	--	9.0	Capacity 75% MeOH/ 25% LPG
High Cost				
Alberta	Pipeline + Ship	--	10.0	6" Pipeline to Vancouver
North Sea	Ship	10.2	9.4	Via Panama
Arab Gulf	Ship	11.4	10.5	Via Panama, Suez
Very High Cost				
Alberta	14" Pipeline	--	13.6	At 50% Capacity
Alberta	Rail + Ship	15.9	13.4	Unit Train 1995+
Alberta	Unit Train	16.9	16.3	
Alberta	Rail	22.0	19.0	

Table 5.3
Ocean Transport and Terminal Costs for Methanol (1988 Cents per U.S. Gallon)

Origin/Destination	Houston	New York	Chicago
British Columbia	21.0	30.4	14.0*
Chile	6.0	8.1	--
North Sea	9.3	7.8	15.5*
Algeria	7.8	7.2	--
Trinidad	7.2	7.2	--
Arab Gulf	11.4	10.8	--

*Via Lake Charles, Louisiana, and then barge to Chicago

Table 5.4
Methanol Transport and Terminal Costs by Rail and Pipe (1988 Cents per U.S. Gallon)

		Cost	
Origin and Destination	Transport Mode	1988	1995
Edmonton to Toronto	Rail + Terminal Costs	20.0	19.9
Edmonton to Chicago	Rail + Terminal Costs	20.1	19.0
Edmonton to Toronto	Batched in LPG Pipeline	--	10.0*
Edmonton to New York	Rail + Ship	40.7	35.0
Edmonton to Houston	Rail + Ship	34.8	30.0

*50 million gallons/year/7¢/gallon at 250,000/year

Table 5.5
Costs of Shipping and Terminaling Methanol in 1988, 8-Million Gallon Shipment from Alberta to Houston, Texas, via Panama Canal

Cost Component	¢/U.S. Gal	% Cost
Rail Tanker Movement to Coast (760 Miles, Alberta to Kitimat)	6.71	39.5
Tank Car Rental	1.41	8.3
Rail/Wharf Throughput and Period Payment	1.46	8.6
Ocean Freight	4.26	25.05
Panama Toll	1.30	7.6
Insurance	0.06	0.35
Receiving Terminal Throughput Cost	1.80	10.6
Subtotal: Cost at Terminal Gate	17.0	100.0
Terminal storage cost	5.0	--
Total Shipping and Terminalling Costs	22.0	--

However, smaller tankships provide an alternative during the transition period when less than 100,000 cars are using methanol in a given urban market. The methanol production facility in Trinidad uses two dedicated tankers of 14,000 tons.

These smaller ships imply somewhat higher delivery costs per gallon. Shipping costs are dominated by crew costs and fuel costs. Crew costs per day are almost independent of ship size, since crew sizes are about the same regardless of the size of the ship (provided the ships being compared are about the same age and degree of automation). As a result, crew costs per ton of cargo decline rapidly as ship size increases. Nevertheless, small ships are better suited to small markets at normal bulk storage prices, which usually will amount to about $0.03 to $0.05 per gallon per month, depending on the local storage situation.

Working out an optimum supply strategy for methanol requires a careful balancing of supplier costs, shipping costs, and storage costs. The strategy will change as the "market" of methanol-consuming cars grows, allowing larger ships and a wider choice of suppliers. However, to illustrate the range of shipping costs that can be achieved for a Los Angeles destination, we estimated total shipping costs including charter rates; crew costs; subsistence costs; ship stores, supplies, and equipment; maintenance and repair; insurance; cargo insurance; fuel for main engines; fuel for auxiliary engines; miscellaneous expenses; general administrative costs; and canal charges and port costs during loading and un-loading. We considered as examples ships of 6, 14, 30, and 60 thousand tons. To confirm our estimates, we obtained letter quotes from shipping brokers. Figure 5.3 shows a plot of estimated optimized shipping costs as a function of market size for various potential supply locations.

With increased methanol demand, the delivered cost of methanol to Los Angeles can be lowered from the very high price of $0.17 per gallon now paid for rail shipment. Shipping costs for a larger market of several hundred thousand cars range between $0.04 per gallon for the short Kitimat voyage and about $0.11 per gallon for the most remote potential suppliers. Clearly, shipping costs vary greatly depending on origin, market volume, and ship size (Table 5.6). Optimum strategies can drive shipping costs to low levels, except perhaps in the earliest years of a buildup to substantial methanol use.

The principal conclusions from the preceding analysis of transport and terminal costs are the following:

For Movements to California:

1. The lowest cost is available through ocean tanker movement of methanol manufactured at the Pacific coast locations in British Columbia, Chile, and Alaska, and then from expanded output from the plants now operating in Algeria and Trinidad.

2. Product shipped by ocean tanker from the North Sea and Persian Gulf locations incur high shipping costs.

3. Rail movement of fuel methanol from Alberta is costly, even when unit trains are employed.

Figure 5.3
Estimated Shipping Costs for Methanol

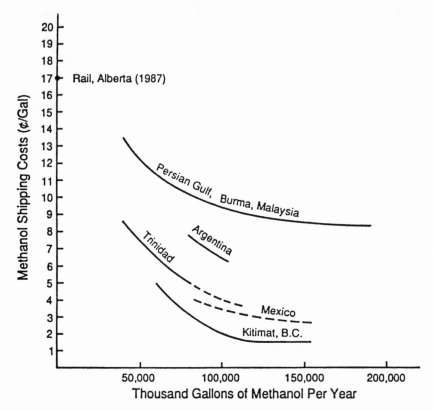

4. Combination of rail movement from Alberta to Canadian Pacific ports and ship to California ports are also not cost effective due to the effects of multiple handling and the high rail costs. Replacement of train movements by pipeline to the coast results in significant cost reductions that permit methanol from Alberta to compete with movements from the North Sea and the Persian Gulf.
5. A new 50,000 barrel per day (10,000 tons per day) methanol pipeline from Alberta to Southern California would provide transport costs similar to the pipeline-ship combination but only at high throughput levels. The 1,400-mile pipeline costs may be reduced by partial use of the line to ship LPG now exported from Canada by rail.

For Movements to Other Locations in North America:

1. In general, shipping costs to the U.S. East Coast are higher than the lowest cost movements to Southern California, but movements from Chile (for the southern seaboard), Trinidad, Algeria, and the North Sea (to New York) are reasonable.
2. Movements to the midwest region (Chicago and Toronto) by rail from Alberta are expensive. Two medium-cost solutions are available: Chicago product can be brought up the Mississippi by barge from Lake Charles which in turn receives product by large

Table 5.6
Methanol Shipping Cost Estimates for California

Origin	Cost ($/gal)	Comment
Canada (Edmonton)	0.17	By rail
Canada (Kitimat)	0.05	By dedicated 6,000 DWT tanker at
		20 million gpy
Canada (Kitimat)	0.04	80 million gpy
Mexico	0.04	80 million gpy
Argentina	0.07	80 million gpy
Brazil, Chile	0.06	80 million gpy
Trinidad	0.05	80 million gpy
Algeria	0.08	120 million gpy
Persian Gulf, Burma, China, Malaysia, and Taiwan	0.10	120 million gpy
India	0.11	120 million gpy

gpy = gallons per year

tanker from Trinidad, Algeria, and British Columbia. Reduced costs are possible when small barges of 1,200 tons capacity are replaced by towed multiples of 20,000 tons capacity. For the Toronto area, in the long term, two pipelines can be converted from LPG-ethane and crude oil-LPG to batch methanol. The former can also serve the Chicago market.

We offer two other observations on the bulk movement of methanol. LPG ocean tankers and the North American LPG pipeline system may provide lower costs than the current chemical-grade methanol system of small tankers and rail shipment. Second, back haul of water may be a means to reduce methanol transport costs from the Persian Gulf. An increasing proportion of the water requirements in this area is met through the desalination of sea water. An investigation is currently underway that estimates the cost of pipelining water from Turkey to be $0.042 per gallon. Back haul of water has already been practiced in bulk wine movements from Algeria to Montreal.

COST OF DISTRIBUTION AND RETAILING OF METHANOL FUEL

About 75 percent of the U.S. fuel market lies within a 100-mile radius of existing public access marine terminals.[10] The largest terminals are located on the Pacific and Atlantic Oceans and the Great Lakes; smaller terminals on the inland waterways are accessible by barge. These terminals can therefore be equipped with fuel methanol storage tanks, the splash-blending equipment required for methanol-gasoline blends such as M85 (15 percent gasoline), and facilities for movement of methanol-containing transportation fuels.

It appears that the majority of the product dispatched from these terminals or similar pipeline terminals will be M85, but bus and truck fleets using 100-percent methanol will also be served. The net effect is that certain volumes of gasoline will be replaced by an 85–15 blend of methanol (M85), and certain gasoline and diesel fuel will be replaced by methanol fuel (M100). The operating costs for expanded facilities at the ocean and pipeline terminals were included in the methanol costs analyzed earlier. The cost of local distribution of large volumes of methanol fuels to public and private refueling stations on a cost per unit volume basis are similar to the cost of distributing gasoline and distillates. In 1988 the average cost for truck transportation of gasoline and diesel fuel could be represented by the equation:

$$\text{Truck distribution cost (¢/gal)} = 0.7 + 0.019 \text{ (miles one way)}^{11} \qquad \textbf{(1)}$$

At an average one-way haul of 40 miles, the distribution cost will be \$0.0146 per gallon; at the maximum preferred range of 100 miles the cost will be \$0.026 per gallon. No economies of scale exist with local truck distribution despite the larger volumes of methanol-gasoline blends that will be hauled compared to the current gasoline volumes.

Our findings regarding the capital costs of upgrading existing equipment at service stations for methanol compatibility and safety and of installing new equipment (storage tanks and dispensers) are similar to those estimated elsewhere.[12] A new methanol tank and associated equipment would cost approximately \$45,000 installed. The *incremental* cost of methanol-compatible materials and components for a new fiberglass tank installation is only about \$2,000 and essentially no added cost for steel tanks; hence proposed rules requiring methanol compatibility for all new tanks would greatly reduce costs associated with a methanol transition.

Retail station profit margins must cover the value of the tank as well as direct operating costs. A number of the cost items are transaction related, unaffected by the increased volume associated with methanol fuel; others are volume related. Our detailed estimates indicate that the average operating cost at those public service stations serving methanol fuels will increase from an average \$0.07 per gallon for gasoline operations to \$0.084 per gallon for methanol service. Costs for private methanol stations serving only fleets will increase from \$0.042 per gallon to $0.046 per gallon. Thus, for example, the average cost of distributing and retailing methanol, including local distribution by truck and the retail margin, for a service station located 40 miles from a rail, ocean, or pipeline terminal is $0.0986 per gallon. In the case of large fleets at private stations the cost of distribution and refueling vehicles with methanol containing fuels is estimated at $0.0606 per gallon.

In the early stages of a transition to methanol, it is likely that M85 will first compete with superpremium and premium gasolines. For these gasolines, the turnover rate of the fuel at the retail stations is longer and hence requires higher than average gross profit margins (retail price minus delivered price minus taxes), typically $0.17 to $0.20 per gallon in California. Because methanol has half the energy density and therefore half the value per unit volume as gasoline, the gross margin on methanol would be about half that of gasoline margin to keep the same gross profit margin, if the station serves as many methanol buyers as it would have been able to serve gasoline buyers. This corresponds to a margin of about $0.09 per gallon of M85, which agrees with the previous detailed estimate for public stations.

We have also allowed a margin for wholesaling. Although it is not clear what commercial relationships for methanol distribution will be established, it is possible that some intermediary will have to take the price risk between the time of bulk purchase and the later retail sale. The California Energy Commission currently fulfills this role in California. In gasoline retailing, the wholesale jobber's margin is several cents per gallon. We will assume a wholesale M85 margin requirement of $0.03 per gallon of M85.

With these profit margins in mind, Table 5.7 shows some example price calculations for M85 delivered at the retail pump, expressed as cents per equivalent gallon of gasoline. The bulk (plant gate) purchase prices are assumed, but are in the range of published calculations taking into account feedstock costs, operating costs, and capital recovery. Shipping cost assumptions are consistent with the data presented earlier. The first column, for a fleet of 500 methanol-using cars, corresponds to the situation in California in 1986 and uses actual costs experienced in the program.

Table 5.7 shows that although M85 has a good chance of competing with superpremium and premium gasolines when the volume is high enough to achieve some economies of scale, during the transition when only a few thousand cars are using methanol it is impossible to control costs for methanol to be price-competitive with gasoline. Transition strategies, therefore, need to focus on establishing an initial flow of methanol to other users less sensitive to price or to establish another mechanism to support the higher initial prices during the early phases of a transition.

SUMMARY

This chapter addressed the costs of distributing CNG and methanol to a transportation fuels market. Natural gas currently has a well-established distribution

Table 5.7
Estimated Methanol Fuel Prices for Los Angeles (Premium Unleaded Rack Price = $ 0.61, Retail Price = $1.023)

	Stage			
	0	1	2	3
No. of Cars Using M85	500	20,000	50,000	200,000
Market Share (Percent)[a]	0	0.2	0.5	2
M85 Volume (1,000 Gal/Year)	500	20,000	50,000	200,000
Gate Price of Methanol (cents/gal)	27.0	35.0	35.0	30.0
Cost of Bulk Delivery (cents/gal)	17.0	17.0	6.0	10.0
Bulk Delivery Method	Rail	Rail	Ship	Ship
Bulk Storage (cents/gal)	5.0	5.0	3.0	2.0
Total (cents/gal)	49.0	57.0	44.0	42.0
Blending Cost for Low Volume (cents/gal)	0.3	0.3	0.0	0.0
Blended M85 Cost (cents/gal)	51.1	57.9	46.6	44.9
Truck Delivery (cents/gal)	2.6	2.6	2.6	2.6
Taxes (U.S. and State) (cents/gal)	9.0	9.0	9.0	9.0
Required Wholesale Margin (cents/gal)	3.0	3.0	3.0	3.0
Required Retail Margin (cents/gal)	9.0	9.0	9.0	9.0
Subtotal (cents/gal)	74.7	81.5	70.2	68.5
Sales Tax (6%) (cents/gal)	4.5	4.9	4.2	4.1
Total (cents/gal)	79.2	86.4	74.4	72.6
Equivalent Gasoline Price[b] (cents/gal)	139.4	152.0	130.9	127.8

Notes: [a]For Ozone-Impacted Areas

[b]Assumes 1.76 Conversion Factor Methanol to Gasoline

network but requires compression to higher pressures to be effectively stored on board vehicles. Methanol currently does not have a fuel infrastructure to support its use as a transportation fuel. Methanol fuel is moved today in very small volumes, which adds substantially to its price.

CNG can compete with gasoline and diesel on price in many regions of the United States, including California, if gas suppliers and utilities provide adequate incentives to fleet and individual vehicle users. This is necessary since compression costs are very capital intensive, about $700 per vehicle for a 350 to 400-vehicle refueling station. To stimulate the use of CNG, the Canadian government has also partially offset these high capital costs with $40,000 grants to station owners.

Since methanol is moved in such low volumes, as a transportation fuel it is currently not competitive with gasoline or diesel even at low production prices of $0.25 to $0.35 per gallon. Volumes of about 50 million gallons per year are needed before more cost-effective ocean ships can be used. This volume roughly corresponds to a market of 50,000 dedicated methanol vehicles.

In comparing CNG and methanol, one of the primary differences between these two fuels is the scale in which each can be implemented in a cost-effective manner. For CNG, only the compression costs have to be amortized over the number of vehicles using a refueling station. Although these costs are not insignificant, CNG can be economically viable in relatively small projects such as 400 cars or 40 or so transit buses. Methanol, on the other hand, must be implemented on a large scale right away in order to be cost competitive with premium or superpremium fuels. This means that during the transition from small to large volumes of methanol, detailed planning needs to occur in order to minimize overall costs.

NOTES

1. Canadian Gas Association, NGV market information, private correspondence.

2. R. W. Duncan, W. D. Jenkins, and R. F. Webb, "The Economics of Propane and Natural Gas Buses," paper presented at APTA meeting, Ottawa, Ontario, May 13, 1987.

3. R. F. Webb Corporation, *Program Evaluation Study of the Propane and Natural Gas Vehicle Grant Program (PVGP and NGVP),* prepared for Energy, Mines, and Resources Canada, Program Evaluation Group, Report #PE 101/1986, March 1987.

4. Duncan, Jenkins, and Webb, "The Economics of Propane and Natural Gas Buses."

5. R. F. Webb Corporation, *Program Evaluation Study of the Propane and Natural Gas Vehicle Grant Program.*

6. Ibid.

7. R. A. Lewis, Hoechst-Celanese Chemical Group, Dallas, Texas. Personal communication with M. D. Jackson, October 1986.

8. Ibid.

9. K. Koyama et al., "Methanol Supply and Demand Issues from a California Perspective," 1987 World Methanol Conference, sponsored by Crocco and Associates, Inc., and the California Methanol Task Force, December 1–3, 1987.

10. B. McNutt, J. Dowd, and J. Holmes, "The Cost of Making Methanol Available to a National Market," *SAE* paper 872063, November 1987.

11. J. Holmes, "Distribution Systems and Costs for a Regional Fuel Methanol Market," *SAE* paper 861573, October 1986.

12. Ibid.

6

MARK A. DeLUCHI

Hydrogen Vehicles

Hydrogen is a very attractive transportation fuel in two important ways: it is the least polluting fuel that can be used in an internal combustion engine, and it is potentially available anywhere there is water and a clean source of power. The prospect of a clean, widely available transportation fuel has motivated much of the research on hydrogen fuels.

Serious work on hydrogen vehicles began in the 1930s, when Rudolph Erren converted over 1,000 vehicles to hydrogen and hydrogen/gasoline operation in England and Germany.[1] However, interest in the fuel waned after World War II. The resurgence of research and experimental activity came in the late 1960s and early 1970s as hydrogen programs began in Europe, Canada, the Soviet Union, Japan and the United States.[2] This chapter addresses the use of hydrogen in highway vehicles.

HYDROGEN PRODUCTION

Elemental hydrogen occurs freely on earth in only negligible quantities. It is chemically very active, however, readily forming compounds with many other elements. Hydrogen is most abundant and accessible in water and the fossil fuels: coal, oil, and natural gas. Of the fossil fuels, coal is by far the most plentiful.

Fossil fuels consist of hydrocarbon molecules that can be reformed, cracked, oxidized, or gasified to produce hydrogen. Natural gas is reformed using water vapor to produce hydrogen and carbon monoxide, with an efficiency of about 72 to 86 percent,[3] or is "cracked" directly into carbon and hydrogen when heated in the presence of a suitable catalyst. Oil, oxygen, and water are combined in a partial oxidation reaction to produce carbon monoxide, carbon dioxide, methane, and hydrogen. Coal is gasified using high-temperature steam to produce carbon monoxide, carbon dioxide, and hydrogen. Coal gasification is about 60 percent efficient[4] and produces hydrogen for about $7 to $13 per mmBtu.[5]

Although hydrogen production today is quite small (the main use of hydrogen as a fuel in the United States is in rocket boosters), these production processes—particularly those using natural gas—are well established and could be expanded to accomodate the increased demand for hydrogen were it to become a major transportation fuel.

Of course, hydrogen derived from fossil fuels is not a clean renewable resource. Nevertheless, it may be easiest to sustain a transition to hydrogen by expanding commercially available and relatively inexpensive production processes, such as the manufacture of H_2 from coal. Coal is relatively abundant and could provide a low-cost feedstock for hydrogen for many decades, although there would be major adverse environmental impacts.

Most of the hydrogen research community agrees that eventually hydrogen should be produced from water.[6] There are several methods for splitting water to produce hydrogen: thermal and thermochemical conversion, photolysis, and electrolysis.

Photolysis, the splitting of water by light with the aid of (biological) photochemical electron transfer reagents analogous to chlorophyll, has been described as "the most elegant solution" to the hydrogen production problem.[7] Unfortunately, these systems typically are very inefficient, utilizing less than 1 percent of the incoming energy, and thus are very costly, although recent developments suggest that improvements are possible.[8] Water also can be decomposed thermally and thermochemically, but these methods are not considered as promising as electrolysis, and are not considered further in this chapter.[9]

In electrolysis, an electrolyte is added to water to make it more conductive; then an electric current is applied to the solution. The water molecule is split, and hydrogen gas is evolved at the negative electrode. In conventional electrolysis, the electrolyte is contained in liquid solution, and all the energy needed to split water is supplied by electricity. A variation of the conventional process is to first heat the water and then electrolyze the steam, the advantage being that lower-cost heat energy, such as waste heat from cogeneration, replaces some of the electrical energy. Another method, which researchers expect to be more efficient, is to use solid compounds as the electrolyte rather than solutions. If solar photovoltaic cells are used to generate the electricity, the cost of hydrogen production by electrolysis is $27 to $72 per mmBtu, depending primarily on assumptions regarding the cost of solar electricity.[10]

When hydrogen is produced from water electrolytically, the ultimate energy source is the feedstock used to generate the electricity. Fossil fuels would not be used because it would be cheaper and more efficient and would generate less carbon dioxide to gasify the coal, oxidize the oil, or reform the natural gas directly to produce hydrogen.[11] Nonfossil feedstocks, such as solar, geothermal, wind, hydro, and nuclear energy, would be used. Of these, solar energy and nuclear energy (from breeder reactors or possibly fusion plants) are available for the long term. Since breeder reactor development in the United States is dormant and the prospects for fusion are very uncertain, the cost analysis in this chapter examines solar hydrogen only.

HYDROGEN DISTRIBUTION

Typically, hydrogen would be transported via pipeline from the site of production to end users as a gas. Ideally, the current natural gas pipeline distribution system would be used for at least the initial stages of a transition to hydrogen.[12] The biggest question is whether current pipelines can be used safely. Pure hydrogen reacts at the surface of certain pipeline steels, embrittling them and accelerating the growth of fatigue cracks.[13] Hydrogen could not be shipped in pipelines made of susceptible steel unless the embrittling reaction were inhibited.

Researchers have recently learned that compounds can be added to hydrogen to inhibit embrittlement, and they are looking for safe, inexpensive, and effective inhibitors (none identified thus far is satisfactory on all three counts).[14] If an acceptable inhibitor is not found, dedicated hydrogen pipelines will have to be built. This would require a greater investment cost, which could delay a transition to hydrogen. However, it would not necessarily increase the unit life-cycle cost of hydrogen.

Hydrogen also could be shipped in liquid form, in 13,000-gallon tank trucks, 35,000-gallon rail cars, or, for short distances, in vacuum-jacketed pipelines.[15] The last option would be feasible only for shipment to large potential end users of liquid hydrogen, such as airports.

VEHICULAR FUEL STORAGE AND REFUELING

Hydrogen may be stored on board a vehicle as a gas bound with certain metals (hydrides), as a liquid in cryogenic containers, or as highly compressed gas (690 atm) in ultra-high-pressure vessels. Most systems are bulky and costly.

Metal Hydrides

Vehicular hydride storage systems usually consist of long thin hollow cylinders filled with granular metal alloy, a heat exchange system, and a casing. They are pressurized with hydrogen to about 34 atmospheres. Exhaust heat from the engine, carried by cooling water, is used to release the hydrogen from the metal lattice during vehicle operation.

The main criteria used to evaluate hydrides are cost, energy density, desorption or hydrogen-release temperature, rate of hydrogen absorption, susceptibility to fouling by impurities in the gas, and volumetric expansion. Researchers have not yet identified a hydride that is outstanding on all counts.

The most serious shortcoming of hydrides is their low mass and volumetric energy density (Table 6.1). Complete storage units, which include the housing for the hydride and the coolant system, are very large—25 to 80 gallons or more—and quite heavy—250 to 1,000 lbs.[16] The low mass energy density of most hydrides means that hydride vehicles are much heavier and less efficient

Table 6.1
Characteristics of Hydrogen Storage Systems

Vehicle	Range[a] miles	Weight lbs	Size gallons	Refill Time minutes
Gasoline	260	66	8	3
LH$_2$	260	88	50	4
Hydride[b]	150	585	35	5-20[c]

Notes: See ref. 6 for details

[a]Assumes a 35 mpg gasoline car, and 35 mpg plus efficiency advantage for
    hydrogen vehicles.

[b]For an iron-titanium hydride.

[c]80% of hydride capacity refilled in under 10 minutes.

than comparable gasoline or LH$_2$ vehicles, and are limited by storage weight to about a 100- to 200-mile range.

Hydride vehicle refueling stations either would be sited at hydrogen pipelines or would receive liquid or gaseous hydrogen by truck or rail and have bulk fuel storage on site.

Refueling hydride vehicles is relatively simple. It appears that hydride vehicles can be refueled in 10 minutes or less, although in most cases a good deal more time is needed to fill the last units of capacity.

Liquid Hydrogen (LH$_2$)

Hydrogen liquefies at about –425°F and 2 atmospheres of pressure. LH$_2$ is stored in double-walled, super-insulated vessels designed to minimize heat transfer and the boil off of LH$_2$. A significant advantage of using LH$_2$ is that LH$_2$ systems are much lighter and often more compact than hydride systems providing an equal range. Moreover, LH$_2$ storage is not significantly heavier than gasoline storage, on an equal-range basis, although it is still about six times bulkier (Table 6.1).[17] The main obstacles to widespread use of LH$_2$ vehicles are the bulkiness of storage systems, the high cost of liquid hydrogen and cryogenic technology, and the evaporation loss of LH$_2$.

A safely-run, easy-to-use, public LH$_2$ refueling station appears to be within the scope of current technology and practices. The latest German LH$_2$ vehicle, a BMW 745i, can be refilled in three minutes at a fully automatic filling station, at which the opening and closing of tank valves is automatic.[18] Researchers also have developed concepts for a mobile LH$_2$ storage and refueling system.[19]

Perhaps the most serious problem with LH$_2$ storage is boil-off. Cryogenic containers are not perfect insulators; gradually the liquid hydrogen warms and evaporates, increasing the pressure of the gas in the tank. If the vehicle is not driven for two to five days, the gas is vented to the atmosphere. Boil-off creates safety problems, economic loss, and inconvenience for the user.

The most important of these is safety. Gas vented into a confined space such as a garage might create a serious fire hazard. The most attractive solution is to react the venting gas with other compounds to create a harmless product. Hot wires, boil-off catalysts, and fuel cells have been used or proposed, but no solution has yet been widely accepted.[20] An important area of research is to determine the cost of reliable boil-off control for LH_2 vehicles. If boil-off control proves to be feasible and relatively inexpensive, then the market potential of LH_2 vehicles will not be limited by safety concerns related to boil-off.

Compressed Gaseous Hydrogen

An appealingly simple way to store hydrogen on board is as a compressed gas in high-pressure aluminum vessels wrapped with carbon.[21] However, to achieve a satisfactory energy density, the storage pressure must approach 690 atm (10,000 psi). At this pressure, the storage system would be about nine times larger than the gasoline tank providing the same range, but if carbon were used, only about three times heavier.[22] However, carbon fiber is very expensive, and refueling at 690 atm would likely be very expensive. The cost and bulk of compressed hydrogen storage must be reduced if the method is to be practical.

Hydrogen can also be stored in glass microcapsules at high pressure,[23] absorbed on cold activated carbon,[24] and stored as a liquid hydride.[25] Of these, the liquid hydride method is most developed. In this system, toluene, a liquid at ambient temperature and pressure, is analagous to the metal in a metal hydride. It reacts with hydrogen gas upon refilling to form methylcyclohexane, and releases hydrogen fuel when exhaust heat is added in the presence of a catalyst. Thus far, however, the cost, bulk, and weight of the total system appear to limit its use to heavy-duty vehicles.

PERFORMANCE OF HYDROGEN VEHICLES

To date, hydrogen vehicles have used modified stock gasoline engines. It is important to recognize that test-vehicle engines have not been designed and optimized for hydrogen combustion and that the present internal combustion engine may not be best for hydrogen. The results achieved thus far are probably only suggestive of what can be achieved with a dedicated hydrogen engine. This section briefly summarizes the available data on the efficiency and power of hydrogen vehicles relative to gasoline vehicles.[26]

Hydrogen fuel is very expensive, and thus it is especially important to maximize the efficiency of hydrogen vehicles. Fortunately, hydrogen can be considerably more efficient than gasoline, primarily because it burns better in excess air and permits the use of a higher compression ratio. Data from engine tests indicate that hydrogen combustion can be 15 percent to 50 percent more thermally efficient than gasoline. The actual value depends on the air/fuel ratio, how well the engine is optimized for hydrogen combustion, whether gaseous or liquid hydrogen is used, the fuel-injection scheme, and other factors.

The power of hydrogen vehicles relative to gasoline vehicles depends on the air/fuel ratio, the temperature of the charge, and the compression ratio. Lean operation increases efficiency, as noted above, but reduces power. And, at near-ambient temperatures, gaseous hydrogen displaces much more air in the combustion chamber than does atomized gasoline, reducing the density of the fuel charge and significantly decreasing power. The power loss from lean operation can be avoided, without losing most of the efficiency advantage of lean operation, by enriching the fuel charge only at maximum load. The loss due to operation at ambient temperature can be reduced or eliminated if the charge is condensed by using ultra-cold cryogenic fuel.[27] Several vehicles using ultra-cold (cryogenic) LH_2 have even shown increases in power with respect to gasoline operation.[28]

The use of LH_2 offers the best performance. Direct injection of cyrogenic hydrogen late in the compression stroke increases power output and efficiency, eliminates backfiring, and reduces NO_x formation. For these reasons many hydrogen researchers feel that it is the most desirable form of hydrogen use. However, the technical demands of this method are very exacting.[29] More importantly, though, direct, late injection of cryogenic pressurized fuel requires on-board storage of LH_2. If safe disposal of boil-off gases is not possible, widespread acceptance of LH_2 vehicles, and thus the use of direct LH_2 injection, is unlikely.

Hydrogen and gasoline can be burned together in an internal combustion engine in a wide range of mixtures and generally with good results. Hydrogen also can be used in diesel engines, at least in small quantities, and also with good results. Generally, dual-fuel vehicles and engines are more efficient and have lower emissions than stock gasoline vehicles and engines, but are less efficient and have higher emissions than dedicated hydrogen vehicles.

SAFETY ISSUES

Hydrogen has a largely undeserved reputation as a particularly dangerous fuel. The limited experience with hydrogen and analyses of its physical and chemical properties indicate that hydrogen is not necessarily more dangerous than gasoline and may be safer.

Hydrogen is more hazardous than gasoline in certain ways, however. First, it is invisible and odorless, and therefore requires odorants and colorants to enable detection. Second, hydrogen flames are very hot, yet radiate very little heat and are invisible, which makes them harder to locate and thus harder to extinguish or to avoid. Third, hydrogen can ignite within a rather large range of hydrogen/air densities, from 4 percent to 74 percent (by volume). Compared to methane or gasoline, it needs very little energy to ignite.

On the other hand, hydrogen has several safety advantages relative to gasoline. First, in many situations hydrogen is less likely to explode. Second, hydrogen fires burn very rapidly and radiate very little heat, and thus are relatively short lived. Gasoline fires are much more persistent than LH_2 fires involving an equivalent amount of energy. Third, hydrogen storage systems are relatively safe. In test crashes of LH_2 vehicles, the storage systems have remained intact.[30] If LH_2 does leak in a crash, it will evaporate and disperse exceedingly fast, unlike gasoline, which will puddle and remain a fire hazard for much longer. Similarly, research on hydrides and experience with hydride vehicles have demonstrated that hydride storage is at least as safe as gasoline storage.[31] Hydride systems are fairly strong, and fuel leaks are self-limiting because heat is needed to release the hydride. As with liquid hydrogen, any released gaseous hydrogen will disperse very quickly.

For the reasons cited above, the U.S. National Bureau of Standards,[32] the Stanford Research Institute,[33] and the German "Alternative Fuels for Road Transport" program,[34] have concluded that hydrogen hazards are different from, but not necessarily greater than, those presented by current petroleum fuels.

ENVIRONMENTAL IMPACTS OF HYDROGEN VEHICLES

The great attraction of hydrogen is pollution-free combustion. While many undesirable compounds are emitted from gasoline and diesel fuel vehicles or formed from their emissions, the main combustion product of hydrogen and air is water. Hydrogen vehicles would not produce significant amounts of carbon monoxide (CO), hydrocarbons (HCs), particulates, sulfur oxides (SO_x), sulfuric-acid deposition, ozone and other oxidants, benzene and other carcinogenic aromatic compounds, formaldehyde and other aldehydes, lead and other toxic metals, smoke, or carbon dioxide (CO_2) and other greenhouse gases. The only pollutant of concern would be nitrogen oxides (NO_x). If hydrogen is made from water using a clean power source, then hydrogen production and distribution will be pollution-free. Hydrogen is a truly clean fuel.

Environmental Impacts of Hydrogen Production and Distribution

The environmental impacts of hydrogen production and distribution depend in large part on the feedstock. The production of hydrogen electrolytically, using clean solar power or other forms of renewable energy, and the distribution of hydrogen by pipeline, are essentially pollution free. Solar hydrogen plants would be very land intensive, but many, and perhaps most, of the major metropolitan areas of Europe, North America, Africa, China, the Middle East, India, the Soviet Union, and Australia could be supplied with solar hydrogen via pipeline from desert areas, where solar hydrogen plants would have minimum land cost

Table 6.2
Emissions of Hydrogen Vehicles, Percent Change Relative to Gasoline Vehicles

	RHC	CO	NOx	O_3
H_2	-95	-99	-60^a	-95^b

Notes:

Rough estimates only. Table compares advanced, dedicated H_2 vehicles without catalytic converters to gasoline vehicles with 3-way catalysts.

[a]Using very lean operation with H_2, and EGR or LH_2 for additional cooling, but no catalyst.

[b]Reduction in O_3 assumed to be same as reduction in RHCs.

and ecological impact. No CO_2 would be produced during hydrogen production or distribution if solar or other renewable energy were used as the energy source.

On the other hand, the use of coal to produce hydrogen would not be a desirable long-term option from an environmental standpoint for several reasons. Subsurface coal mining causes ground subsidence, endangering people and property, and produces environmentally harmful wastes. Surface mining is damaging to nearby property values, disrupts land uses, and pollutes surface and ground water with acid run-off and other effluents. Dangerous concentrations of carcinogenic polycyclic aromatic hydrocarbons have been measured in the aqueous waste streams of coal gasification facilities, and potentially harmful levels of toxic trace elements may occur in solid and aqueous waste streams.[35] If emissions from coal-to-hydrogen plants were to be similar to emissions from current coal-to-synthetic natural gas plants (the production processes would be similar), then coal-to-hydrogen plants would emit large amounts of regulated pollutants, such as SO_x and NO_x, and may release harmful levels of unregulated pollutants.[36] Finally, and perhaps most importantly in the long run, the use of coal inevitably releases large amounts of CO_2 and other trace "greenhouse gases." This is discussed further below.

Vehicular Emissions

Data from hydrogen engine and vehicle emissions tests and knowledge of emission-formation processes support several conclusions about emissions from hydrogen vehicles (summarized in Table 6.2).[37] First, if oil combustion is kept within normal limits, emissions of HC and CO from hydrogen vehicles are roughly an order of magnitude lower than HC and CO from properly maintained, relatively new, catalyst-equipped gasoline vehicles, and in fact are near zero.

However, if the hydrogen engine burns excess oil, HC and CO emissions become significant. In this case, though, the proper comparison is with an older gasoline vehicle whose emission control equipment has deteriorated, and the hydrogen vehicle still emits about an order of magnitude less CO and HC.

Second, emissions from hydrogen vehicles would not contribute to the formation of ozone and other oxidants that require hydrocarbons as a precursor.

Third, because hydrogen contains no sulfur, carbon, or fuel additives, hydrogen vehicles would emit essentially zero unregulated pollutants, such as benzene, a carcinogen, or formaldehyde, a toxin.

The only pollutant of concern, therefore, is NO_x, which is formed by the reaction of atmospheric nitrogen and oxygen in any internal combustion engine. NO_x emissions can be reduced by reducing the temperature of combustion. In hydrogen vehicles this is accomplished by running the engine very lean or cooling the combustion environment by adding water or exhaust gases, or using cryogenic fuel.

It appears that an optimized hydrogen vehicle could meet the current U.S. NO_x standard and probably have lower lifetime average NO_x emissions than a current-model catalyst-equipped gasoline vehicle without any catalytic control equipment on the engine. Lean operation, combined with some form of combustion cooling, such as EGR, water injection, or the use of very cold fuel, would be required. (One of these might also be used to prevent preignition.) Part of the reason for potentially lower lifetime NO_x emissions with hydrogen use is that the NO_x emissions deterioration rate probably would be much lower than for gasoline vehicles, since a significant portion of the increase in NO_x emissions over the life of current gasoline vehicles is due ultimately to deterioration of the reduction catalyst.

Dual-fuel operation with hydrogen and gasoline or diesel fuel—as little as 5 to 10 percent hydrogen by mass—can substantially reduce emissions of CO, HC, and NO_x.

Hydrogen Vehicles and the Greenhouse Effect

Carbon dioxide (CO_2), methane (CH_4), nitrous oxide (N_2O), ozone (O_3), chlorofluorocarbons (CFCs), and other trace gases absorb and reradiate outgoing infrared energy from the earth. Over the past 100 years, anthropogenic emissions of these gases have measurably increased their atmospheric concentrations. A continuation of this trend is expected to result in a substantial increase in the mean global temperature. This could have serious consequences for agricultural areas and coastal cities, as precipitation patterns shift, oceans expand thermally, and, eventually, polar ice sheets melt. It is widely accepted that there is no economically and environmentally sound method of scrubbing CO_2 emission from the stack and disposing of it, countering climate change, or sequestering carbon in forests.[38] Thus, it is important that emissions of these greenhouse gases be reduced.

The use of hydrogen made from nonfossil electricity and water is one of the most effective ways to reduce anthropogenic emissions of greenhouse gases. Highway vehicles burning hydrogen would emit essentially no CO_2 or CH_4, and because they would emit no reactive hydrocarbons—a necessary precursor to ozone formation in the troposphere—would not contribute to the increase in ozone levels. It has been shown that vehicles without catalytic converters emit very little N_2O, and thus advanced hydrogen vehicles, which would not require a NO_x catalyst, would emit little N_2O.

Table 6.3 shows emissions of greenhouse gases from a wide variety of alternative transportation fuels. It is assumed in Table 6.3 that each fuel supplies the entire U.S. highway fleet. All emissions from feedstock extraction to end use are considered. As shown, the only effective long-term options for eliminating emissions of greenhouse gases from the transportation sector are hydrogen from nonfossil power or a combination of electric vehicles and ICE vehicles using biofuels.

If, however, coal is used to make hydrogen, greenhouse gas emissions would increase 100 percent (hydride vehicles) and 143 percent (LH_2 vehicles) relative to current emissions of greenhouse gases from petroleum use. It thus turns out that hydrogen is either the best or the worst fuel from a greenhouse perspective, depending on the feedstock.

LIFE-CYCLE COST

Hydrogen's environmental advantages generally are thought to be outweighed by the very high cost of hydrogen fuel and hydrogen vehicles. But few, if any, detailed life-cycle cost comparisons between gasoline and hydrogen vehicles have been done. This section presents the results of a life-cycle cost analysis of hydride and LH_2 vehicles.

Costs are calculated with respect to a gasoline base case for which all relevant, baseline, life-cycle cost items have been estimated. The summary cost statistic used in this analysis is the break-even gasoline price. This price statistic is used because it allows the incorporation of not only fuel production costs but also differences in vehicle attributes such as engine life and vehicle maintenance costs. The break-even price of gasoline is that retail price of gasoline, including taxes of $0.20 per gallon, at which the total cost-per-mile of the gasoline vehicle is equal to the total cost-per-mile of the hydrogen vehicle. Costs are estimated for the near term, through the year 2020, in which hydrogen is assumed to be made from coal, and for the middle term (the years beyond), in which hydrogen is assumed to be made by water electrolysis using solar power. Both these scenarios assume mass production and advanced technology but no major technical breakthroughs. Uncertainty in cost projections is handled by using high and low cost estimates; the lower bounds represent very optimistic assessments of what is attainable in the future, and the higher bounds more conservative assessments.

Table 6.3
Emissions of Greenhouse Gases from Alternative Vehicular Fuels, Including Production, Distribution, and End-Use of Fuel

FUEL AND FEEDSTOCK	TOTAL CO_2-equivalent emissions, GT/yr	% CHANGE per km, rel. to petroleum
Hydrogen from nonfossil electricity	0	-100
EVs from nonfossil electricity	0	-100
CNG/LNG/methanol from biomass	0	-100
Doubled U.S. fleet efficiency	0.668	-50
CNG from natural gas	1.081	-19
Methanol from natural gas	1.293[a]	-3
EVs from current power mix (U.S.)		-1[a]
Gasoline and diesel from crude oil	1.336	baseline
Methanol from coal	2.639	+98
Hydride from coal	2.677	+100
LH_2 from coal	3.240	+143

Notes: From ref. 44.

[a] Aggregate CO_2-equivalent emissions would vary depending on the level of penetration assumed possible for EVs and the operating characteristics of the vehicle.

All cost figures are in 1985 dollars. Full details of this analysis and a brief discussion of costs for CH_2 vehicles are given elsewhere.[39]

Several general conclusions can be drawn from the cost-per-mile results shown in Table 6.4 and the break-even gasoline prices shown in Table 6.5. First, there is not much difference between the total life-cycle cost of hydride and LH_2 vehicles. Second, for both types of hydrogen vehicles there are big differences between the high and low break-even prices, which reflect the great uncertainty in estimating some of the key cost parameters such as the cost of fuel production. Third, in all the high-cost cases, both types of hydrogen vehicles are prohibitively expensive. In the middle term, for example, the break-even gasoline price for solar-hydrogen vehicles soars to over $10 per gallon. If coal is the feedstock, the break-even price is around $5 per gallon of gasoline, in the high-cost case. These costs would appear to make hydrogen vehicles uneconomical in any plausible oil-price scenario. The main reason for this, as shown in Table 6.4, is the extremely high cost of hydrogen fuel, under pessimistic assumptions.

In the low-cost case in the middle term (which assumes, for example, $0.06 per kwh for solar electricity on site, very efficient vehicles, and low-cost storage systems), the break-even price is on the order of $3.50 per gallon. Even this cost is higher than gasoline prices are likely to be. Thus it appears that hydrogen vehicles will be cost competitive in the middle term only if the most optimistic cost projections are realized and the price of gasoline at least triples. In the near term, with coal as a hydrogen feedstock, the break-even price is about $1.50 per gallon. In 1987 most gasoline in the United States sold for under $1 per gallon.

The large range of values in Tables 6.4 and 6.5 shows that there is considerable uncertainty in the total life-cycle cost of hydrogen vehicles. This is attributable primarily to uncertainty in production costs. Uncertainty regarding the relative fuel efficiency of hydrogen vehicles, and the cost and salvage value of hydrides and LH_2 tanks also contributes significantly to uncertainty in the vehicle-related costs. Better data on these factors will enable more precise estimates of the overall life-cycle cost of hydrogen vehicles.

The life-cycle cost of the hydride vehicle is quite sensitive to the assumed range of the vehicle because the cost of the storage material (the metal) increases linearly with required range. If the range is assumed to be 260 miles rather than 130 (as in the base case), the near-term, lower-bound break-even price for the hydride vehicle increases by $0.70 per gallon. The break-even gasoline prices are not as sensitive to changes in other base-case parameters. For example, if the reference gasoline vehicle is assumed to cost $15,000 rather than $9,500 (the base case), and the interest rate is assumed to be 5 percent rather than 9 percent (the base case), the lower-bound, break-even gasoline prices decline by about 10 percent for both LH_2 and hydride vehicles.

SUMMARY AND CONCLUSION

The attractiveness of hydrogen vehicles hinges on progress in three areas besides cost. First, hydrides with high mass energy density, low dissociation

Table 6.4
Comparison of Important Components of the Total Life-Cycle Cost, Gasoline and Hydrogen Vehicles, 1985 U.S. Cents per Mile

Cost Component	Hydride Vehicle	LH$_2$ Vehicle	Gasoline
Fuel cost, near term[a]	3.44 - 7.37	4.75 - 11.04	2.70
Fuel cost, middle term[b]	9.44 - 28.79	10.36 - 34.56	3.52
Vehicle initial cost[c]	12.30 - 15.46	11.54 - 14.49	12.07

Notes: See ref. 6 for details.

[a]Assumes hydrogen made from coal,, and gasoline at $1.15/gallon incl. taxes. Gasoline vehicle efficiency assumed to be 35 mpg.

[b]Assumes hydrogen made by solar electrolysis, and gasoline at $1.50/gallon incl. taxes. Gasoline vehicle efficiency assumed to be 35 mpg.

[c]Gasoline vehicle assumed to cost $9,500; hydrogen vehicles $9,500 plus extra cost of hydrogen storage. 9% real interest rate.

Table 6.5
Gasoline Prices at Which Hydrogen and Gasoline Vehicle Life-Cycle Costs Are Equal, 1985 U.S. $/Gallon, with U.S. Taxes

Vehicle	Near Term		Middle Term	
	Low	High	Low	High
LH_2	$1.60 -	$5.34	$3.48 -	$13.55
Hydride	$1.47 -	$4.54	$3.48 -	$23.03

See note 5 for details.

temperature, rapid adsorption, and relatively low susceptibility to degradation by impurities in the gas must be found in order to increase hydride vehicle range and performance. At present, the probability of hydride vehicles achieving performance and range parity with gasoline vehicles seems low. Second, the loss of trunk space to bulky hydrogen storage systems needs to be minimimzed. Hydride, LH_2, and CH_2 tanks are many times larger than gasoline tanks of equal range. Barring virtually unimaginable advances in technology (such as the development of thin, single-walled, effective cryogenic containers, or safe, lightweight, high-pressure vessels for hydrogen compressed to 1,000 atmospheres), this disparity is not likely to change. Therefore, to minimize the loss of trunk space, all vehicle components must be cleverly repackaged, or the vehicle made larger. Third, reliable, low-cost boil-off control devices should be developed for LH_2 vehicles. These will make the difference between LH_2 being potentially acceptable to virtually everyone and being potentially acceptable only to a minority of drivers.

The possibility of small-scale liquefaction of hydrogen at the vehicle refueling station should be investigated. Small liquefiers are available for natural gas. The considerable advantages of on-site liquefaction would be the elimination of two LH_2 transfers—from LH_2 plant to delivery truck and truck to bulk-storage at the station—and the associated fuel loss, and the elimination of the accident costs associated with truck distribution.

Safety should not be a problem. It is now reasonably well established that the hazards of hydrogen use in transportation would be different from, but not necessarily worse than, the hazards of using petroleum fuels. Refueling either hydride, LH_2, or CH_2 vehicles would be sufficiently safe and convenient. Perhaps the most important research issue is to determine under what conditions hydrogen can be shipped in the existing natural gas network. If safe metal embrittlement-inhibiting compounds are found, it probably will be safe to transport hydrogen in most existing pipelines.

It is important that these conclusions about the safety of hydrogen, widely accepted by the alternative fuels technical community, be recognized in the broader policy-making arenas, so that hydrogen R&D is not slighted on account of ungrounded fears.

The most attractive feature of hydrogen is very low emissions of pollutants, including greenhouse gases. Hydrogen vehicles will have at least order-of-magnitude lower emissions of HC and CO than catalyst-equipped gasoline vehicles of the same age and condition. The only significant pollutant would be NO_x, but even in that case hydrogen vehicles should easily meet the current U.S. NO_x standard and emit less than gasoline vehicles, given lean operation and some form of combustion chamber cooling. Hydrogen use would cause serious environmental damage only if coal were used as the feedstock. Even if "clean coal" technologies are commercialized, the use of coal still entails damages from coal mining, and a large increase in emissions of greenhouse gases.

The most fundamental barrier to hydrogen use is cost. Optimistic projections of the cost of hydrogen vehicles and hydrogen fuel must be realized if hydrogen vehicles are to even begin to be cost competitive with gasoline vehicles on a private life-cycle cost basis. This requires that hydrogen vehicles have only a slightly to moderately higher initial cost than gasoline vehicles; that the storage equipment have a high salvage value, and the vehicle a longer life than a gasoline vehicle; that hydrogen vehicles be considerably more fuel efficient than gasoline vehicles; and, most importantly, that solar energy or coal-based hydrogen be as inexpensive as in the most optimistic projections. However, achieving low-cost hydrogen vehicles and hydrogen fuel still may not make hydrogen vehicles competitive with gasoline vehicles on a private-cost basis. Cost competitiveness will depend on consideration of full social cost. For hydrogen to be "economical," it must turn out that the external costs of gasoline use are relatively high—that relatively high value is placed on reducing air pollution, avoiding a greenhouse warming, and reducing dependence on an insecure and diminishing resource.

In conclusion, hydrogen vehicles are not strictly an exotic, distant future possibility. Although all hydrogen vehicles have serious shortcomings, none of the problems is necessarily insuperable. With key advances, any of the three types of hydrogen vehicles considered here, or others, could become attractive. With a strong R&D effort, normal technological progress, and the hoped-for reduction in the price of solar electricity, hydrogen vehicles could be cost-competitive on a social cost basis within a relatively short period of time—perhaps 30 years or less. This prospect of developing a socially cost-effective, non-CO_2 producing, long-run alternative to gasoline, while still very uncertain, is attractive enough to warrant a much stronger R&D commitment to the improvement of hydrogen production and hydrogen vehicle technology.

NOTES

1. P. Hoffmann, *The Forever Fuel, the Story of Hydrogen* (Boulder, Colorado: Westview Press, 1981).

2. W. F. Stewart, "Hydrogen As a Vehicular Fuel," *Recent Developments in Hydrogen Technology, Vol. II,* ed. K. D. Williamson, Jr., and Frederick J. Edeskuty (Boca Raton, Florida: CRC Press, Inc., 1986), 69–145.

3. D. van Velzen, "Cost Comparisons of Different Methods of Producing Hydrogen," *Hydrogen: Energy Vector of the Future* (London: Graham and Trotman, 1983), 106–26.

4. Ibid; K. H. van Heek, "Coal Gasification for the Production of Hydrogen," *Hydrogen: Energy Vector of the Future* (London: Graham and Trotman, 1983), 27–54.

5. M. DeLuchi, "Hydrogen Vehicles: An Evaluation of Fuel Storage, Performance, Safety, Environmental Impacts, and Cost," *Int. J. Hydrogen Energy,* 14, no. 2 (1989): 81–130.

6. J. O'M. Bockris et al. "On the Splitting of Water," *Int. J. Hydrogen Energy* 10, no. 3 (1985): 179–201.

7. J. Gretz, "Conversion of Solar Energy into Hydrogen and Other Fuels," *Hydrogen: Energy Vector of the Future* (London: Graham and Trotman, 1983), 89.

8. J. Miyake and S. Kawamura, "Efficiency of Light Energy Conversion to Hydrogen by the Photosynthetic Bacterium Rhodobacter Sphaeroides," *Int. J. Hydrogen Energy,* 12, no. 3 (1987): 147–49.

9. J. W. Warner and R. S. Berry, "Hydrogen Separation and the Direct High-Temperature Splitting of Water," *Int. J. Hydrogen Energy,* 11, no. 2 (1986): 91–100; T. Ohta et al., "Hydrogen Production from Water: Summary of Recent Research and Development Presented at the 5th World Hydrogen Energy Conference," *Int. J. Hydrogen Energy,* 10, no. 9 (1985): 571–76.

10. DeLuchi, "Hydrogen Vehicles." If recent cost projections for amorphous silicon photovoltaic technology prove correct, hydrogen could be produced for as little as $15 per mmBtu.

11. van Velzen, "Cost Comparisons."

12. G. F. Steinmetz, "Review Presentation for Transmission, Distribution, and Bulk Storage of Hydrogen Relative to Natural Gas Supplementation," *Hydrogen Energy Progress V,* Proc. 5th WHEC, ed. by T. N. Veziroglu and J. B. Taylor, (New York: Pergamon Press, 1984), 1187–1200.

13. T. N. Veziroglu, ed., *Metal-Hydrogen Systems,* Proc. Miami International Symposium on Metal Hydrogen Systems (Oxford, England: Pergamon Press, 1981).

14. Battelle, Columbus Laboratories, *Hydrogen Degradation of Pipeline Steels,* Executive Summary, contract BNL 550722S, for Brookhaven National Laboratory, Upton, New York, January (1986).

15. W. E. Timmicke, "Storage and Distribution of Cryogenic Fuels," *Utilization of Alternative Fuels for Transportation,* ed. by Martin Newman and Jerry Grey (New York: American Institute of Aeronautics and Astronautics, 1979), 59–69.

16. R. Povel et al., "Hydrogen Drive in Field Testing," *Hydrogen Energy Progress V,* Proc. 5th WHEC, ed. by T. N. Veziroglu and J. B. Taylor (New York: Pergamon Press, 1984), 1563–77; G. C. Carter and F. L. Carter, "Metal Hydrides for Hydrogen: A Review of Theoretical and Experimental Research, and Critically Compiled Data," *Metal-Hydrogen Systems,* Proc. Miami International Symposium on Metal Hydrogen Systems, ed. by T. N. Veziroglu (Oxford, England: Pergamon Press, 1981), 503–29; DeLuchi, "Hydrogen Vehicles."

17. Ibid.

18. W. Strobl and W. Peschka, "Liquid Hydrogen as a Fuel of the Future for Individual Transport," *Hydrogen Energy Progress VI,* Proc. 6th WHEC, ed. by T. N. Veziroglu, N. Getoff, and P. Weinzierl (New York: Pergamon Press, 1986), 1161–73.

19. W. F. Stewart, "Refueling Considerations for Liquid-Hydrogen Fueled Vehicles," LA–UR–84–1490, Los Alamos National Laboratory, New Mexico (1984).

20. W. Peschka, "The Status of Handling and Storage Techniques for Liquid Hydrogen in Motor Vehicles," *Int. J. Hydrogen Energy,* 12, no. 11 (1987): 753–64; J. J. Donnelly et al., *Study of Hydrogen-Powered versus Battery-Powered Automobiles,* ATR–79(7759)–1, vols. 2, 3, prepared for the U.S. Department of Energy, May (1979).

21. E. E. Morris, M. Segimoto, and V. Lynn, "Lighter Weight Fiber/Metal Pressure Vessels Using Carbon Overwrap," AIAA–86–1504, AIAA/ASME/SAE/ASEE 22nd Joint Propulsion Conference, June 16–18, Huntsville, Alabama (1986).

22. DeLuchi, "Hydrogen Vehicles."

23. K. L. Yan et al., "Storage of Hydrogen by High Pressure Microencapsulation in Glass," *Int. J. Hydrogen Energy,* 10, no. 7/8 (1985): 517–22.

24. J. S. Noh et al., "Hydrogen Storage Systems Using Activated Carbon," *Int. J. Hydrogen Energy,* 12, no. 10 (1987): 693–700.

25. M. Taube et al., "A System of Hydrogen-Powered Vehicles With Liquid Organic Hydrides," *Int. J. Hydrogen Energy,* 8, no. 3 (1983): 213–25.

26. A detailed review is provided in DeLuchi, "Hydrogen Vehicles."

27. Strobl and Peschka, "Liquid Hydrogen."

28. Stewart, "Refueling Considerations."

29. W. Peschka, "Hydrogen Combustion in Tomorrow's Energy Technology," *Hydrogen Energy Progress VI,* Proc. 6th WHEC, edited by T. N. Veziroglu, N. Getoff, and P. Weinzierl (New York: Pergamon Press, 1986), 1019–36.

30. W. Peschka, "Liquid Hydrogen Fueled Automotive Vehicles in Germany—Status and Development," *Int. J. Hydrogen Energy,* 11, no. 11 (1986): 721–28.

31. H. Quadflieg, "From Research to Market Application? Experience with the German Hydrogen Fuel Project," *Hydrogen Energy Progress VI,* Proc. 6th WHEC, ed. by T. N. Veziroglu, N. Getoff, and P. Weinzierl, (New York: Pergamon Press, 1986), 65–78; G. Strickland, "Hydrogen Storage Technology for Metal Hydrides," *Hydrogen for Energy Distribution, Symposium Papers* (Chicago: Institute of Gas Technology, 1978): 509–39.

32. S. Hord, "How Safe is Hydrogen?" *Hydrogen for Energy Distribution, Symposium Papers* (Chicago: Institute of Gas Technology, 1978): 613–43.

33. Hoffman, *The Forever Fuel.*

34. Quadflieg, "From Research to Market Application."

35. M. J. Chadwick et al., eds. *Environmental Impacts of Coal Mining and Utilization* (New York: Pergamon Press, 1987).

36. For a discussion of new "clean coal" technologies, see D. F. Spencer et al., "Cool Water: Demonstration of a Clean and Efficient New Coal Technology," *Science,* 232 (May 2, 1986): 609–12. These new technologies do not reduce emissions of greenhouse gases significantly.

37. Reviewed in DeLuchi, "Hydrogen Vehicles."

38. M. A. DeLuchi, R. A. Johnston, and D. Sperling, "Transportation Fuels and the Greenhouse Effect," Universitywide Energy Research Group, UER–182, December, University of California at Berkeley (1987). Forthcoming in abridged form in *Transportation Research Record* (1989), no. 1175.

39. DeLuchi, "Hydrogen Vehicles."

7

DICK RUSSELL

Alternative Fuels as a Solution to the Air Quality Problem: An Environmentalist's Perspective

A logical place to begin an examination of urban air quality problems is in the Los Angeles basin, where one out of every twenty Americans resides. Population growth has resulted in a 50 percent driving increase over the past decade. Here, according to congressional testimony by James Lents, executive director of the area's South Coast Air Quality Management District (AQMD), "continuation of present air quality levels will expose over nine million people to air pollution 200 to 300 percent [above] the health standards each year." Lents adds, "If you took the air pollution problems in Houston, Denver, Albuquerque and New York and added them all together, it wouldn't equal the problem in L.A." Ozone levels here, some of the highest on the planet, are the biggest dilemma. Formed in sunlight, ozone is a blessing in the upper atmosphere because it absorbs harmful ultraviolet rays, but it is a curse at ground level where it results from a reaction between nitrogen oxides (a product of combustion) and volatile organic compounds.

The L.A. basin has been described as a veritable amphitheater for ozone— 12.5 million people with 8 million cars, 10 million mobile sources of air pollution, and thousands more stationary sources including a dozen oil refineries. On the smoggiest days, ozone readings are three times the allowable federal level. Despite a substantial reduction in evaporative and tailpipe emissions from automobiles through more stringent state controls, the increase in cars and driving time still finds the basin also well in excess of federal carbon monoxide regulations. For two out of every three days the region violates the EPA's air quality standards.

A recent state study indicated that smog contributes to over $330 million a year in crop damage in California, wiping out 20 percent of Riverside County's oranges and grapes. The health effects, besides the expected respiratory ones, include reproductive problems and state estimates that chemically contaminated air could cause cancer in nearly 30,000 of California's 27 million-plus current residents, with the toll rising to almost 44,000 when the population reaches 40 million early in the next century.

In the L.A. basin, that population figure is expected to soar past 18 million people, with new Southern Californians making 3 million more trips to work each day. Those commuting polluting vehicles are the first line of attack for the South Coast AQMD. Starting in January 1988, all companies with more than 500 workers were required to offer tangible rewards aimed at persuading employees to join car pools or ride buses or bicycles. By 1990 the same requirement will apply to firms with over 100 employees. Companies failing to submit incentive plans by specified dates will face fines of at least $1,000 per day. "If you want to live 50 miles from your job and still have air worth breathing, you're just going to have to learn to share your car," says the AQMD's new board chairman Norton Younglove.

Lents's 20-year game plan also calls for more freeway car-pool lanes and restrictions on how and where future businesses and housing developments can locate, with industrial growth being forced to occur near affordable housing. He notes that without such changes the cost of building new freeways over the next decade will reach $100 billion, with near-gridlock traffic during rush hours that are already three hours long. In March 1988, the AQMD also ordered a new crackdown on excessively smoking vehicles. After over a decade of laxity, the expanded "smog patrols" expect to issue 200 citations per day (compared to 3,800 in all of 1986).

Gasoline stations will face similar scrutiny. In April 1988, a Long Beach judge issued the first jail term ever meted out for violating Southern California air quality rules after a local gas station owner pleaded no-contest to using defective vapor recovery nozzles on his pumps. The district has estimated that if these systems, first mandated in 1976, were in proper working order at all 7,000 service stations in the basin, 140 fewer tons of hydrocarbons per day would enter the atmosphere. "Other polluters should now be on notice," Lents warns.

There will be stakeouts at maintenance yards of the region's Rapid Transit District, which operates the largest all-bus fleet in the United States, and the district is about to ban all heavy-duty diesel trucks from the freeways during rush hours, forcing more night deliveries. "Some trucking firms have actually said they'd prefer this because they waste money sitting in traffic jams," says Lents. "Most say they'll go along if we can convince businesses to keep receiving and shipping open. Well, people repairing highways have adjusted to doing it at nights and on weekends." (The 850,000 diesel trucks and buses in California produce almost as much nitrogen as all passenger cars combined.)

But perhaps most significant of all is Lents's new "clean fuels" program. It will require all transit buses, rental cars, and other fleet vehicles purchased after 1993 to run on something other than gasoline. The goal is that within 20 years half of all the vehicles on the road will be powered by either methanol fuel or, more preferably, electricity. Lents has already established an AQMD Office of Technology Assessment, hired a chief scientist, and allotted a $2 million dollar budget for 1988, which is anticipated to increase by $3 to $4 million each year. Getting matching funds from Southern California Edison and the Metropolitan Water District, both of which would profit from a move to electric vehicles, the AQMD is studying not only electric and methanol options but also fuel cell and superconductor technologies. Sixteen demonstration projects have been earmarked to test the advantages and limitations of alternative fuels, with the district seeking a $1 surcharge on annual vehicle registration fees to help fund the $30.4 million program.

At the moment, the major problem with electric vehicles (besides opposition from the oil and gas industry) is the lack of battery staying power. Under current technology, standard lead-acid batteries in experimental vans have about a 60-mile range in city driving before they need recharging, which takes about eight hours. But Chrysler will have a van available next year with up to a 100-mile range, with the batteries placed underneath so as not to affect cargo space. And batteries are in development that could more than double that while requiring less recharging time.

Southern California Edison spokesperson Michael Hertel says that electric vehicles may be particularly adaptable to Southern California because the fuel supply is cheap and readily available—up to a million or more of the vehicles could hit the road without any demand for new electric power plants. Homeowners with 220-volt wiring could recharge overnight during non-peak electric hours, and fuel costs would be about half that of a similar van running on gasoline.

In November 1987, an innovator named Kenneth Kurtz displayed his sleek, three-wheel, open-cockpit electric car to AQMD officials and members of California's Senate Transportation Subcommittee. Made out of a $200 wrecked Volkswagen, Kurtz spent over five years developing the low, cone-shaped "Solar Wind" which can reach 70 mile-per-hour speeds on its batteries. Kurtz says "it'll store energy just sitting in the parking lot," due to its dashboard lined with solar cells.

That same day in Washington, D.C., the Senate Commerce Committee endorsed legislation encouraging manufacturers to produce more methanol-powered automobiles. By April 1988, this had passed both houses of Congress. California congressman Henry Waxman's House bill revising the Clean Air Act, while giving lagging areas like the South Coast basin another ten years to meet the federal air standards, would require 30 percent of all new cars to run on cleaner fuels like methanol by 1998.

In September 1988, Congress approved legislation providing incentives for manufacturers to turn to methanol, ethanol, or natural gas-powered vehicles. For each car produced and sold using an alternative fuel, the manufacturer would get large credits making it easier to meet the government's minimum fuel economy requirements for the manufacturer's entire fleet. Additionally, the Department of Energy would be authorized to start buying and testing alternative fuel cars, trucks, and buses offered by manufacturers in late 1989 at a cost of up to $18 million.

Methanol already powers Indianapolis race cars at 200 mile-per-hour speeds, and Ford has about 500 experimental methanol cars in California, providing as much or more horsepower as gasoline and costing about the same. The California Energy Commission has an agreement with Atlantic Richfield to begin marketing methanol at 25 service stations in the basin in 1988, with Chevron expected to follow suit. Ford, General Motors, and others are also working on prototype cars built to run on combinations of methanol, ethanol, or gasoline.

Potential pollution problems also exists with methanol, however. Senators Robert Stafford of Vermont and John Chafee of Rhode Island have expressed concerns that methanol, produced from coal or natural gas, would have to burn twice as much carbon and therefore add to the climatic "greenhouse effect." In addition, they warn it could bring a tenfold increase in tailpipe emissions of formaldehyde.

But Freeman Allen, a chemistry professor at California's Claremont College and chairman of the state's Sierra Club chapters, believes that methanol would not add significantly to the total amount of atmospheric carbon dioxide nor would it pose the same evaporative problems as hydrocarbon gasoline. "Its biggest benefit would be in replacement of diesel fuels," says Allen, "because it reduces the amount of particulate matter and nitrogen oxides." As for formaldehyde, Allen and other experts say that this can be minimized with the proper kind of catalytic converter placed on the back end of vehicles. Says Lents, "It just may take a different design to achieve this. The state is committed to set specific standards so new cars being built won't emit more formaldehyde. The methanol formaldehyde problem has really been blown out of proportion by [the oil and auto] industry. It takes a lot of gall to take benzene, a known human carcinogen, and purposely put it in gasoline to increase the octane and then to go out and knock methanol."

The nation's first large-scale experiment to control urban pollution with high-oxygen fuels was in the Denver area. Residents were restricted to buying only corn-based ethanol or ether-based MTBE for the first two months of 1988, an effort that ended without the epidemic of clogged fuel injectors and stalled cars predicted by opponents from the oil industry. The alternative fuels measure will expand to cover a four-month winter period, when air pollution trapped by the Rocky Mountains is the worst, starting November 1988.

Yet another possibility is hydrogen power. The California basin area of Riverside is studying plans to use such vehicles in its city fleet and may manufacture the fuel itself through an inexpensive electrolysis process. This much is certain: one or more of these alternatives is going to be on the front burner in Los Angeles and elsewhere. And should the movement toward different sources bog down, Lents says that gasoline rationing "certainly can't be discounted."

Changes in peoples' habits and lifestyles are imminent, with staggered business working hours an eventual likelihood as well as limits on such common pollution-producing household items as hair spray and deodorants. Lents has his technology office working on electric and natural-gas powered lawn mowers, and is pushing his own 800 employees to set an example by car pooling or riding bicycles and, as he's done, returning to the old push mower for small yards.

Lents has also created an Office of Public Affairs for the district, with $2.5 million—about 5 percent of the agency's annual budget—earmarked for a massive public relations blitz in 1988. Television, radio, and newspaper ads are encouraging car pooling. District officials are setting up community meetings to demonstrate methanol and electric vehicles. An April 1988 conference was held to sell the new smog-control program to 4,000 elected officials. "The District is moving ahead in a remarkable way," says the Sierra Club's Allen. "You have to gulp a little at a 20-year attainment plan for clean air, but they couldn't realistically do it a whole lot faster. But the aggressive approach will bring some backlash, especially from industries that are just getting over the shock of what's being proposed for them."

Backlash from industry has indeed begun in earnest. For years, the 30,000-plus industrial and commercial polluters in the basin—mostly refineries, heavy industries, and chemical manufacturers—pretty much had their way with the AQMD. The district had only 100 inspectors to monitor 60,000 pieces of equipment and buildings on 32,000 sites. They did issue 5,000 industrial pollution citations in 1986, but fines averaged about $260 per violation, "a slap on the wrist," as Waxman puts it. Although all polluting industries must have a permit, 90 percent of those issued by the AQMD in 1986 failed to supply sufficiently detailed restrictions on the quantity and quality of emissions allowed, or imposed no restrictions at all. And 60 percent of industrial expansions were taking place without even bothering to get a permit. Most offensive of all, a "smog credit" program set up by the EPA allowed companies to buy and sell the right to pollute above safe standards. Some firms, selling their unused credits when they were going out of business for example, were literally making millions on the deals. Brokers began listing "Emission Reduction Credits" as matter-of-factly as real estate agents list houses.

Last October, the AQMD board voted to eliminate credits stockpiled by companies that have shut down or ceased certain operations, establishing a "bank" for such credits to be controlled by the agency. Lents substantially beefed up the inspection and permit-processing programs. But those moves are only the beginning of a crackdown that has some corporations crying potential bankruptcy.

About 29 percent of the basin's nitrogen dioxide pollution—the key ingredient in smog, acid rain, particle pollution, and poor visibility—comes from industry (the rest emanating from cars and trucks). New rules promulgated by the AQMD will require a 90 percent reduction in nitrogen dioxides emitted from boilers and heaters used to refine oil, and a similar reduction from gas turbines that make electricity. This could cost Southern California's oil industry as much as a billion dollars according to several spokespeople, and drive some refineries out of business. Southern California Edison is opposing the AQMD on this one, citing estimates that it will cost them $20 million annually for 20 years to install the required controls on turbines at its electric generating station in Long Beach. The utility wanted the plan put off for 9 to 18 months, but Lents refused to consider this.

Industry claims that controls on car exhaust, for example, should happen before they're asked to share the burden. But the technology exists for them to make the changes, the best option being a process known as selective catalytic reduction (similar in concept to catalytic converters now required in California's automobiles), which has been used successfully for a decade in Japan. The AQMD has 13 projects underway to demonstrate the viability of clean-burning fuels at industrial sites.

How Los Angeles, with the worst air quality in the country, is trying to cope with the problem is, of course, only a microcosm of what's become a global dilemma. Close to 400 million vehicles clog the world's streets today, according to the Worldwatch Institute, with a record 125,000 cars rolling off the assembly lines every working day in 1987. Compare this with the period immediately after the Second World War, when only 50 million vehicles were on the road.

And the health and environmental costs everywhere are escalating, too. Researchers at the University of California estimate that the use of gasoline and diesel fuel in the United States alone may contribute to 30,000 deaths every year. Some 59 American cities still don't meet federal carbon monoxide standards. Auto emissions also contribute to acid rain, which is destroying aquatic life and forests.

Most ominous of all, there looms the "greenhouse effect." In 1985, the United States contributed nearly one-quarter of the total world production of carbon dioxide from fossil fuels, with fuel burned on highways contributing almost one-quarter of that. It is critical, if methanol is utilized on a large scale in the future, that it not be derived from coal, which would only exacerbate the global warming trend. Two fuel feedstock options that don't produce net CO_2 are biomass, which can be converted to alcohol and used to fuel methanol or natural gas vehicles, or non-fossil-fuel electricity. At current fleet efficiency, however, biofuels are prohibitively expensive and could not supply even half of the total demand for highway fuels.

Clean-burning hydrogen, captured from water, could become an energy currency that will store solar-generated power for nighttime use and power automobiles. A California company, Pacific Energy Resources, is seeking to make hydrogen competitive with other fuels. It is currently negotiating with a major plywood manufacturer in Oregon that has expressed interest in revamping a plant to meet its energy needs with hydrogen.

Tom Bathe, of the Solar Energy Research Institute in Golden, Colorado, notes accelerated research into the use of photovoltaic cells, man-made silicon chips that chemically convert natural light to electricity to replace fossil-fuel power plants. And research continues into possibly using thermal systems, which convert heat captured directly from the sun, to replace fossil-fuel heat sources.

There is no time to lose. Efficiency improvements—cars with mileage ratings of at least 50 miles per gallon—and less wasteful factories could probably eliminate 10 percent of all carbon dioxide emissions by the year 2005. That, at least, would be a start. Coupled with a "climate protection tax" on all fossil-fuel use, development of alternatives would doubtless move faster. (A 50 percent reduction in CO_2 is considered necessary to stem an otherwise disastrous temperature rise by the middle of the next century.)

Ultimately, it's up to people to make the choices necessary for a sustainable future. As California senator Tom Hayden commented recently about Los Angeles in a statement with far broader ramifications, "It's a race against time. L.A.'s air pollution has turned an environmentalist's issue into everybody's issue." In March 1989, Los Angeles regulatory officials adopted the nation's most sweeping air-quality plan. It is proposed that by 1998, 40 percent of all cars and 70 percent of trucks and buses would run on methanol or other "clean" fuels. By 2008, the plan calls for auto makers to be selling vehicles that run only on electricity or other very clean alternative fuels. In coming years many companies will have to alter their products and manufacturing processes, or cease operations. For example gasoline engines on lawn mowers will eventually be banned. "The technological developments that come out of this will have enormous worldwide implications," says AQMD board member Larry Berg.

NOTE

This chapter was adapted from an article prepared for the summer 1988 issue of *Amicus Journal*, the magazine of the Natural Resources Defense Council.

Editor's note: The author is a journalist who prepared this chapter with the objective of expressing how environmentalists view the urban air quality problem in the United States. Its intent is to demonstrate the nature and urgency of the pressure being applied to political leaders to solve urban air quality problems. Unlike all others in this book, this chapter was not reviewed for accuracy by technical experts.

8 JEFFREY A. ALSON, JONATHAN M. ADLER and THOMAS M. BAINES

Motor Vehicle Emission Characteristics and Air Quality Impacts of Methanol and Compressed Natural Gas

Consideration of alternative transportation fuels in the United States has come full circle. Research into nonpetroleum fuels in the late 1960s was motivated by the realization that combustion of gasoline and diesel fuels was responsible for a large portion of urban air pollution. In the 1970s, as emissions from new gasoline-fueled vehicles were reduced and as oil prices soared, the focus for alternative fuels changed from environmental to energy policy. Currently, air quality concerns are again the primary motivation for alternative fuels.

The present interest in alternative fuels has been manifested in activity at all levels of government. The Environmental Protection Agency (EPA), under the auspices of the Alternative Fuels Working Group that includes representatives from agencies and departments throughout the executive branch, recently released two documents that evaluate the potential of alternative fuels to improve urban air quality.[1] The Department of Energy (DOE) is currently undertaking a comprehensive 18-month study of alternative transportation fuels.[2] The states of California, Colorado, Arizona, Nevada, and New York either have or are considering alternative fuels programs in selected urban areas. Pending before Congress are several pieces of legislation relating to alternative fuels, ranging from corporate average fuel economy incentives for automotive manufacturers that sell vehicles capable of operating on nonpetroleum fuels to mandates that centralized vehicle fleets in areas not attaining air quality standards begin purchasing vehicles that operate on alternative fuels.

Alternative fuels can be divided into two distinct groups: those that could completely replace gasoline and those that can be low-level additives to gasoline. Much of the current near-term interest, particularly in urban areas with very high levels of carbon monoxide pollution, is in gasoline additives such as ethanol, methanol, and methyl tertiary butyl ether (MTBE), which can be added to the current gasoline pool and provide immediate carbon monoxide reductions. These are not addressed further.

This chapter will address two alternative fuels—methanol and compressed natural gas (CNG)—that EPA believes have the potential in the near term to completely replace gasoline and/or diesel fuel, at least in certain new vehicle applications. Liquefied petroleum gas (LPG) and ethanol are not addressed here because of long-term supply constraints. There are, of course, other fuels that could provide very significant urban emission reductions, such as electricity and hydrogen, but there appears to be little likelihood that these fuels will be feasible in the near and medium term (except for extremely limited applications).

The primary purpose of this chapter is to identify the potential emissions reductions available through the use of methanol and CNG in motor vehicle applications. In order to better understand the context for such a discussion, it is necessary to first consider the magnitude of our current air quality problems and the contribution of motor vehicles to these problems. These issues are addressed in the first section. The second section will address the potential emission reductions available from the use of methanol and CNG in light-duty vehicles (i.e., passenger cars and light trucks), by presenting test data from prototype vehicles and comparing these data to emission factors for comparable gasoline vehicles. The following section will consider the use of methanol and CNG in heavy-duty engines (most work to date has focused on bus engines), and will compare emission results from these fuels with those of diesel fuel. The relationship of vehicle emission reductions to urban air quality improvements will be addressed briefly in the final section. A secondary purpose of this chapter is to outline for the reader some of the nonemissions issues that are relevant in any overall evaluation of methanol and CNG as motor vehicle fuels.

MOTOR VEHICLES AND AIR QUALITY

EPA has established National Ambient Air Quality Standards (NAAQS) for six pollutants: lead, sulfur dioxide, nitrogen dioxide, ozone, carbon monoxide, and particulate matter. Historically, motor vehicles have been a significant contributor to urban levels of each of these pollutants with the exception of sulfur dioxide. The motor vehicle contribution to nationwide lead emissions has dropped dramatically recently, due both to the use of unleaded gasoline in new vehicles and the lead phase-down program, but vehicles continue to contribute significantly to urban levels of ozone, carbon monoxide, particulate matter, and nitrogen dioxide.

Figure 8.1 shows the number of persons, based on the 1980 census, living in counties with air quality levels above the NAAQS in 1986.[3] It can be seen that a large number of areas continue to be in violation of air quality standards for ozone, carbon monoxide, and particulate matter. Motor vehicles are a major source of all these pollutants, as well as of nitrogen dioxide, a principal precursor of ozone formation. Thus, the remainder of this chapter will focus on four

Figure 8.1
Number of Persons Living in Counties with Air Quality Levels Above National Ambient Air Quality Standards in 1986 (Based on 1980 population data)

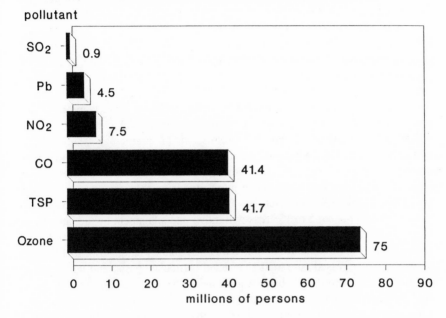

pollutant

millions of persons

pollutants: ozone, carbon monoxide (CO), nitrogen dioxide (NO$_2$), and particulate matter.

Ozone is our most serious long-term urban air quality problem. The ozone NAAQS is a maximum one-hour level of 0.12 ppm, not to be exceeded more than three times in a three-year period. Ozone is not emitted directly but is a product of a series of complex atmospheric processes involving hydrocarbons (HC), nitrogen oxides (NO$_x$), and sunlight. EPA believes that for most urban areas HC control is generally the most promising strategy for reducing ozone levels. Motor vehicles are typically responsible for 30 to 50 percent of urban HC emissions. Based on data through 1987, there are 68 areas not attaining ozone standards. As new, cleaner vehicles continue to displace older, more polluting vehicles, and as EPA and state agencies implement other controls, we expect many nonattainment areas to move into compliance. Still, 20 to 30 of our largest cities will require major HC emission reductions (on the order of 40 percent or more) to reach attainment, and such reductions will be very difficult to achieve, especially since HC sources are a very diverse group. Recent studies have suggested that ozone levels near the NAAQS level of 0.12 ppm can affect otherwise healthy adults, and the agency is currently reviewing these new studies to see if the standard ought to be lowered.

The NAAQS for NO$_2$ is an annual mean of 0.053 ppm. Motor vehicles generally emit about 40 to 60 percent of urban NO_x emissions. While Los Angeles is currently the only city in nonattainment for NO_2, NO_x control is still very much a priority for two reasons. First, because NO_x emissions have not been as tightly controlled as other emissions, continued economic growth will likely mean that overall NO_2 levels will begin to grow at some point in the future and some areas currently in attainment could be threatened with nonattainment. Second, in certain ozone nonattainment areas, EPA encourages NO_x reductions as an ozone control strategy.

CO is the one urban pollutant that is almost exclusively a motor vehicle problem. CO results from incomplete fuel combustion, a more common characteristic of automotive engines than stationary engines. Vehicles are generally responsible for 80 to 90 percent of CO emissions in urban areas. The NAAQS for CO is 9 ppm, eight-hour average, and 35 ppm, one-hour average, not to be exceeded more than once per year. Currently, 59 areas are in nonattainment for CO, but the future situation is much more promising than for ozone. Because new gasoline vehicles emit much less CO than older vehicles, EPA projections show that all but about 5 to 15 areas will move into attainment by the late 1990s simply with the improvements brought about by existing motor vehicle standards.

The final NAAQS pollutant of concern is particulate matter. Prior to 1987 the standard was expressed on a total suspended particulate basis, and the majority of EPA's data is still on this basis (including the data used in Figure 8.1). Now the standard is expressed on an inhalable particulate basis and considers only those particles less than 10 micrometers in diameter. The new standards of 50 micrograms per cubic meter (micrograms/m^3) as an annual mean and 150 micrograms/m^3 for a 24-hour average are projected to be approximately equivalent in stringency to the older total suspended particulate standards.[4] Thus, while the exact number of nonattainment areas for the new standards is not known at this time, it is expected to be significant. Vehicles that use unleaded gasoline emit very low levels of particulate, but diesel trucks and buses are important sources of inhalable particulate, particularly in central city areas. Diesel particulate is a special concern because of both its small size and hazardous composition.

Thus, as we look toward the future, the emission characteristics of greatest concern for light-duty vehicle applications are HC and CO emissions, and for heavy-duty applications, particulate emissions.

ALTERNATIVE FUELS IN LIGHT-DUTY APPLICATIONS

Light-Duty Gasoline Vehicle Emissions

Over 95 percent of all fuel consumed in passenger cars and light trucks in the United States is gasoline. Accordingly, gasoline-fueled applications are the appropriate passenger vehicle target for alternative fuels. Since most light trucks do not differ significantly from passenger cars, and in order to simplify the types

Table 8.1
Average Zero-Mile Emissions from Gasoline Cars (Grams per Mile over the EPA Federal Test Procedure)

Model Year	Exhaust HC	Evap. HC	Total HC	CO	NO_x
1966	7.2	4.5	11.7	78	3.4
1986	0.23	0.55	0.78	1.2	0.54
Reduction	97%	88%	93%	98%	84%

Table 8.2
Average 50,000-Mile Emissions from Gasoline Cars (Grams per Mile over the EPA Federal Test Procedure)

Model Year	Exhaust HC	Evap. HC	Total HC	CO	NO_x
1966	8.1	4.5	12.6	89	3.4
1986	1.0	0.6	1.6	13	1.3
Reduction	88%	87%	87%	85%	62%

of comparisons that need to be made, this section will focus on gasoline-fueled passenger cars.

Any projection of the future potential of alternative fuels to reduce gasoline passenger-car emissions must recognize that very large emission reductions have occurred over the last two decades. Congress and EPA established progressively more stringent passenger-car emission regulations that culminated in 1981 with the following standards: 0.41 grams per mile (gpm) exhaust HC, 3.4 gpm exhaust CO, 1.0 gpm exhaust NO_x, and 2.0 grams per test evaporative HC (these standards must be met by all new gasoline and diesel passenger cars sold in the United States, except for those sold in California, which has somewhat more stringent standards).

These standards have provided the impetus for the development of sophisticated emission control technologies that have greatly reduced passenger car emissions as shown in Tables 8.1 and 8.2 (all data in this section are from the EPA Federal Test Procedure urban driving cycle).[5] Table 8.1 shows that zero-mile emissions from new gasoline passenger cars have been reduced by 84 to 98 percent between the years 1966 and 1986. Emissions tend to increase with vehicle age, and of course in-use emissions are the relevant measurement. In-use emissions reductions have not been quite as great as zero-mile (new-vehicle) emissions reductions (see Tables 8.1 and 8.2), but they are still very impressive. These large per-mile emission reductions have resulted in lower overall motor vehicle pollutant burdens in the late 1970s and 1980s even with the growth in the number of vehicles and number of vehicle miles traveled.

The regulatory program currently in place, with very stringent emission standards for gasoline passenger cars, provides a challenging context in which to project emission reductions for alternative fuels. When gasoline engines were basically uncontrolled in the 1960s, emissions were almost exclusively a function of fuel type. The substitution of fuels with inherently more benign properties such as methanol and CNG would have clearly reduced vehicle emissions. Today, however, emissions are a primary design concern and are a function not only of the fuel being utilized but of many other variables such as the level of applicable emissions standards, the type of engine design, the specific calibration of various design parameters, the emission control system design, and others. And, of course, emissions are just one of many items of interest to automotive engineers and must always be viewed in terms of their relationship to other important characteristics such as power, driveability, reliability, fuel economy, and cost. The relevant issue is not just whether emissions reductions are possible with a given fuel. The answer to that question is almost always *yes*. The more critical issue is whether alternative fuels have properties that will inherently reduce emissions of certain pollutants while at the same time obviating the need for more complex emission controls or sacrifices in other vehicle characteristics.

Relative Reactivity of Hydrocarbon Compounds

The potential for methanol and CNG vehicles to reduce urban ozone levels is related to the photochemical reactivity of organic compounds (*organic* is used here to refer to all unburned and partially combusted fuel compounds that contain carbon, i.e., HC and oxygenated HC). Gasoline itself is a mixture of HC compounds; the combustion of gasoline results in a large number of individual organic products. Although many of these organic compounds in gasoline exhaust are considered to be toxic, the primary justification for their regulation is their role in the formation of ozone. Because there are far too many organic compounds in gasoline exhaust to regulate individually, and because almost all organic compounds participate in ozone photochemistry, EPA regulates gasoline vehicle organics by setting standards for exhaust and evaporative HC emissions.

It has long been recognized that different organic compounds have different photochemical reactivities, i.e., each compound has a unique rate at which it reacts in the complex photochemical reactions that lead to ozone formation. Our present exhaust and evaporative HC emission standards implicitly assume that the mix of individual HC constituents remains fairly similar from one gasoline vehicle to the next, which is probably a reasonable assumption. But for fuels that are considerably different than gasoline, it is no longer valid to simply

Table 8.3
Reaction Rates with Hydroxyl Radicals Relative to Butane

Compound	Relative Reaction Rate
Methane	0.003
Methanol	0.40
Benzene	0.51
n-Butane	1.0
Toluene	2.4
Ethylene	3.4
Formaldehyde	3.6
Acetaldehyde	6.4
m-Xylene	9.7
Propylene	10.4

assume that unburned fuel-related emissions will have the same overall photo-chemical reactivities as gasoline vehicle HC emissions.

One important characteristic of ozone formation is the reaction of organics with the hydroxyl radical (OH). Table 8.3 lists the OH-rate constants for a number of organics, all of which, except methanol, are present in gasoline vehicle emissions; the rates are normalized so that the reaction rate of butane, a common gasoline constituent, is unity.[6] This table can be used as one indicator of relative reactivity, while acknowledging that other parameters such as maximum ozone yield, NO_2 oxidation rate, and HC/NO_x ratio are also important.

Methane, the primary constituent of natural gas and the dominant HC constituent in CNG vehicle exhaust (and present in gasoline and methanol exhaust as well), is the simplest compound to address in this regard. As reflected by the relative reaction rates shown in Table 8.3, it is considered to have such a negligible photochemical reactivity that EPA recommends that methane be excluded from regulatory controls.[7] EPA's current motor vehicle HC standards do in fact include methane, but EPA proposed a nonmethane HC standard in the early 1980s and would likely reconsider this issue if certification of CNG vehicles appeared imminent. Past practice has been to assume that the methane component of CNG emissions has zero reactivity while the remaining HC has an overall reactivity similar to those of gasoline vehicle HC. This chapter will follow that practice and thus will focus on nonmethane HC from CNG vehicles.

Methanol vehicle organics are a somewhat more difficult issue to address. The existing data base suggests that when pure methanol is used as a fuel, the organic emissions are largely unburned methanol with a small amount of formaldehyde and only trace amounts of a small number of HC compounds. When gasoline is added to methanol fuel to aid cold starting, then there is a higher percentage of HC emissions. As with CNG vehicles, it has been commonplace in the past to assume that methanol vehicle HC has reactivity profiles similar to those of gasoline vehicle HC. But what about the relative reactivity of the unburned methanol and formaldehyde emissions?

Table 8.3 shows that methanol itself tends to have a relatively low reaction rate with the hydroxyl radical while formaldehyde has a relatively high reaction rate. In order to assess the overall ozone impact of substituting methanol vehicle organics for gasoline vehicle organics, a number of computer simulation studies have been performed. These studies simulated air chemistry and transport within certain urban areas and accounted for dispersion of pollutants. Based on these studies, EPA has developed a model that provides reactivity factors for methanol and formaldehyde relative to typical nonoxygenated HC (i.e., HC without oxygen atoms) from gasoline vehicles. Based on this model, the average reactivity factors are projected to be 0.43 for methanol and 4.8 for formaldehyde.[8] That is, on an equivalent carbon basis, the methanol molecule has only 43 percent of the potential to form ozone as the typical gasoline HC molecule, while the formaldehyde molecule has a 4.8 times higher potential. On a gram-per-mile basis, the reactivity factors are 0.19 for methanol and 2.2 for formaldehyde (these compounds have higher mass-to-carbon ratios than typical gasoline compounds). It must be emphasized that this model simplifies a very complex process that is best simulated by detailed computer programs. There are a number of caveats pertaining to the studies upon which the model was based as well as the model itself, which limit the scope of its applicability. While EPA believes it is useful as an analytical tool, EPA does not consider it to be appropriate for use in formulating standards or making other binding regulatory decisions.

Light-Duty Methanol Vehicle Emissions

Methanol has long been considered to be an excellent motor vehicle fuel. Its simple molecular structure, high octane, wide flammability limits, high flame speed, and low flame temperature result in a fuel that can potentially be burned in a very clean and efficient way relative to petroleum fuels. Because methanol is such a different fuel than gasoline, it is helpful to distinguish between two types of methanol vehicles—current-technology and advanced-technology methanol vehicles. These two types of methanol vehicles would be expected to differ

Table 8.4
Summary of Exhaust Emissions from Current-Technology Methanol Vehicle Data Base (Grams per Mile over the EPA Federal Test Procedure)

Pollutant	Average Emissions	Emissions Range
Methanol	0.47	0.12 - 1.00
Formaldehyde	0.035	0.00 - 0.17
CO	1.7	0.43 - 3.2
NO_x	0.61	0.04 - 0.88

*Data are from the 40 vehicle configurations that met current federal CO and NO_x emission standards.

with respect to both engine design and vehicle emissions. Methanol is not considered to be a good fuel for retrofitted vehicle programs because of its corrosive effect on many materials used in older vehicles.

Current-Technology Methanol Vehicles. Current-technology methanol vehicles utilize engines that are very similar to engines used in today's gasoline vehicles, with modifications to allow the engine to operate well, but not optimally, on a blend of 85 percent methanol and 15 percent gasoline (M85). These are the types of methanol vehicles currently involved in demonstration programs and have emissions and efficiency characteristics similar to flexible fuel vehicles (FFVs) or variable fuel vehicles (VFVs) operating on M85.

A large amount of emissions data has been generated from current-technology methanol vehicles over the last few years; EPA has summarized it in a computerized data base.[9] The data base currently includes results assembled from 13 different studies conducted by EPA, the California Air Resources Board,.and other organizations; it includes data from 10 different vehicle models and 64 different engine and vehicle configurations. Since many of these first-generation methanol vehicle prototypes were not designed to meet any specific emission requirements, some vehicles failed either the CO or NO_x federal emissions standards. The data base includes exhaust emission test data for 40 vehicle configurations that met the proposed methanol vehicle standards; Table 8.4 gives the average and range for exhaust emissions over the federal test procedure for these vehicles. The average mileage of these vehicles was around 10,000 miles, although individual vehicle mileage ranged from zero to over 100,000 miles.

With respect to CO and NO_x emissions, the data in Tables 8.1, 8.2, and 8.4 are very instructive. Average CO and NO_x emissions from current-technology methanol vehicles are somewhat higher than the zero-mile emissions for current gasoline vehicles shown in Table 8.1, but considerably lower than the 50,000-

Table 8.5
Projected Total Organic Emissions from Current-Technology Methanol Vehicles
(Grams per Mile over the EPA Federal Test Procedure)

Mileage	Methanol	Formaldehyde	Hydrocarbon
Zero-Mile	0.71	0.048	0.21
50,000 miles	1.57	0.106	0.48

mile emissions for these gasoline vehicles given in Table 8.2. Since the methanol vehicles had, on average, accumulated around 10,000 miles, CO and NO_x emissions appear to be about the same for today's gasoline and methanol vehicles. This is to be expected for CO emissions, as CO levels are strongly correlated to air/fuel ratio, and current gasoline and methanol vehicles have all generally been designed to operate at stoichiometric air/fuel ratios.

Because methanol has a relatively low flame temperature, it has been hypothesized that methanol vehicles should yield lower NO_x levels. With current NO_x standards, however, we believe that manufacturers will likely trade off methanol's low-NO_x characteristic to gain other benefits such as fuel economy, performance, or a less expensive catalytic converter.

The analysis of the ozone impact of organic emissions from current-technology methanol vehicles is more complex. Instead of using organic emission factors from the data base (the data are much more limited for organics than for CO and NO_x because of differences in measuring and reporting these emissions among various organizations), EPA has assumed that the organic emission levels of current-technology methanol vehicles would be the maximum levels permitted under the new methanol vehicle standards (which essentially require that methanol vehicles emit no more than the amount of carbon allowed from gasoline vehicles). The data base was utilized, however, to project the proper methanol-to-hydrocarbon-to-formaldehyde ratios. The projected total organic emissions (exhaust plus evaporative), both at zero miles and at 50,000 miles, are shown in Table 8.5.[10] Utilizing the reactivity factors (on a gram-per-mile basis) given earlier in the chapter of 0.19 for methanol and 2.2 for formaldehyde and assuming that the types of HC from methanol vehicles are similar to the HC from gasoline vehicles, the projected methanol vehicle emissions in Table 8.5 would result in a 42 percent reduction in ozone-formation potential at zero miles and a 37 percent reduction in ozone potential at 50,000 miles, relative to current gasoline vehicles emitting at the levels given in Tables 8.1 and 8.2. These reductions apply only to the light-duty vehicle contribution to ozone formation; the net impact on ozone would be much less, as addressed later, due to the presence of other polluting sources.

The EPA model used to generate the relative reactivities is a simple generalization of the complex modeling that must be performed to project ambient ozone impacts. Nevertheless, the many studies that have been performed to date support the contention that methanol is less reactive than typical gasoline HC, and that therefore, as long as formaldehyde emissions are not excessive, current-technology methanol vehicles will provide some ozone benefits.

Advanced-Technology Methanol Vehicles. There are reasons to believe that future engine and vehicle designs optimized to take full advantage of the combustion properties of methanol fuel could provide much larger emission benefits than those discussed above for current methanol vehicles. Current methanol vehicle prototypes are basically gasoline vehicles with only the most rudimentary modifications to permit methanol combustion. Because of the perception that there is no near-term market for methanol vehicles, manufacturers have not attempted to optimize vehicle designs for methanol utilization. Methanol's combustion properties suggest that a high-compression lean-burn engine could be developed that has significant efficiency and emission improvements.

Gasoline cannot be combusted at a high air-to-fuel ratio because of engine misfire and the fact that stoichiometric air-to-fuel ratios are necessary in order to allow the NO_x reduction function of the catalytic converter to operate effectively. But methanol's wide flammability limits, high octane, and higher flame speed allow it to maintain stable combustion at much leaner air/fuel ratios than gasoline. And methanol's low combustion temperature inherently produces low engine NO_x levels. These properties suggest the possibility of an optimized lean-burn high-compression methanol engine with good driveability that would be able to comply with NO_x emission standards without the need for a reduction catalyst. Such an engine would likely provide very significant benefits in terms of low CO emissions and improved energy efficiency. Once again, we would not expect NO_x emission reductions from such an engine design relative to a catalyst-equipped gasoline vehicle, although methanol's low-NO_x characteristic facilitates the application of lean-burn technology.

One important issue with optimized methanol vehicles is fuel specification. All of the potential improvements discussed above could be achieved through the use of either M85 or pure methanol (M100) fuel. Almost all prototype and demonstration vehicles to date have utilized M85 both to improve cold startability and to provide a more luminous flame in case of a fire. But there are several tradeoffs associated with choosing between M85 and M100, including several with safety ramifications (for example, the addition of high-volatility gasoline makes M85 more likely to ignite and burn more severely than M100). But the most important consideration for purposes of this chapter is that the use of M85 could limit the potential ozone benefits available from the use of methanol fuel. The use of M85 would increase total evaporative HC emissions and would increase the proportion of reactive HC (as opposed to less reactive methanol) in both exhaust and evaporative emissions. Thus, from an environmental perspective, M100 is the preferred fuel. In order to utilize M100 a breakthrough in cold starting will be necessary, which has not been achieved to date, but is the focus of considerable research.

Table 8.6

Emissions from Low-Mileage Methanol Toyota Carinas (Grams per Mile over the EPA Federal Test Procedure)

Test Site	Fuel	Methanol[*]	Formaldehyde	HC[*]	CO	NO_x
Toyota	M85	-	-	-	0.93	0.69
CARB	M85	-	0.014	-	0.70	1.00
EPA	M85	0.23	0.007	0.07	1.07	0.75
EPA	M100[**]	0.33	0.011	0.02	0.74	0.76

[*]Methanol and HC emissions are total (exhaust plus evaporative) emissions. Methanol and HC were not measured separately, but were projected from FID readings based on ratios determined from the EPA methanol vehicle emission data base for both M85 and M100.

[**]This vehicle cannot be started on M100 at low ambient temperatures.

Toyota is the first automotive manufacturer to make available vehicles utilizing many of the concepts discussed above. Toyota has recently commercialized the lean-burn concept on selected gasoline vehicle models in the Japanese market. They adapted the lean-burn concept to methanol by increasing the engine's compression ratio and making other changes to the intake system, the control system, and the exhaust catalyst. These modifications resulted in a methanol engine that yielded stable combustion at leaner air/fuel ratios with better energy efficiency, higher torque, improved driveability, and lower engine-out NO_x emissions than the gasoline engine counterpart already available in the Japanese market.[11]

The 1.6-liter 4-cylinder engine was installed in two Toyota Carina vehicles with inertia weights of 2,250 lbs. The cars can operate under federal test procedure conditions on both M85 and M100 (though they cannot start on the latter at low temperatures). They have been loaned to EPA and the California Air Resources Board (CARB), and all of the available emission data on them are given in Table 8.6.[12]

The methanol and HC data combine both the evaporative and exhaust components, and so can be compared to the total HC values given in earlier tables. Since all of the Carina testing has been performed at low mileage, it is appropriate to compare Table 8.6 and Table 8.1. As with the current-technology methanol vehicle data, CO and NO_x emissions are fairly similar. The trend of somewhat lower CO emissions and somewhat higher NO_x emissions (though still below the EPA standard) is to be expected with the lean-burn calibration.

Once again, the largest emissions benefits would be with respect to ozone-producing potential. The data for methanol, formaldehyde, and HC emissions in Table 8.6 include both exhaust and evaporative emissions expressed on a gram-per-mile basis. The Carinas exhibited much lower levels of these pollutants compared to the zero-mile projections for current technology vehicles given in Table 8.5. Since we do not have emissions data from the methanol Carina at high mileage, the only calculation that can be done is to compare the ozone-producing potential of the low-mileage Carina to typical zero-mile gasoline vehicles as shown in Table 8.1. Using the reactivity factors developed earlier, the emissions from the advanced-technology Carina fueled with M85 would yield an 83-percent reduction in ozone-producing potential, while the emissions from the M100 testing suggest a reduction of 86 percent compared to current zero-mile gasoline vehicle emissions.

It must be emphasized that even this Toyota Carina must be viewed as only a partial step toward an optimized methanol vehicle. It would be expected that future, more comprehensive research programs will provide vehicles with more favorable emission characteristics than those of the Carina. At this time, EPA projects that future advanced-technology methanol vehicles could provide as much as a 80 to 90 percent reduction in ozone-producing potential and significantly lower CO emissions while simultaneously maintaining NO_x emissions at levels comparable to current gasoline vehicles.

Related Concerns with Methanol Vehicles. Are the types of potential emission reductions discussed in the previous sections realistic? Will the automotive industry and consumers accept methanol vehicles? These questions are critical in any evaluation of the potential of alternative fuels to reduce motor vehicle emissions absent a legislative fuel-use mandate.

Methanol is considered to be an excellent vehicle fuel by the automotive industry. It is a very efficient fuel, with current-technology vehicles generally exhibiting slightly higher efficiencies and advanced-technology vehicles projected to be considerably more efficient. Methanol almost always provides a boost in power output, and this increased performance can sometimes be traded off to further improve efficiency. And being a liquid fuel that can be produced from natural gas and coal, there are no major noneconomic barriers to widespread supply and distribution of methanol.

Potential drawbacks associated with methanol include formaldehyde emissions, vehicle range, and cold startability. Formaldehyde's role in ozone formation has been addressed above, but it is also of concern as a human toxin and carcinogen. EPA has analyzed formaldehyde exposure in great detail and has concluded that formaldehyde levels from current-technology vehicles do not appear to present a serious public health concern.[13] Formaldehyde emissions from advanced-technology vehicles should be much lower, possibly approaching the levels from gasoline vehicles. Formaldehyde emission levels will clearly continue to be a top research priority for industry, EPA, and other regulatory agencies.

Vehicle range is a serious concern because methanol has only half the volumetric energy content of gasoline. This debit is partially offset by methanol's increased efficiency (a small factor for current vehicles but more significant for advanced vehicles) and also by the blending of gasoline in the case of M85. Carrying 50 to 75 percent more fuel in traditional tanks seems unlikely, so unless fuel storage tank designs are modified, range will be reduced. The application of plastic and/or bladder tanks might permit greater volumes of fuel to be carried with acceptable safety characteristics.

The last related problem, primarily associated with M100, is ability to start in cold weather. A breakthrough is thus necessary in order to be able to reap the maximum ozone benefits of M100, but is not necessary in order to achieve the significant benefits available from an advanced technology vehicle fueled with M85.

Light-Duty Compressed Natural Gas Vehicle Emissions

CNG consists primarily of methane but also contains smaller quantities of other compounds such as ethane and propane. Its fuel characteristics are dominated by methane. Methane shares many of the same beneficial fuel characteristics of methanol: simple molecular structure, high octane, ability to combust under lean conditions, among others. The most obvious difference between methane and methanol (and current automotive fuels) is that methane is a gas, not a liquid. This is an advantage in terms of cold startability and cold-start emissions. It is a disadvantage with respect to on-board fuel storage and maximum engine power output.

Unlike methanol, CNG does not pose problems with corrosion and can be used in existing gasoline vehicles retrofitted with CNG conversion kits. These kits typically permit the vehicle owner to fuel with either CNG or gasoline, and are referred to as "dual fuel" conversion kits. The following section will discuss emissions from such vehicles. The succeeding section will address the performance of dedicated and optimized CNG vehicles.

CNG Dual-Fuel Retrofit Vehicles. According to the American Gas Association there are approximately 30,000 dual-fuel vehicles operating in the United States. It has been problematic for EPA to estimate the emissions impact of such vehicles for several reasons: (1) there has been little reliable emission testing performed, particularly on conversions of recent computer-controlled vehicles; (2) the performance of conversion kits can vary greatly depending on a number of factors, including kit manufacturer, the expertise of the installer, and the quality of maintenance; (3) the fact that the vehicle can operate on either CNG or gasoline means that overall emissions can vary greatly depending upon which fuel is actually used; and (4) the conversion process itself sometimes interferes with the gasoline combustion process and can lead to increased emissions on gasoline. It should also be obvious that an engine that must be able to operate on fuels as different as CNG and gasoline cannot be optimized for either fuel (this is true with methanol flexible or variable fuel vehicles as well).

There has been considerable emissions testing of CNG dual-fuel vehicles over the years by EPA, California Air Resources Board, Colorado Department of

Health, various Canadian agencies, and others. Unfortunately, the great majority of this work has involved gasoline vehicles that were not designed to meet the emission standards in effect in the United States since 1981. The evidence is clear that the bulk of CNG dual-fuel conversions of pre–1981 gasoline vehicles resulted in reduced emissions. But as noted earlier, the relevant baseline is now a gasoline vehicle with a more sophisticated emission control system meeting much more stringent emission standards. Consequently, the pre–1981 vehicle data base is of little value in projecting future potential emission reductions.

EPA has performed a comprehensive literature search in order to compile a list of CNG dual-fuel retrofits involving 1981 and later model-year vehicles that have been emission tested over the federal test procedure on both gasoline and CNG at recognized test laboratories. At this time we have only identified five vehicles that fulfill these criteria (listed in Table 8.7). Two of these vehicles were tested by the California Air Resources Board.[14] The other vehicles were tested by EPA as part of a cooperative test program with the American Gas Association and retrofit conversion kit research and marketing companies.[15]

The emissions data in Table 8.7 suggest two clear trends. The first is that CNG dual-fuel vehicles offer the potential for very significant CO emissions benefits. CO emissions from each of the five vehicles in Table 8.7 were significantly reduced with CNG, with three of the vehicles emitting near-zero CO. These results confirm both theoretical expectations (better mixing of gaseous fuel, lean operation, lack of fuel enrichment for starting, etc.) and data from programs involving pre–1981 vehicles. One important lesson from the EPA test program, however, is that CO emission reductions with CNG are not " automatic," i.e., as with any fuel, state-of-tune is an important determinant of CNG vehicle emissions. While all three CNG dual-fuel vehicles tested by EPA ultimately gave very low CO values, two of the three vehicles yielded much higher CO levels when originally tested by EPA in the as-received condition and had to be recalibrated in order to bring the CO levels down. CO emissions from the Crown Victoria before and after recalibration were 4.3 and 0.5 gpm, respectively, and for the Celebrity were 2.6 and 0.1 gpm, respectively. While the Celebrity exhibited some driveability difficulties in the high-emissions mode, the Crown Victoria did not, raising the possibility that such increases could go undetected in the field.

The second clear emissions trend indicated in Table 8.7 is that NO_x emissions can be significantly increased on CNG relative to gasoline. While two of the vehicles showed slight improvements with NO_x on CNG, three of the vehicles suffered large increases in NO_x. With two of the three dual-fuel vehicles tested by EPA, the increases were large enough to cause the vehicles to exceed the 1.0 gpm NO_x passenger car standard on CNG. NO_x emissions from CNG vehicles are a serious concern for three reasons: methane's relatively high flame temperature; the desirability of burning CNG fuel at high air-to-fuel ratios; and because advanced spark timing is often used to compensate for methane's lower flame speed and thereby improve performance. All three conditions in principle

Table 8.7
Exhaust Emissions from CNG/Gasoline Dual-Fuel Cars (Grams per Mile over the EPA Federal Test Procedure)

Test Site	Vehicle	Fuel	NMHC	CO	NO$_x$	Eff. 1[**]	Eff. 2[***]	Accel[***]
EPA	CNG Fuel Sys. 1984 Delta 88	Gasoline	0.30[*]	9.8	0.40	-15%	Base	Base
		CNG	0.25[*]	1.7	1.18	-7%	+10%	-30%
EPA	Total Fuels 1987 Crown Victoria	Gasoline	0.27[*]	1.4	1.07	-15%	Base	Base
		CNG	0.36[*]	0.5	0.93	-13%	+4%	-25%
EPA	Wisconsin Gas 1987 Celebrity	Gasoline	0.20[*]	1.3	0.60	-15%	Base	Base
		CNG	0.16[*]	0.1	1.19	-8%	+8%	-35%
CARB	Dual Fuel Sys. 1983 Ford LTD	Gasoline	0.36	3.3	0.56	NA	Base	NA
		CNG	0.35	0.1	0.47	NA	-20%	NA
CARB	Pacific Light. 1985 GMC Pickup	Gasoline	0.26	7.0	0.70	NA	Base	NA
		CNG	0.05	0.2	1.06	NA	-20%	NA

[*]
Non-Methane HC was not measured, but was calculated assuming that methane was equal to 25% of gasoline HC emissions and 90% of CNG HC emissions.

[**]
Post-Conversion energy efficiency over the FTP relative to EPA certification fuel economy data for the specific gasoline vehicle model.

[***]
These final two columns use the post-conversion gasoline fuel mode as a baseline for comparison with the CNG fuel mode (negative numbers mean lower efficiency or less acceleration with CNG).

will cause higher NO_x emissions. CNG Fuel Systems provided data to EPA that suggested that NO_x emissions would be increased by 55 percent on the Delta 88 if timing were advanced by a typical margin.[16]

The nonmethane hydrocarbon (NMHC) results were much more inconclusive. NMHC emissions were higher in one case, slightly lower in three cases, and significantly reduced in the final case with CNG relative to gasoline. Conclusions about NMHC are particularly difficult given that measurement techniques for HC from CNG vehicles are still being optimized (the values in Table 8.7 for the vehicles tested by EPA are not direct measurements but rather are based on assumptions of the methane fraction in gasoline and CNG vehicle HC emissions). Formaldehyde emissions were detected from the CNG vehicles, and were found to be comparable to gasoline levels.

These recent test data suggest the need for further work in this area. Clearly, dual-fuel vehicles have the potential to provide very large CO emission reductions when operated on CNG. But the specific magnitude of CO emission reduction that can be achieved will depend on the ability of in-use vehicles to maintain low-CO calibrations as well as the need for any design tradeoffs to provide acceptable NO_x levels, which are a serious concern on CNG operation. Based on the current state of dual-fuel kits and NMHC measurement, it is not possible to state whether dual-fuel vehicles will reduce NMHC. Related issues include the impact of the conversion on emissions when the vehicle is operated on gasoline, and the emission impacts of calibration changes that may be attractive to vehicle owners for improved efficiency and/or performance.

Advanced-Technology CNG Vehicles. CNG is such a different fuel than gasoline that, as with methanol, there is every reason to expect that the optimum engine for CNG will be much different than today's CNG dual-fuel engine, and that such an engine would likely provide greater emission reductions and better performance and efficiency than are available from dual-fuel engines.

For purposes of efficiency and CO emissions, the optimum CNG engine should be a high-compression lean-burn engine. CNG may have a slightly more difficult challenge in this regard than methanol because of its relatively higher flame temperature and lower flame speed. The issue is whether it will be possible to reap the efficiency and CO benefits of a high compression lean-burn design while maintaining NO_x emissions within acceptable levels.

The most complete attempt to date to design, build, and evaluate an optimized CNG vehicle was undertaken by Ford in 1983 and 1984. Ford built and leased 27 dedicated CNG Ranger pickup trucks in cooperation with the American Gas Association and member utilities. These vehicles have been in service since that time.

The 2.3 liter gasoline engine normally used in the Ford Ranger was modified in several ways to improve it for CNG utilization including higher compression ratio and advanced timing. The final engine provided efficiency and performance very similar to that of gasoline Rangers, though with a greatly reduced range. Emissions data from three low-mileage Rangers, two fueled with CNG and one

Table 8.8

Exhaust Emissions from Low-Mileage CNG and Gasoline Ford Rangers (Grams per Mile over the EPA Federal Test Procedure)

Test Site	Fuel	NMHC	CO	NOx
Ford-1984	Gasoline	0.20	3.2	1.1
Ford-1984	CNG	0.14**	0.03	1.9
EPA-1988	CNG	0.14**	0.04	2.0

*The CNG Ranger was designed to have approximately the same efficiency and acceleration characteristics as the gasoline Ranger.

**Non-Methane HC was not measured, but was calculated assuming that methane was 90% of CNG HC emissions.

fueled with gasoline, are given in Table 8.8.[17] These data correlate fairly well with those for dual-fuel vehicles discussed above. CO emissions for the two CNG Rangers represented a 99 percent reduction relative to the gasoline Ranger. NMHC emissions were 30 percent lower with CNG for both Rangers. NO_x emissions with CNG were higher in both cases, still below the NO_x standard of 2.3 gpm that was in effect for light trucks certified in 1984 but higher than the 1.2 gpm NO_x standard that took effect beginning in 1988.

The development of advanced-technology CNG vehicles is in its infancy and the Ford Ranger was simply the first attempt to optimize a CNG vehicle. It is very likely that future research will yield improvements in emissions, efficiency, and performance, but the question is whether all can be improved simultaneously. At this time EPA projects that future advanced-technology CNG vehicles will likely be able to provide reductions of over 90 percent for CO emissions and should be able to provide large reductions in NMHC emissions as well. The magnitude of the NMHC reductions and of any possible deleterious effects on NO_x emissions are not clear at this time, however.

Related Concerns with CNG Vehicles. Cold startability and formaldehyde emissions, two of the major concerns with pure methanol as a fuel, should not pose problems for CNG. As a gaseous fuel it is generally considered to have good cold-start characteristics (though regulator freezing has been a problem in some cases), and emission testing has shown that formaldehyde levels from CNG vehicles are generally equivalent to or less than formaldehyde levels from gasoline vehicles.

Efficiency, performance, and range, often interrelated, are significant concerns with CNG vehicles, particularly with respect to dual-fuel retrofits. Each of the five dual-fuel vehicles shown in Table 8.7 suffered a major penalty either in terms of efficiency or acceleration performance. Table 8.7 shows that all three of the vehicles tested by EPA had higher efficiencies on CNG than on gasoline after conversion, but in all cases the CNG energy economy was lower than the

EPA certification fuel economy data for the preconversion gasoline vehicle (i.e., the conversions appear to have reduced efficiency on gasoline). All of the vehicles tested by EPA also yielded large decreases in acceleration performance, measured by 5-to-60 mph and 30-to-60 mph times. Finally, each of these dual-fuel vehicles had a much lower range on CNG than on gasoline, typically at least a 50 percent reduction. It has been estimated that current CNG storage tanks only provide one-sixth of the range of an equivalent size gasoline tank. This is an extremely serious problem for dual-fuel vehicles, since the existing gasoline tank is not removed. These trends for efficiency, performance, and range of dual-fuel vehicles are also confirmed by evaluations of pre–1981 vehicle conversions as well. Decreases in efficiency, performance, and/or range are especially relevant with a dual-fuel vehicle because of the potential for the user to be motivated to use gasoline fuel, or to tamper with the CNG control system, both of which would likely increase emissions. These concerns make reliance on CNG dual-fuel vehicles as an air quality strategy very problematic.

Efficiency, performance, and range characteristics can be improved with dedicated and optimized CNG vehicles, but it is unclear whether they can begin to approach levels of current gasoline vehicles. With passenger cars there may simply be insufficient space to store enough fuel to travel hundreds of miles. The best case for fuel storage is a light-duty truck, such as the Ford Ranger discussed above. The Ranger did have efficiency and performance similar to its gasoline counterpart, but its range was only approximately one-third that of the gasoline model. It would be possible to install additional fuel tanks in the truck bed, but that would not only reduce usable cargo area but would also add considerable weight to the vehicle which would, in turn, reduce efficiency and performance. Using accepted relationships between weight, acceleration, and fuel economy, we estimate that a CNG Ranger with range and power equivalent to the gasoline model would be 25 percent less efficient. This tradeoff between efficiency, performance, and range is one of the reasons why many experts believe CNG is better suited for centralized urban fleet applications than for general public use.

Conclusions—Light-Duty Applications

Vehicle and engine designs available today that operate on methanol and CNG fuels can provide emission benefits. Current technology methanol vehicles (which operate only on M85 fuel and meet proposed EPA emissions standards) are projected to reduce the peak one-day ozone-producing potential of a motor vehicle by approximately 30 to 40 percent relative to that of a current gasoline vehicle. CO and NO_x emissions would not be expected to be affected by the use of current-technology methanol vehicles. CNG dual-fuel retrofit vehicles could provide very large CO reductions on the order of 80 to 95 percent compared to current gasoline vehicles. The NMHC and NO_x emission impacts can vary greatly depending on the conversion. While emission benefits are possible, there are practical concerns with current-technology methanol vehicles and CNG dual-fuel vehicles that may limit their effectiveness as air quality control strategies.

The emission benefits available from both methanol and CNG would be greater in dedicated vehicles optimized for the individual alternative fuels. From an environmental perspective, both fuels would be best utilized in high-compression lean-burn designs that should yield very large NMHC/ozone and CO benefits. But with only very preliminary designs and data at this time, it is impossible to project specific emission reductions with any certainty. EPA believes reductions of up to 90 percent for both pollutants with both fuels may be achievable, although the validity of this conclusion is dependent on the resolution of specific engine design issues for both fuels.

ALTERNATIVE FUELS IN HEAVY-DUTY APPLICATIONS

EPA regulations define a heavy-duty vehicle as one with a gross vehicle weight rating (GVWR is empty vehicle weight plus cargo carrying capacity) of 8,500 pounds or greater. The heavy-duty vehicle class includes a very broad spectrum of vehicles, ranging from an 8,500-pound GVWR gasoline-fueled panel truck to an 80,000-pound GVWR diesel-fueled tractor trailer. To date, the development of heavy-duty engines that operate on alternative fuels has focused on one very small but important segment of this market—transit buses. Accordingly, this section will focus on transit bus engines. It should be noted, however, that the emission trends identified herein would also be expected to apply to other heavy-duty applications where methanol or CNG would be substituted for diesel fuel.

Transit Buses as an Initial Application for Alternative Fuels

Historically, transit buses have not been considered to be a significant environmental problem primarily because there are so few of them. But a number of factors has forced EPA, state, and local environmental agencies to reassess this evaluation. It has become clear that public exposure to transit bus pollution is much higher than previously thought. Vehicle emissions are strongly correlated to the type of vehicle duty cycle; the stop-and-go nature of transit bus operation has been found to yield much higher levels of emissions than the more steady-state type of operation that is more common with long-haul trucks. Transit buses are operated exclusively in urban areas, typically in very busy urban corridors with maximum population exposure, and their pollution is emitted at ground level directly into the human breathing zone. Accordingly, per unit mass, transit bus pollution probably affects a greater number of people than any other pollution source. Consistent with this is the fact that environmental agencies receive more complaints about diesel transit bus pollution than any other motor vehicle pollution problem.

EPA has had emission regulations for diesel-fueled heavy-duty engines since the early 1970s. While emissions from light-duty vehicles are measured using a chassis dynamometer and reported in grams per mile, emissions from heavy-duty vehicles are measured while the engine is operated on an engine dynamometer and are expressed in grams per horsepower-hour (g/hp-hr). Present HC and CO standards for new heavy-duty diesel engines are 1.3 and 15.5 g/hp-hr, respectively, and these standards are not expected to change in the future. Emission standards for NO_x and particulate matter are in a state of transition, however. The NO_x standard for engines produced in 1988 and 1989 is 10.7 g/hp-hr; in 1990, 6.0 g/hp-hr; and in 1991, 5.0 g/hp-hr. The particulate standard is 0.60 g/hp-hr until 1991, when it drops to 0.10 g/hp-hr for transit bus engines and 0.25 g/hp-hr for truck engines. In 1994 all heavy-duty diesel engines will be subject to the 0.10 g/hp-hr particulate standard. Thus, for the three-year period from 1991 through 1993, engines in new transit buses must meet a more stringent particulate standard than engines used in other heavy-duty applications.

When EPA established these NO_x and particulate emission standards in March 1985, it was believed that diesel trap-oxidizers would be available by 1991 and that all transit bus engines and most truck engines would need such an aftertreatment device to comply with the standards. Instead, it now appears that all truck engine manufacturers will be able to meet the 0.25 g/hp-hr particulate standard without trap-oxidizers. Some observers have concluded that the bus engine market is too small to justify large-scale trap-oxidizer research and development programs and so these programs are now geared toward the 1994 truck engine market rather than the 1991 bus engine market. Accordingly, it is possible that trap oxidizers will not be available in 1991 and that diesel bus engines will not be able to comply with the 0.10 g/hp-hr bus engine particulate standard that takes effect in 1991. This has led to serious interest in alternative fuels such as methanol and CNG.

In addition to the EPA emission standards, there are other reasons why transit buses are considered a promising initial application for alternative fuels. Transit buses are typically centrally fueled and maintained, which means that they could utilize new fuels without the need for massive changes in the current fuel distribution and engine maintenance infrastructures. Since transit agencies receive a large portion of their funding from government agencies, they should be extremely sensitive to environmental concerns, and there could be an effective governmental lever for the application of new technology. Finally, if the public could be convinced that transit buses are part of the urban air quality solution and not part of the problem, then state and local planning agencies would have greater support for programs that encourage urban travelers to switch from cars to buses.

Diesel Transit Bus Engine Emissions

The development of an emissions data base for bus engines utilizing alternative fuels is much more complex than for light-duty vehicles, largely due to the procedures used to obtain emissions information.

Table 8.9
Diesel Transit Bus Engine vs. Chassis Emissions

Pollutant	Low Mileage Bus Engines (g/hp-hr)	In-Use Bus Chassis (gpm)
PM	0.57	5.5
NOx	6.3	26.1
CO	3.2	51.9
HC	1.5	3.4
Aldehydes	0.10	NA

EPA's heavy-duty federal test procedure, commonly referred to as the *heavy-duty transient* test to distinguish it from the older *steady-state* test, is used for gasoline and diesel-fueled heavy-duty engines regardless of engine size or vehicle application, and will likely be adapted for use with heavy-duty engines that use alternative fuels as well. Two aspects of the test procedure are most relevant for the discussion here. First, the test involves the engine only, removed from the vehicle and operated over a duty cycle on an engine dynamometer. Testing a "bare" engine instead of an entire chassis greatly simplifies the process since one engine may be used in a large number of vehicle applications, and engine dynamometer test facilities can be both smaller and cheaper than a facility necessary to test an 80,000-pound GVWR vehicle. Second, the transient test duty cycle consists of engine speed and load transients that were selected to simulate intracity *truck* operation. This was done because trucks greatly outnumber buses in urban areas. As mentioned above, engine testing yields emissions in units of grams per horsepower-hour (g/hp-hr), or pollution per unit of work output of the engine. Since air quality modelers utilize grams-per-mile (gpm) emissions factors, EPA has developed a "conversion factor" methodology to convert g/hp-hr to gpm. Historically, this conversion factor has been approximately 3, i.e., multiplying a g/hp-hr value by 3 yields a projected gpm value.

Recent testing by EPA and other organizations has suggested that the use of the EPA transient test with transit bus engines may underestimate actual bus chassis emissions. Table 8.9 illustrates this point. The first column gives average engine emissions data for three low-mileage diesel bus engines that EPA tested in the early 1980s. The second column gives average chassis emissions data for seven diesel buses pulled directly from operating service and tested on a chassis dynamometer over driving schedules designed to simulate transit bus operation.[18] These buses had been built between 1980 and 1983 and had accumulated between 50,000 and 250,000 miles prior to testing. It can be seen that multiplying the engine data by a conversion factor of 3 is fairly reliable for HC and NO$_x$, but the methodology does not hold for particulate and CO emissions. We do not

Table 8.10
Low-Mileage Exhaust Emissions from Diesel Bus Engines Calibrated to Meet 5 g/ hp-hr NO$_x$ (Grams per Horsepower-Hour over the EPA Transient Test)

Pollutant	Current DDC Engine	Current Cummins Engine
PM	0.33	0.49
NOx	5.1	4.2
CO	1.3	3.4
HC	0.74	0.74
Aldehydes	NA	NA

believe this finding calls into question the validity of using engine testing in general, but rather that there are specific factors that make it much more difficult to utilize engine testing to simulate transit bus emissions, e.g., the unique stop-and-go nature of transit bus operation, the fact that buses typically utilize automatic transmissions while trucks use manual transmissions, and other factors.

We believe that chassis testing over driving cycles designed to simulate transit operation is the most accurate laboratory means of projecting actual transit bus emissions, and that, accordingly, the in-use chassis data in Table 8.9 are the most realistic emission factors for diesel transit buses manufactured in the early 1980s. In addition, work is ongoing to evaluate other engine test cycles that might better simulate transit operation. However, the only type of heavy-duty emissions test cycle over which there is sufficient data to be able to compare diesel, methanol, and CNG bus engine emissions is the EPA transient engine cycle. It must also be emphasized that all of the data in this section with the exception of the second column in Table 8.9 are from testing of engines with zero or near-zero miles. The important issue of the relative emissions deterioration rates for the different fuels cannot be addressed at this time since methanol and CNG engines have only recently been developed, and no testing of very high mileage engines or vehicles has yet been possible. However, data relevant to this issue will be generated by the methanol and CNG bus demonstrations now underway.

The U.S. transit bus engine market is a small one (typically 2,500 to 3,500 units per year) and has historically been dominated by Detroit Diesel Corporation (DDC, previously known as Detroit Diesel Allison Division of General Motors). Today's standard DDC engine is a 6V–92TAD two-stroke engine. The Cummins L10, a four-stroke engine similar to those used in truck applications, has recently achieved some market share. Table 8.10 gives 1988 certification emissions data from the diesel bus engines currently being produced by these two companies. Both of these engines were targeted at 5.0 g/hp-hr NO$_x$ which is appropriate for comparative purposes since that will be the federal standard in 1991. Compared to the emissions from engines produced in the early 1980s (see Table 8.9), today's designs have reduced emissions of all pollutants. The one huge emissions obstacle is the 0.10 g/hp-hr particulate standard that is still far below the levels achieved by current engines.

Methanol Transit Bus Engine Emissions

Methanol was traditionally considered a poor fuel for diesel cycle engines because of its very low cetane number. Serious investigation of methanol's potential as a diesel fuel substitute began in the early 1980s in response to two factors. First, with diesel fuel becoming more expensive and future cost and availability becoming ever more uncertain, transit operators and engine manufacturers alike became interested in alternatives. Second, it was clear that in response to the 1977 Clean Air Act amendments, EPA would be establishing more stringent particulate and NO_x emission standards, and methanol was known to be an inherently low emitter of both pollutants.

All major heavy-duty diesel engine manufacturers now have ongoing methanol truck engine research programs. With respect to bus engines, however, only two companies merit discussion at this time: DDC and M.A.N., a West German Company with an assembly plant in the United States.

M.A.N.'s involvement in the German Alcohol Fuels Project led them to modify its four-stroke, naturally aspirated diesel engine for methanol combustion. The two key aspects of the modification were the addition of spark ignition and an increase in volumetric flow rate of the fuel injection system.[19] M.A.N. has also included a catalytic converter as an essential feature of its methanol bus engine. M.A.N. has had methanol buses in several cities around the world since the early 1980s. It delivered its first U.S. methanol bus to San Rafael, California (a suburb of San Francisco) in 1984, and the bus has accumulated approximately 60,000 miles. Maintenance and fuel economy of this bus have been very impressive.[20] An engine similar to the one installed in the methanol bus in San Rafael was tested by EPA; results are shown in Table 8.11. This engine was outfitted with an oxidation catalyst. It can be seen that particulate emissions were low, as expected, and that the catalyst greatly reduced CO and methanol emissions. Particularly impressive was the extremely low formaldehyde level, indicating that the catalyst was effective at nearly all operating conditions. The NO_x level was surprisingly high for a methanol engine, above the 1991 standard. In 1987, M.A.N. delivered 10 methanol buses to Seattle, Washington, at the time the largest number of methanol buses ever delivered to one city. The engines in these buses are turbocharged versions of the basic M.A.N. methanol engine. The turbocharged engine has not been emissions tested over the EPA transient test to date.

DDC's methanol research program has centered on its 6V–92TAD bus engine.[21] It was found to be possible to autoignite methanol in the two-stroke engine at normal operating temperatures by controlling the exhaust gas scav-

Table 8.11
Low-Mileage Exhaust Emissions from Methanol Bus Engines (Grams per Horsepower-Hour over the EPA Transient Test)

Pollutant	DDC Engine w/o Catalyst	DDC Engine w/Ag Catalyst	DDC Engine w/Pt Catalyst	M.A.N. Engine w/Catalyst
PM	0.08	0.08	NA	0.04
NO_x	1.6	1.8	1.5	6.6
CO	8.8	7.8	5.3	0.31
Methanol	3.4	2.1	1.4	0.68
Formaldehyde	0.19	0.05	0.14	0.001

enging process to produce the requisite in-cylinder conditions at the time of fuel injection. In effect, exhaust gases are maintained in the cylinder providing sufficient temperature for methanol ignition. Glow plugs were added for use in cold starting and light-load operation. The first bus with a DDC methanol engine was also delivered to San Rafael, in 1983. This bus, which has the first methanol engine ever built by DDC, had many problems at first, but has since operated well and has accumulated approximately 70,000 miles to date.

DDC now has a major methanol research and development effort and has announced that it intends to offer a methanol bus engine to transit operators beginning in 1991.[22] As such, the DDC methanol engine has undergone extensive development and emissions testing. DDC now has approximately 15 methanol engines in daily service in San Francisco, Riverside, New York City, and Canada, and by early 1989 will have delivered 30 engines for use in buses in Los Angeles and five more in Denver.

The most recent transient test emissions data from DDC methanol engines are also shown in Table 8.11 (all data reported by DDC). The data in the second column represent the stage of development of the DDC engine without catalytic after-treatment as of January 1989. This basic engine will be provided in the aforementioned demonstrations in Los Angeles and Denver (except for experimental injectors which are not widely available at this time), and existing engines in the field will be upgraded to configurations as similar as possible. It can be seen that the noncatalyst engine yields low particulate and NO_x levels, but CO, methanol, and formaldehyde are all higher than desired. The development of an effective oxidation catalytic converter is one of the remaining tasks for DDC. This will be a significant challenge given the low exhaust-gas temperatures and high exhaust volumes of the two-stroke methanol engine and the potential for some catalysts to convert a portion of the methanol in the exhaust to formaldehyde (at low temperatures). The data in the third and fourth columns of Table 8.11 show emissions for the base DDC methanol bus engine with two different catalytic converters. The third column gives data with a silver catalyst located "downstream" in the typical muffler location. This catalyst reduces formaldehyde emissions quite well, but is not very effective at lowering CO and methanol. The following column shows data for a platinum catalyst located "upstream," between the exhaust manifold and the turbocharger. The platinum catalyst was more effective on CO and methanol, but did not reduce formaldehyde emissions nearly as well. DDC has now given top priority to development of a catalyst that can reduce CO and formaldehyde emissions simultaneously. Such a catalyst, which would also be expected to reduce particulate (largely HC from lubricating oil) and methanol emissions, could make the DDC methanol engine a very clean engine in all respects.

Taken as a whole, the data in Table 8.11 suggest that, at a minimum, methanol bus engines would provide significant reductions in particulate and NO_x emissions, the two greatest emissions concerns associated with current transit buses. It seems quite likely that with an effective catalytic converter, CO, HC/methanol, and formaldehyde emissions will be equivalent to or less than emissions from controlled diesel engines.

CNG Transit Bus Engine Emissions

The low cetane number of CNG, like methanol, makes CNG use in diesel engines problematic. Its use is especially unattractive in the two-stroke diesel engine manufactured by DDC, and DDC has therefore done very little work with CNG to date. CNG is better suited to the four-stroke diesel engines built by most other diesel engine manufacturers and is best suited, again like methanol, to Otto-cycle (gasoline-type) engines. In any case, CNG promises reduced emissions relative to diesel fuel and therefore there has been a great deal of interest in CNG as a transit fuel by the gas industry, diesel engine manufacturers, and companies promoting aftermarket conversions.

Several heavy-duty diesel engine manufacturers now have CNG development programs underway. For the transit market, the most relevant program involves Cummins and its four-stroke L10 diesel engine. To date, however, Cummins has not provided any CNG engines for demonstration buses nor has it publicly released any transient test emissions data. There are two CNG buses presently operating in Tacoma, Washington, but these are aftermarket conversions that utilize a (fumigated) mixture of diesel and CNG; no transient emissions data have been reported.

At this time, the only ongoing dedicated CNG transit bus program in the United States is a two-bus demonstration sponsored by Brooklyn Union Gas (BUG) in New York City. The buses were inaugurated in July 1988 and began revenue service operation in September. The performance of the buses will be monitored by local officials, and this will permit a comparison with the six methanol buses that began operation in New York City in April 1988.

When BUG initiated the demonstration program, no heavy-duty engine man-

Table 8.12
Low-Mileage Exhaust Emissions from Brooklyn Union Gas CNG Bus Engine
(Grams per Horsepower-Hour over the EPA Transient Test)

Pollutant	IMPCO Open Loop w/o Catalyst	IMPCO Open Loop w/Catalyst	TNO Closed Loop w/Catalyst
PM	0.01	0.01	0.01
NOx	6.6	1.3	1.2
CO	31.9	10.8	6.6
HC	3.6	1.0	1.0
NMHC	0.82	0.15	0.17
Formaldehyde	0.032	0.001	0.001

ufacturer was offering a dedicated CNG engine for transit bus application. Accordingly, the utility chose to convert a Chevrolet 454-cubic-inch Otto-cycle engine normally used in light and medium-duty gasoline-fueled trucks. The most significant engine modifications included the addition of a CNG mixer, the use of high-compression turbo pistons and a turbocharger that raised the effective compression ratio to nearly 13:1 at rated conditions, and some materials changes to accommodate the higher combustion temperatures of CNG. The engine was equipped with a three-way catalyst that utilized both rhodium and platinum.

After extensive adjustment of the engine, EPA and BUG agreed to evaluate three CNG gas mixer configurations over the transient test at EPA's Motor Vehicle Emission Laboratory. The IMPCO open-loop system is based on a traditional mixer design and uses mechanical adjustments to maintain operation near stoichiometric levels. The IMPCO closed-loop and TNO closed-loop systems both utilize oxygen sensors and microprocessors to maintain tight air/fuel ratio control near stoichiometric levels. EPA evaluated each of these systems with and without the three-way catalyst. Table 8.12 gives a representative summary of the results of this test program. As was discussed in the light-duty section, other things being equal, one would prefer to operate CNG engines at very high air/fuel ratios (i.e., very lean) in order to take advantage of CNG's wide flammability limits and thereby improve efficiency and reduce CO emissions. Previous experience with large CNG engines, however, has suggested that it may not be possible to achieve low NO_x levels without catalytic reduction, and three-way catalysts only provide effective NO_x emission reduction under stoichiometric air-fuel conditions. Given the NO_x and CO levels from the non-catalyst open-loop data in Table 8.12, it is clear that the engine was indeed operating at or near stoichiometry. Comparing these data with those of the IMPCO open loop with catalyst configuration, it is apparent that the catalyst was quite effective for the full range of pollutants. In fact, both catalyst configurations shown in Table 8.12, IMPCO open loop and TNO closed loop, gave low emissions well under the 1991 bus engine standards.

Comparing the CNG bus engine data in Table 8.12 with the diesel bus engine data in Table 8.10, it is apparent that the catalyst-equipped Otto-cycle CNG bus engine offers very large emission reductions for particulate and NO_x emissions. Total HC emissions from the CNG engine with catalyst are comparable to those from the diesel engine, but NMHC and formaldehyde emissions are lower with CNG. The one possible emissions increase is with CO, due to the need to operate the engine at or near stoichiometry. Given the fact that the NO_x levels are well below the 1991 standard, however, there may well be the opportunity to trade off somewhat higher NO_x emissions to achieve lower CO emissions comparable to current diesel CO levels.

A comparison of the CNG data with the methanol bus engine data in Table 8.11 clearly shows that the two fuels offer similar emission benefits. At the present time the CNG engine with catalyst gives lower particulate and formaldehyde levels, but application of an effective oxidation catalyst to the DDC LA engine configuration is expected to provide similar results as well as lower CO emissions.

Related Concerns with Methanol and CNG Buses

It was noted earlier that methanol and CNG have distinct advantages and disadvantages with respect to the displacement of gasoline as a light-vehicle fuel. Many of the concerns that exist with respect to the use of these fuels in light vehicles are not relevant to transit buses. For example, vehicle range does not appear to be as serious of a problem for either fuel in a transit application because of the increased space for fuel tanks and the ability to refuel at a central location (although the added weight due to additional tankage, especially with CNG, could increase axle weights above legal limits). Cold starting with methanol in a diesel-cycle engine is achievable with either glow plugs or spark plugs in concert with direct cylinder injection. Larger CNG engines are expected to provide adequate power output. Of the concerns discussed previously, the only ones that could present difficulties for bus applications are higher formaldehyde emissions from methanol engines with improperly functioning catalysts and lower energy efficiency from CNG engines calibrated to operate near stoichiometry, but these concerns are likely to be overcome with further engine and catalyst development.

Nevertheless, methanol and CNG face a number of barriers as they seek to replace diesel as the primary transit bus engine fuel. Some of these barriers are due to differences in the fuels, others because the transit industry tends to be a follower and not a leader with respect to engine technology. While methanol engines are ready for production at one major bus engine manufacturer (DDC), a catalytic converter still needs to be developed that can reduce formaldehyde, methanol, and CO emissions simultaneously.

The challenge is more significant for CNG, since no diesel engine manufacturer has committed itself to a full-scale CNG engine development program. It is uncertain whether the transit industry, which currently relies exclusively on diesel engines, would adopt CNG if the only option were to utilize it in Otto-cycle engines. In order to best utilize CNG's full emissions and energy-efficiency potential in a diesel-cycle engine, it would need to be operated very lean. Such engines are under development, and if they prove successful, the design should be available within a few years.

Both methanol and CNG must prove that they can provide the high level of durability that is characteristic of current diesel engines. Specific concerns include the operation of glow plugs and fuel injectors for the DDC methanol engine, spark plugs and catalysts for the M.A.N. methanol engine, and fuel system calibrations, engine block durability, and catalysts for the BUG CNG engine.

Finally, the primary nontechnology barrier to use of these fuels by transit operators is fuel cost. Transit properties generally operate with very tight budgets, and any increased operating cost is a serious concern. It will be impossible for alternative fuels to compete with $15 to $20-per-barrel oil, purely on the basis of economics, although CNG is probably more competitive. Of course, the economics of methanol and CNG would improve in the future if diesel fuel prices rise due to regulatory controls and/or changes in world oil prices.

Conclusions—Heavy-Duty Applications

Emissions data indicate that the use of either methanol or CNG as a transit bus engine fuel could significantly reduce the two problematic pollutants, particulate matter and NO_x emissions.

At this time the most promising alternative fuel option for transit buses appears to be the DDC methanol engine, as the company has committed to a production effort with a base engine design that has been the industry standard for many years. The energy efficiency of the methanol engine has already proven to be equivalent to that of diesel engines calibrated to meet future NO_x standards. At the current time, CO and formaldehyde emissions are higher from the DDC methanol engine, but engine-out levels have been significantly improved, and it seems certain that application of an effective catalytic converter will reduce CO and formaldehyde emissions to levels at or below those of diesel engines.

The BUG bus engine suggests that CNG can be utilized in a very clean manner as well. Energy efficiency and CO emissions are presently somewhat worse than diesel, but continued development work with other engine designs will likely yield improvements. The critical question with the BUG engine is whether transit operators will accept a converted Otto-cycle engine for transit applications.

Probably the two most important issues to the transit industry with alternative fuels are fuel cost and engine reliability and durability. These will continue to be evaluated in ongoing and future demonstration programs.

AMBIENT AIR QUALITY BENEFITS OF POSSIBLE IMPLEMENTATION SCENARIOS

The previous sections have concluded that the use of methanol and CNG in engines and vehicles optimized for these fuels can result in significant vehicle emission reductions. The roles that these fuels can play in reducing ambient urban air pollution levels are, of course, very dependent on the rate at which vehicles utilizing these cleaner fuels are actually introduced into urban areas. The economic, political, and social issues that will govern the introduction and utilization of nonpetroleum fuels in the future cannot be addressed here, but it would be instructive to determine the air quality benefits that could accrue based on some possible alternative fuels implementation scenarios. It must be emphasized that the selection of these scenarios should not be taken to imply that the scenarios are projections of what will actually occur, only that they are possible. The actual introduction of methanol and CNG vehicles may be faster or slower than indicated in these hypothetical scenarios.

The following analysis only considers the use of methanol and CNG in passenger cars and light-duty trucks. As such, the emissions of interest are NMHC and CO, and the ambient pollutants of interest are ozone and CO. While it is clear that the use of alternative fuels could provide significant reductions in bus emissions of particulate matter and NO_x, these emission reductions would result in significant air quality improvements only in localized areas such as central city locations where buses congregate and the immediate areas near bus routes. Modeling these localized areas would require a different type of analysis.

The basis for the following scenario analysis is our MOBILE3 computer program, which projects fleetwide emission reductions for NMHC, CO, and NO_x, and does not include particulate. The MOBILE3 model uses EPA's data base of many thousands of emission tests to estimate the effects of altitude, temperature, speed, vehicle wear, and other factors on emissions. For purposes of this analysis, the most important facet of MOBILE3 is that it calculates fleetwide emissions based on calendar year emission factors and the age distribution of the fleet.

MOBILE3 lists emissions by model year and vehicle type. Vehicle type is a classification that separates light-duty gasoline-fueled vehicles (mostly passenger vehicles) from light-duty trucks, heavy-duty trucks, and others. For each model year within each vehicle class it is possible to designate what portion of the vehicles will be able to use alternative fuels. To find the effect on the entire group of vehicles, the effect of the alternative fuel is weighted by the sales fraction of vehicles able to use it and added back to the rest of the group. For example, if one were to model a scenario in which 25 percent of the 1999 model year of light-duty gasoline-fueled vehicles were actually dedicated methanol vehicles and that the dedicated vehicles emit 50 percent of the CO that gasoline-fueled vehicles emit, then the actual emissions of the group will be 12.5 percent (.25 x .50) less than they would have been if the whole group were designed to use gasoline. The reductions calculated in this way are applied to the emissions as listed by MOBILE3, to find the emissions that would occur under the modeled scenario. When these emissions have been calculated for each model year and vehicle type, they are weighted together by the proportion of the miles traveled by the group with the population to arrive at a final emission level, in grams per mile, for the population.

We utilized three implementation scenarios described below. Particular emphasis was placed on fleets, because fleet vehicles "turn over" in a short time relative to vehicles in the population at large, because they travel more miles per year and because they are often centrally fueled. Since these vehicles are replaced so often, it is possible that most fleet vehicles in a particular area could be displaced by vehicles that use alternative fuels over the course of a few years. Conversion or replacement of vehicles in private use would take much longer. We assumed that fleet vehicles make up 5 percent of all urban vehicles and are responsible for 10 percent of all miles traveled in urban areas.

Based on the analyses earlier in this chapter, the following assumptions were utilized with respect to vehicle emission reductions: methanol FFVs would yield a 35 percent reduction in NMHC emissions and no reduction in CO emissions when operated on methanol; dedicated methanol vehicles would reduce NMHC emissions by 80 percent and CO emissions by 50 percent; CNG dual-fuel vehicles would reduce NMHC emissions by 20 percent and CO emissions by 50 percent when operated on CNG; and dedicated CNG vehicles would give 70 and 80 percent reductions in NMHC and CO emissions, respectively. Furthermore, we assumed that methanol FFV and CNG dual-fuel vehicles would use gasoline half of the time (with emissions equal to those of standard gasoline vehicles) and the alternative fuel half the time. With respect to ambient air quality projections, we assumed that passenger cars and light-duty trucks would be responsible for approximately 30 percent of NMHC emissions and 90 percent of CO emissions in a typical city (the NMHC contribution is somewhat higher today in some cities but is expected to lessen in the future). We also assumed that ambient ozone reductions are proportional to reductions in NMHC emissions, which is a major simplification of a complex issue.

We chose three scenarios to analyze. Each scenario assumes that alternative fuels are increasingly utilized in new vehicles. The first scenario assumes that vehicles in fleets begin converting to FFVs in 1992 and that by 1996 all new fleet vehicles are either FFVs or dedicated to methanol. By 2000, all new fleet vehicles are dedicated methanol users. We assume that the general population of vehicles moves toward methanol use at a much slower rate due to slower public acceptance of the product and the need for commercial refueling stations to begin carrying methanol. In this scenario new FFVs begin to appear in the general population in 1996 and steadily increase in market share, peaking at 50 percent in 2004. No dedicated methanol vehicles are present among the general population in this scenario.

In the second scenario, we assume that no FFVs are introduced and that dedicated methanol vehicles are introduced at a faster rate than in the first scenario. These vehicles make up 25 percent of new fleet vehicles in 1993 and 100 percent by 1996. Dedicated methanol vehicles appear in the general population beginning in 1997 and peak at a 20 percent market share in 2002.

The third scenario describes a switch from gasoline fuel to CNG for a portion of the fleet. It is analogous to the first methanol scenario, with some changes. Because there has been less development of CNG technology than the technology needed to use methanol, we assume that the introduction of CNG vehicles is delayed relative to our assumed introduction of methanol vehicles. This delay is one year for fleet vehicles and three years for general population vehicles. Because of the difference between liquid and gaseous fuels and since liquid fuels are considered more appropriate for some applications, we assume in the CNG scenario that half of those vehicles that we modeled in the first methanol scenario as dedicated are actually dual-fuel CNG vehicles instead of dedicated CNG vehicles. In this CNG scenario, dual-fuel vehicles begin entering fleets in 1993 and dedicated CNG vehicles begin fleet penetration in 1997. By 2001, all new fleet vehicles are dedicated CNG vehicles. With general public vehicles, dual-fuel vehicles enter the market in 1999 and reach 25 percent of the new vehicle market by 2005.

Table 8.13 summarizes the results of this analysis. The first two lines in each of the scenarios give alternative vehicle sales for both fleet and general public vehicles for the years 1995, 2000, and 2005. The third and fourth lines show the percent reduction in motor vehicles emissions of NMHC and CO, and the final two lines project the ambient urban improvements in ozone and CO for a typical U.S. city.

The most significant result of this analysis is simply that the ambient ozone and CO improvements are not huge for the scenarios analyzed. The largest improvement in ambient ozone is 2.6 percent with the second scenario, where dedicated methanol vehicles represent all new fleet vehicles by 1996 and 20 percent of all new general public vehicles by 2002. This scenario would reduce CO levels by almost 5 percent, which is the largest CO impact among the scenarios.

There are several reasons why the impacts on air quality are rather modest. First, with respect to ozone, the motor vehicle contribution to overall ozone is decreasing over time; it is assumed to be 30 percent in this analysis. Table 8.13 shows that the reductions in NMHC vehicle emissions are considerably larger than the projected reductions in ozone. Second, since we ended the analysis in the year 2005, we will not see the full benefits of the vehicles sold prior to and in 2005. Even if new vehicle sales simply continued at the rates assumed in 2005, increasing fractions of the overall fleet vehicle miles traveled would be with alternative fuels, and this would yield larger reductions in future years. If sales of vehicles utilizing alternative fuels increased, then the impacts would be larger yet.

Finally, it must be recognized that at any time the motor vehicle emission

Table 8.13
Projected Ambient Urban Ozone and Carbon Monoxide Reductions Based on Three Scenarios

	1995	2000	2005
Scenario 1: Methanol			
Fleet Vehicle Sales	90% FFV	100% DED	100% DED
Public Vehicle Sales	0	25% FFV	50% FFV
Vehicle NMHC Reduction	0.6%	4.3%	7.0%
Vehicle CO Reduction	0%	2.1%	2.9%
Ambient Ozone Reduction	0.2%	1.3%	2.1%
Ambient CO Reduction	0%	1.9%	2.6%
Scenario 2: Methanol			
Fleet Vehicle Sales	75% DED	100% DED	100% DED
Public Vehicle Sales	0	5% DED	20% DED
Vehicle NMHC Reduction	1.4%	5.1%	8.5%
Vehicle CO Reduction	0.7%	3.1%	5.4%
Ambient Ozone Reduction	0.4%	1.5%	2.6%
Ambient CO Reduction	0.6%	2.8%	4.9%
Scenario 3: CNG			
Fleet Vehicle Sales	75% DF	55% DF	50% DF
		45% DED	50% DED
Public Vehicle Sales	0	10% DF	25% DF
Vehicle NMHC Reduction	0.7%	2.4%	4.8%
Vehicle CO Reduction	0%	2.3%	4.5%
Ambient Ozone Reduction	0.2%	0.7%	1.4%
Ambient CO Reduction	0%	2.1%	4.1%

inventory is dominated by older gasoline vehicles. The methanol and CNG vehicles are assumed to displace newer, cleaner gasoline vehicles (that would otherwise have been bought), which are responsible for less than a proportional share of the overall emissions inventory because their emissions are less deteriorated with regard to their emissions standards and because they may have been certified to tighter emission standards than older gasoline vehicles. Thus, at any given time a significant portion of the fleet is still composed of older gasoline vehicles that account for a greater fraction of total emissions than their VMT percentage would suggest. As a result, the impact of alternative-fueled vehicles on total fleet emissions is significantly less than on the emissions of each model year's fleet into which such vehicles have penetrated. By extension, it is also significantly less than it would be on the emissions of an entire fleet in which methanol and CNG vehicles are given an age distribution equivalent to gasoline vehicles.

Carrying forward an implementation scenario until the entire gasoline fleet is able to turn over would provide a demonstration of the maximum potential impact of alternative fuels on mobile source emissions. In a true steady-state condition, when all gasoline vehicles have the same base emission levels and the age distribution of gasoline and alternative-fueled vehicles are equivalent, the impacts would be very straightforward. If dedicated methanol or CNG vehicles gave an 80 percent NMHC emission reduction and accounted for 50 percent of all vehicle miles traveled and motor vehicles contributed 30 percent of all NMHC, then the ozone reduction would be 12 percent (the product of the three values). It does not appear that a steady-state condition is appropriate, however, until well after the year 2000.

NOTES

1. Environmental Protection Agency, Office of Mobile Sources, "Guidance on Estimating Motor Vehicle Emission Reductions from the Use of Alternative Fuels and Fuel Blends," EPA-AA-TSS-PA–87–4, January 1988; Environmental Protection Agency, Office of Mobile Sources, "Air Quality Benefits of Alternative Fuels," July 1987.

2. Department of Energy, "Assessment of Costs and Benefits of Flexible and Alternative Fuel Use in the U.S. Transportation Sector—Progress Report One," DOE/PE–0080, January 1988.

3. Environmental Protection Agency, Office of Air Quality Planning and Standards, "National Air Quality and Emissions Trends Report, 1986," EPA–450/4–88–001, February 1988.

4. 52 *Federal Register* 24634, July 1, 1987.

5. Environmental Protection Agency, Office of Mobile Sources, "Compilation of Air Pollutant Emission Factors, Volume II: Mobile Sources," AP–42, September 1985.

6. R. Atkinson, "Kinetics and mechanisms of the gas-phase reactions of the hydroxyl radical with organic compounds under atmospheric conditions," *Chemical Reviews* 85, no. 1 (1985): 69.

7. 42 *Federal Register* 35314, July 8, 1977.

8. Michael Gold and Charles Moulis, "Effects of Emission Standards on Methanol Vehicle-Related Ozone, Formaldehyde, and Methanol Exposure," Environmental Protection Agency, Air Pollution Control Association Paper 88–41.4, 1988.

9. Michael Gold and Charles Moulis, "Emission Factor Data Base for Prototype Light-Duty Methanol Vehicles," Environmental Protection Agency, *Society of Automotive Engineers* Paper 872055, 1987.

10. Ibid.

11. Kenji Katoh, Yoshihiko Imamura, and Tokuta Inoue (Toyota Motor Corporation), "Development of Methanol Lean Burn System," *Society of Automotive Engineers* Paper 860247, 1986.

12. J. D. Murrell and G. K. Piotrowski, "Fuel Economy and Emissions of a Toyota T-LCS-M Methanol Prototype Vehicle," Environmental Protection Agency, *Society of Automotive Engineers* Paper 871090, 1987; California Air Resources Board, "Alcohol-Fueled Fleet Test Program—Seventh Interim Report," MS–87–04, 1987.

13. Michael Gold and Charles Moulis, "Effects of Emission Standards on Methanol Vehicle-Related Ozone, Formaldehyde, and Methanol Exposure."

14. California Air Resources Board, "Evaluation of Dual Fuel Systems, Inc.'s Compressed Natural Gas/Gasoline Dual Fuel Conversion System," 1983; Memorandum from Rod Summerfield, Chief, Standards Development and Support Branch, to K. D. Drachand, Chief, Mobile Source Division, California Air Resources Board, October 29, 1986.

15. Robert Bruetsch, "Emissions, Fuel Economy, and Performance of Light-Duty CNG and Dual-Fuel Vehicles," Environmental Protection Agency, EPA/AA/CTAB–88–05, June 1988.

16. Letter from Stephen Carter, vice-president of CNG Fuel Systems, to Richard Polich, Consumers Power, December 3, 1987.

17. Tim Adams (Ford Motor Company), "The Development of Ford's Natural Gas Powered Ranger," *Society of Automotive Engineers,* Paper number 852277, March 1985.

18. EPA Office of Research and Development, "Characterization of Heavy-Duty Motor Vehicle Emissions Under Transient Driving Conditions," 1984; EPA Office of Mobile Sources, "Emissions Characterization of Heavy-Duty Diesel and Gasoline Engines and Vehicles," EPA 460/3–85–001, 1985.

19. A. Neitz and F. Chmela, "The M.A.N. Methanol Engine Powering City Buses," Fifth International Alcohol Fuel Technology Symposium, May 1982.

20. Michael Jackson, Stefan Unnasch, Cindy Sullivan, and Roy Renner, "Transit Bus Operation with Methanol Fuel," *Society of Automotive Engineers,* Paper number 850216, 1985.

21. R. R. Toepel, J. E. Bennethum, and R. E. Heruth (Detroit Diesel Allison Division), "Development of Detroit Diesel Allison 6V–92TAD Methanol Fueled Coach Engine," *Society of Automotive Engineers,* Paper number 831744, November 1983.

22. J. E. Bennethum (Detroit Diesel Corporation), "A Strategic Review of Heavy-Duty Emission Regulations and Alternate Fuels," *Transportation Research Record,* 1988.

9

ROBERT SAUVÉ

Compressed Natural Gas and Propane in the Canadian Transportation Energy Market

If one is permitted to oversimplify a little, it is possible to characterize the North American transportation response to energy matters as profligacy in the use of universally available oil-based fuels before 1974, followed by conservation until about 1980, with attempts to develop or promote the use of alternative fuels ever since. This chapter deals with the last phase in Canada, especially as regards propane and compressed natural gas (CNG). Particular attention will be accorded the evolving use of these gaseous fuels from the point of view of the vehicle owner or operator. The emphasis will be on a review of the evidence in order to contribute to informed debate on this subject. With this focus, the analysis will attempt to reveal the thrust of policies by the government of Canada in this area.

I shall argue first that this issue has to be viewed at two levels, the psychological and the material. As regards the psychological, it is my view that, at the currently low levels of crude oil prices and consumption taxes, gasoline prices to Canadians (and Americans?) are so low that most motorists do not even consider alternative fuels, no matter how economical they may be in fact. Therefore, until there is a real increase in the retail price of gasoline, the actual financial advantages of converting to alternative fuels will tend to remain hidden to all but the most discriminating buyers.

The material level has two elements: supply and financial constraints. As regards supply, I shall argue that whatever supply assumptions one cherishes these finite resources will dry up sooner rather than later. Although it may appear heretical to some, it is possible that supplies of gaseous fuels will start to decline within the next 50 years and effectively disappear before the end of the next century. In Canada, natural gas reserve depletions have exceeded new reserve additions in the last few years. Furthermore, estimates of proven natural gas reserves can be seductively misleading.

To declare that we have, say, 50 years of proven reserves is to state only that current reserves will satisfy our normal demand for that period. However, since natural gas currently supplies but a very small share of the transportation energy market, this 50-year estimate clearly does not apply to it. In fact, if natural gas had also to supply the whole transportation energy market in Canada, this 50-year estimate would quickly drop below 25. Therefore, we are dealing with a fairly limited planning horizon.

As supplies of gaseous fuels decline, how will markets or governments allocate them between the transportation, heating and industrial sectors? Since gaseous fuels can compete without subsidies in all sectors except transportation at the present time, this future allocation is not likely to favor transportation. One must therefore conclude that gaseous automotive fuels are supply constrained.

With respect to the financial level, I shall demonstrate that at current market prices for fuels and conversion equipment only high fuel-consumption vehicles are able to amortize the cost of converting to gaseous fuels. The data suggest that, while this represents a theoretically healthy share of the national vehicle stock, the practical share is relatively modest. Furthermore, it will be shown that the currently favorable competitive position of gaseous fuel retail prices relative to gasoline and/or diesel fuel is a function of the level of consumption tax abatements accorded these fuels by all governments in Canada. While it is possible to assume that governments will continue these tax abatements for a reasonably small volume for strategic reasons, it would appear unreasonable to assume that financial demands on governments will permit this to be extended to large shares of the transportation market unless compensating taxes are levied elsewhere. And, with the potentially adverse political consequences of such a levy known to all, this is not a workable option except possibly during periods of oil supply shortages. This confirms a reasonably modest short-term market share for gaseous fuels.

The strategic conclusions of this chapter will be:

1. that gaseous fuels have a role to play as transitional fuels between the oil-based era of today and future transportation markets that will be supplied by an unknown range of environmentally benign fuels;

2. that gaseous fuels will enjoy a reasonably modest share of the total transportation energy market in this period; and

3. that this share will be dominated by high fuel-consumption vehicles in specialized niche markets.

BACKGROUND

Before dealing with the developing use of compressed natural gas and propane as transportation fuels in Canada, it would be useful to circumscribe the subject.

Table 9.1
Canadian Energy Demand Structure

	1969	1978	1987
Growth in energy demand		54 %	13 %
Oil share of energy demand	49 %		35 %
Transportation share of oil demand	46 %		55 %

National Energy Board, Ottawa.

Since reliable market data are difficult to acquire in this area, the reader is invited to scrutinize carefully the assumptions that necessarily have had to be made.

The structure and trend of energy demand in Canada have changed quite significantly in the last two decades. As indicated in Table 9.1 this applies particularly to the transportation sector which remains over 98-percent reliant on oil-based fuels.[1] The reader will recognize 1978 as the year prior to the last crude-oil price shock in 1979; the first was in 1974. It would seem reasonable to conclude from these data that energy conservation and efficiency measures will continue to ride on the shoulders of oil and that the transportation sector will still drive oil.

When it comes to oil, Canada is not different from other countries: its objective is not only to use less but also to encourage others to do likewise, Canada's position as a net exporter of crude oil notwithstanding. Whether this goal is supported by supply-security imperatives, environmental concerns, or economic motives is not critical. As a matter of fact, the various reasons advanced for conserving oil could be mutually contradictory. For instance, a decision to use relatively more expensive domestic crude rather than less costly and less reliable imports would clearly frustrate the objective of improving the export potential of our industrial sector: it would instead weaken it by increasing its energy costs. Also, concerns over security of supply wax and wane over time, making it very difficult to base long-term policy on this factor. Finally, policy dictated by oil depletion assumptions may be dangerously inappropriate since there is a developing body of opinion that holds that environmental disaster may occur as a result of the continued burning of fossil fuels at current consumption rates long before their economic supply actually dries up. Canada wishes to use less energy, especially oil, no matter what motive is advanced.

Figure 9.1 represents a stylized model of the supply/demand balance for gasoline (61 percent of the total demand for transportation fuels in Canada) or its equivalent for some indeterminate forecast period. The reader should pause here to become aware of its limits. It obviously uses hyperbole and artistic license to make certain points; at no time does it purport to represent actual

Figure 9.1
Whither Transportation Fuels in Canada?

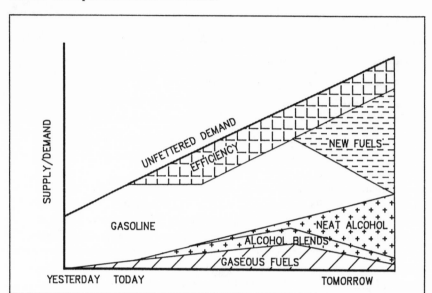

supply/demand trends. For instance, the unfettered demand line is clearly not representative of any forecast known to this author; its slope was chosen simply to allow the efficient graphical depiction of the theoretical supply elements required to meet any given demand.

With this in mind, one should now recognize the unfettered demand line as typical of the trend in the pre–1974 period. This demand trend was depressed after 1974 by various efficiency measures, from technological improvements in vehicles to changes in personal driving habits. As a result, in the decade following 1975, new-car fuel efficiency improved by over 50 percent. Although Figure 9.1 shows efficiency as growing to an absolute maximum, its true contribution probably will increase continuously, although at a diminishing rate.

The decline in the demand for gasoline, for alcohol blends, and gaseous fuels is shown taking place at a time—as yet unknown—when oil supply constraints become critical or when pollution concerns can no longer be ignored. The date of this turning point depends very much on the forecaster consulted. I belong to the school that feels that environmental pressures will force a turn before economic supply constraints become effective; this pressure will certainly mount from now on. As well, if the turning point for oil turns out to be environmentally determined, the turning point for gaseous fuels will obviously follow and will not coincide with it as shown in Figure 9.1.

To put these alternative fuels into some type of relative perspective, one should note: (1) since alcohol in gasoline blends currently does not exceed 10 percent of gasoline demand wherever blends are available, it is not likely to displace a greater share if blends ever become universally available; (2) as will be shown later, gaseous fuels are currently, in theory at least, financially attractive to between 15 percent (CNG) and 30 percent (propane) of the Canadian fleet of automobiles and light trucks at current market prices for transportation fuels; (3) neat alcohols such as methanol and ethanol are promising though uncertain replacements for gasoline and/or diesel fuel in the short to medium term (some consider them too hazardous or too costly for current use in gasoline engines; technical problems are preventing methanol from proceeding past the demon- stration stage in diesel engines at the present time); (4) new fuels such as hy- drogen, cost-effective electric batteries, and fuel cells (methanol?) on which our long-term transportation energy future may depend are not currently on anyone's medium-term economic horizon. Therefore, as far as the subject matter of this chapter is concerned, the data suggest that CNG and propane will continue to form part of the matrix of transitional fuels available to Canadians. But there is little evidence to suggest either that they will dominate this transitional matrix or that they can can survive into the long term. On the other hand, should new supplies of natural gas become available, this could extend the life of gaseous fuels past the transition phase.

Finally, before addressing the particular subject matter of this chapter, one should take particular note of the financial facts facing typical automobile owners as they become embroiled in the second most costly purchase decision of their lives—buying a car. Figure 9.2 depicts the typical distribution of the costs experienced by anyone owning and operating an average passenger car today.[2] With depreciation and financing charges accounting for about 70 percent of all first-year costs, it is clear that other cost elements pale by comparison. On the other hand, if this graph had been drawn at European prices, the capital cost share could have been quite different. In either case, because of the obscure nature of depreciation and financing charges, the average car owner usually remains oblivious to their true significance. And there is a corollary: without a reduction in capital cost, decreases in other operating costs would have a limited impact on total ownership and operating costs.

But, alas, decisions are sometimes more influenced by illusions and emotions than by reality. For example, whereas it is not uncommon for someone to select the more-expensive gas guzzler over the more fuel-efficient and less-expensive vehicle, it is commonplace to see this "irrational" buyer chasing "rationally" all over town after the lowest gasoline price!

As can be seen in Figure 9.2, the gasoline share of total costs increases as the vehicle ages. Therefore, to the extent that new-car buyers are at all motivated by fuel economy—an unproven hypothesis at low energy prices— any policy directed at influencing fuel choice should be most effective if di- rected at the owners of older vehicles, the largest pool of potential new car buyers.

Figure 9.2
Car Costs in Canada, 1987

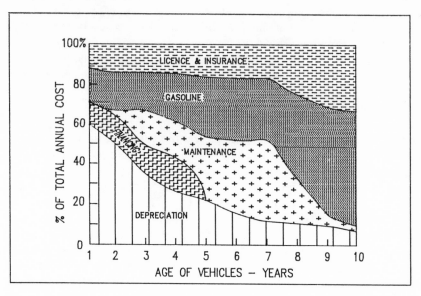

GASEOUS FUELS POLICY DEVELOPMENT

As thinking matured at Energy, Mines and Resources (EMR) Canada with respect to gaseous fuels in the transportation sector, it became increasingly difficult to segregate policy planning for propane from that for CNG. They were, after all, competing for the same gasoline and diesel fuel conversion candidates. Nevertheless, human nature being what it is, we initially found ourselves attempting to plan for one independent of the other. This was complicated by the fact that each fuel had its separate industry supporters: propane was in the hands of private-sector producers and distributors; CNG was effectively dependent on public-and private-sector natural gas utilities. Each fuel was and remains subject to different financial imperatives. Another factor is that whereas auto-propane accounts for about 30 percent of the domestic business of propane distributors, CNG probably accounts for less than 1 percent of the income of natural gas utilities.[3]

Nevertheless, we gradually came to realize that although propane and CNG were in fact competing for the same market they were nevertheless complementary, not competing, wings of a single gaseous fuel policy: to encourage consumers to use either fuel wherever it made financial sense to do so. Our intent was and remains to promote both fuels with the resources required to

Table 9.2
Gaseous Fuel Conversion Grants and Subsidies, 1988

```
CNG        : Federal Government - conversion grant of $500.
                                - elimination of sales and excise
                                  taxes on CNG.
             Ontario            - up to $1,000 sales tax rebate
                                  on conversion equipment.
             All Provinces      - 0 to 100% reduction of fuel
                                  (road) taxes.
             Gas Utilities      - conversion grants of $200-$900.

Propane    : Federal Government - elimination of excise tax,
                                  minimal sales tax.
             Ontario            - up to $750 sales tax rebate
                                  on conversion equipment.
             All Provinces      - 0 to 100% reduction of fuel
                                  (road) taxes.
             Distributors       - various conversion subsidies.
```

develop a self-sufficient industry capable of fostering the long-term growth of each gaseous fuel. So far, it would appear that the propane industry has become self-sufficient (at current subsidy levels) and is growing while CNG continues to struggle.

Government and industry support of gaseous fuels in Canada has followed two complementary paths: a federal government conversion grant, sometimes partially matched by provincial governments, fuel distributors and utilities, sometimes augmented by relaxation of provincial sales taxes on the purchase of the conversion equipment; and general, but not universal, reduction of federal and provincial government fuel consumption taxes. Table 9.2 provides an overview of these elements.

Since Canadian vehicle manufacturers do not currently market vehicles equipped with gaseous fuel capability as original equipment, those wishing to use propane or CNG must make a supplementary decision to spend the additional time and money to convert their new vehicles from gasoline or diesel fuel. It presently costs about $3,000 (Canadian) per vehicle for a two-tank conversion for CNG, and $1,500 (Canadian) for propane.[4] Since this represents a sizeable increase in the effective price paid for a new vehicle, consumers are reluctant to convert without capital assistance in the form of grants of one type or another, their purchase of other expensive nonproductive options notwithstanding. In fact, the natural gas utilities consider the federal government grant of $500 (Canadian) for each CNG vehicle conversion critical to the success of their efforts to increase their share of the transportation fuel market. On the other hand, the propane industry has reached a level of penetration where it appears to be able to carry on without the benefit of government conversion grants. Nevertheless, as was shown in Table 9.2, the propane industry considers conversion grants important enough to provide its own subsidies in some areas.

Table 9.3
Retail Price Structure: Toronto, April 1988 (Cents Canadian Per Liter)

	Gasoline	Diesel	Propane	CNG
Product Cost	20	20	16-18	12-15
Federal Taxes	10.0	6.7	0.2	
Ontario Tax	8.3	9.9		
Fuel.Stat.Costs - Full	8	8		
- Incr.			3	9
Fuel.Stat. Margin	4	4	4	7
	-----	-----	-----	----
Current Pump Prices	50	49	24	29
Implied Retail Price Differential		1	26	21

Source: Petroleum Products Marketing Division, EMR Canada

In addition to this capital assistance, the consumer also benefits from reduction of federal and provincial consumption taxes on gaseous fuels. Table 9.3 outlines the major cost elements in the retail price (pump price paid by the motorist) of the most common transportation fuels. Note that the indicated fueling station costs for gasoline and diesel represent the full costs of the station charged to these fuels, while the incremental costs for propane and CNG represent only the additional costs of the gaseous fuel portion of the station: excluded are such items as real estate and some capital costs, taxes, and administration costs that are borne only by gasoline and diesel fuel. Therefore, if equivalent grass-roots stations had to be built for gaseous fuels (they now generally share premises with gasoline and diesel fuel), their operating costs (and perhaps their retail prices) would have to rise dramatically.

The competitive effect of fueling station costs may be lost on those who attempt to explain market dynamics as a function of product costs alone. In this case, propane has a higher product cost than CNG but can be retailed at a lower pump price than CNG simply because of its lower fueling station costs. Also, the prices shown are as the consumer sees them, not as an energy analyst would consider them. The analyst would adjust them for energy content, a practice that would favor diesel relative to gasoline, and both relative to gaseous fuels. But we are concerned here with the consumer's perception, not with energy analysis.

As will be shown later, gaseous fuels appear to require pump price advantages of about $0.25 per liter to compete effectively with gasoline and diesel fuel. This means that while propane in Toronto seems to be competitive, CNG may be in trouble with respect to all fuels, particularly propane. Further, as mentioned previously, the current price advantage enjoyed by gaseous fuels is primarily a function of relief from government consumption taxes. The importance of this tax relief becomes even more evident when we consider that, while the incremental fueling station costs of gaseous fuels will decline as volumes rise, their retail pump prices will nevertheless have to rise as they start building and have

Figure 9.3
Kilometers Traveled and Fuel Used by Light Vehicles in Canada

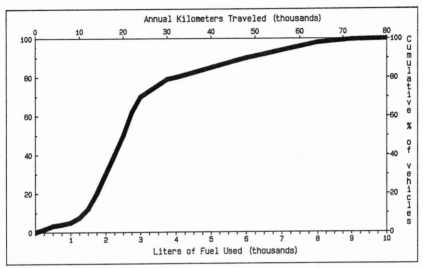

to pay the full costs of their own stations. It is also likely that the product-cost differential between oil-based and gaseous fuels will widen in favor of gaseous fuels only as crude oil prices escalate significantly, an unlikely prospect in the short term. Therefore, continued relief from government consumption taxes is critical to the survival of gaseous transportation fuels in Canada over the short to medium term.

Figures 9.3 through 9.5 attempt to put these financial considerations into further perspective for light vehicles.[5] The two horizontal scales are common to all curves in those figures; the lower scale converts the upper at the average fuel efficiency of the Canadian fleet of light vehicles. The thick vertical scale on the right in the three figures applies only to the thick curve.

The thick solid curve in Figure 9.3 is a rough approximation of the distribution of automobiles and light trucks in Canada. It shows, for example, that approximately 80 percent of vehicles travel less than about 30,000 kilometers per year, consuming less than 3,800 liters of gasoline in the process. Since this curve was developed from raw data that are still in need of further analysis, it should not be taken as anything more than the probable distribution of the Canadian fleet. Bearing this caveat in mind, we might still conclude that a very large share of the fleet (80 percent or so) travels less than about 33,000 kilometers per year and utilizes less than 4,000 liters of fuel. This is not unexpected since normal

Figure 9.4
Gaseous Fuel Vehicle Conversion Costs

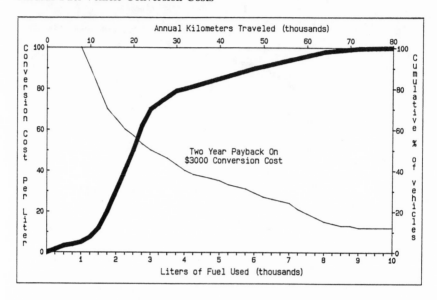

Figure 9.5
Gaseous Fuel Vehicle Conversion Costs and Paybacks

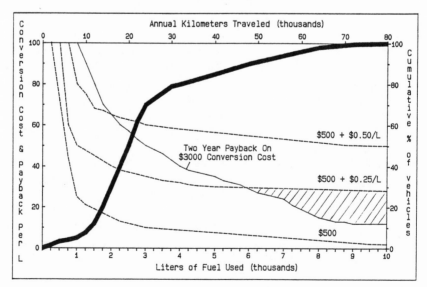

passenger cars, the bulk of the fleet, are located at the lower end of the spectrum. Taxis and other commercial vehicles account for much of the balance.

In Figure 9.4, the new curve represents the $3,000 (Canadian) conversion cost of a typical two-tank CNG conversion—allocated over the volume of fuel consumed in a two-year payback period. If one were to assume a longer payback, this cost curve would move down and suggest an apparently lower cost per unit of fuel consumed. However, it would be misleading since most consumers would not consider a payback in excess of two years; they would simply become disinterested as potential rewards became more remote.

Figure 9.5 incorporates consumer income equivalents into Figure 9.4. The lowest income line indicates the prorated value of the $500 (Canadian) federal government CNG conversion grant. Although conversion grants may loom large in the consumer's eyes at the time of new vehicle purchase, they cannot cover the complete cost of conversion. In fact, it could be argued that their value lies as much in their attention-getting marketing value as in their pure financial value. Therefore, lower retail prices for gaseous fuels are required to fill the breach. How much lower than gasoline they must be is an open question; it really depends on one's assumptions concerning the impact of this price spread on the consumer. One large Canadian propane supplier has said that the retail price of propane must not exceed 60 percent of the retail price of gasoline; this would indicate a propane price of about $0.30 per liter versus a current gasoline price of just under $0.50 per liter in Toronto. Others suggest a minimum price-differential for either propane or CNG of $0.25 per liter.[6] The evidence suggests that both points of view are valid. For example, it is reasonable to expect consumers to recognize readily and understand the advantage of saving $0.25 on a commodity that normally sells for $0.50 as opposed to another where the same absolute saving of $0.25 applies to an article costing $50. On the other hand, no matter how much the relative saving may be, it must still be sufficient to amortize completely the remaining conversion costs within the required payback period. The next higher income line in Figure 9.5 represents the combined value to the consumer of the conversion grant and a $0.25 per liter price advantage over gasoline. As it turns out, this appears to be close to the price differential typically available today in the Canadian market. The third income line ($500 conversion grant plus a $0.50 price advantage) is shown for reference only.

Based on where these cost and revenue curves intersect in Figure 9.5 (hatched area) with respect to the vehicle distribution curve, it would appear that about 85 percent of vehicle owners (those who travel less than 40,000 kilometers per year and consume less than 4,700 liters of fuel in the process) do not have a financial incentive to convert to CNG with the combination of a $500 (Canadian) conversion grant and a $0.25 fuel price advantage. Since lower conversion costs for propane vehicles would shift the break-even point downward, propane conversions would be unattractive to about 70 percent of vehicle owners at the same price differential. Some feel that manufacturers of vehicles dedicated to CNG (or propane) would be able to reduce substantially these high conversion costs.

But this has yet to be demonstrated. In any event, assuming reasonable accuracy in our data for the distribution of the Canadian fleet, this theoretically represents a potential market of about 2 million vehicles for CNG and 4 million for propane out of a national fleet of about 13 million cars and light trucks. This could require something in the order of 18,000 fueling stations for gaseous fuels. In Canada, there are currently about 20,000 gasoline service stations serving over 10 million vehicles, 120 CNG fueling stations serving 16,000 vehicles, and 4,000 propane stations serving 120,000 vehicles. The New Zealand experience and other independent analyses suggest that a CNG fueling station should accommodate a minimum of 300 vehicles. Hence, a 2-million vehicle market would theoretically require a maximum of about 6,000 CNG stations. The same assumptions would indicate a need for a maximum of about 12,000 propane fueling stations. Since these 18,000 stations would normally be simple adjuncts to service stations in the 20,000-outlet gasoline network, these data suggest that the indicated number of candidates for conversion to gaseous fuels is at least plausible in that the base of the required fueling station network is in place. However, should gaseous fuels be so successful as to penetrate this far into the gasoline market, they could no longer be considered as incremental additions at established gasoline outlets: competition would require them to bear their share of real estate and other fixed costs that currently are borne only by gasoline and diesel fuel. Also, the ability or desire of governments to continue current policies on consumption tax abatements on gaseous fuels at these high penetration levels is arguable. Therefore, it seems reasonable to designate much lower penetrations as the normal limits for either gaseous fuel; how much lower has yet to be determined. However, a combined target of 5 to 10 percent of gasoline demand might be practical. A more precise determination of the appropriate penetration level awaits further analyses concerning the effect on gaseous fuel profitability of increased sales volume per station, the competitive reaction to higher market penetrations, and government attitudes to continued relaxation of consumption taxes on gaseous fuels as their revenue bases erode.

GOVERNMENT OF CANADA PROGRAMS

Propane

When the EMR Propane Vehicle Program (PVP) began in June 1981, there were some 2,000 propane-powered vehicles in Canada. Before the PVP ended in March 1985, about 71,000 vehicles had been converted to propane as a result of the $400 (Canadian) per vehicle conversion incentive provided in the program. Another 30,000 or so, who were not initially eligible for this subsidy, converted on their own because of the financial attractiveness of propane. Further, since there appeared to be a plentiful supply of propane fueling stations at the outset of the program, no financial incentives were required for this purpose. At the end of 1987, two years after the PVP ended, there were an estimated 140,000 propane-powered vehicles in Canada supported by some 4,000 propane fueling stations. Although there are propane-powered vehicles in all provinces, the majority are in Ontario, Alberta, and British Columbia. Finally, analysis of the data in the last year of the program indicates that 82 percent of the conversions were on commercial (i.e., high fuel-consumption) vehicles, an important datum to be compared with the lower commercial share achieved in the CNG conversion program.[7]

Our efforts since 1985 have been directed toward working with provincial governments and the propane industry itself in order to foster continued growth in the demand for propane as an automotive fuel. This has included: (a) an annual grant to the industry association for market development; (b) an R & D partnership with the government of Ontario and Chrysler Canada to provide propane technology as original equipment on vehicles manufactured in Canada (this may be the ultimate long-term solution to the conversion cost hurdle); (c) a major demonstration of propane in public-transit buses in Ottawa; (d) promotion of the use of propane (and CNG) within the Canadian federal fleet; and (e) various public-information campaigns.

The steady annual increase in the demand for propane suggests that the industry marketing this fuel is healthy and should continue to grow as long as current retail fuel price differentials do not deteriorate. Central to its success so far has been the industry's concentration on market segments where fuel consumption is high: taxis and urban truck fleets. This market also requires the lowest investment in fueling station infrastructure. It is encouraging to note that the propane industry appears to be continuing its focus on these important market segments.

CNG

The companion program to promote the use of CNG has two wings: The Natural Gas Vehicle Program (NGVP) to encourage the conversion of vehicles to CNG, and the Natural Gas Fueling Station Contribution Program (NGFSCP) to assist in the establishment of the related CNG fueling station network. This program was announced by EMR in February 1983. It consisted of a direct contribution of $500 (Canadian) for each vehicle converted to CNG, and a maximum grant of $50,000 (Canadian) per fueling station to help defray the estimated $300,000 (Canadian) required to provide incremental CNG fueling capability at existing gasoline service stations. It must be noted that a considerably higher investment would be required to construct a grass-roots CNG fueling station.

By the end of 1987, EMR had partially funded the conversions of about 16,000 vehicles (virtually all of the CNG conversions in this period) and the construction of 108 CNG fueling stations. There were, in total, 118 CNG fueling stations. The NGVP was most successful in Quebec, Ontario, and British Columbia; it was not offered in the Atlantic provinces since natural gas is not available there.

Table 9.4
CNG Experience

		Vehicle Conversions		New Fueling Stations	
		Canada	NZ		Canada	NZ
1980			6,369			22
1981			10,494			36
1982		700	15,572		6	55
1983		1,000	20,235		6	58
1984		3,000	29,524		18	55
1985		4,350	23,683		34	89
1986		4,300	4,424		25	35
1987	June =	2,800	581	= March	36	15
		16,150	110,882		125	365

Source: CGA NGV Cost Investigation, Feb. 1988

These data clearly demonstrate the strategic importance of these two federal government programs. Not surprisingly, we have been asked to extend our efforts in this area beyond the 1988–89 fiscal year. Although these programs are administered by the federal government (through EMR), they are actually funded by the government of Alberta under a program to promote the use of natural gas in provinces to the east. In British Columbia (BC), the program is funded by EMR, the BC government, and the natural gas utilities of that province. In April 1988 a change was made in the conversion-grant delivery mechanism in order to produce a more cost-effective operation. Where it had been the practice for the consumer to receive grants directly from EMR, it will now be channeled through the utilities.

As indicated in Table 9.4, Canadian and New Zealand CNG conversions peaked in 1985 and 1984 respectively. It should be noted that these turning points coincide roughly with the period when international oil prices started to drop dramatically. However, although no one denies that falling oil prices helped to diminish interest in gaseous fuels generally and CNG in particular, opinion is not unanimous regarding the nature of this cause-effect relationship. Did falling oil prices narrow the retail price differentials between gasoline and CNG to a point where consumers no longer saw any financial advantage in CNG? Although differentials did narrow somewhat, they did not eliminate the incentive to convert. Did oil prices drop so much that, price differentials aside, motor gasoline prices became so low that fuel costs no longer attracted the consumer's attention? I am inclined to the latter view.

Where commercial vehicles accounted for about 82 percent of all propane conversions in the past few years, they provided approximately 66 percent in the CNG sector.[8] These ratios may explain why propane has been so successful to date while CNG has struggled. To begin with, since propane conversions were located largely in the high fuel-consumption commercial-end of the market, propane had a better chance to build a stronger financial base than did CNG with a lower commercial share. Other factors such as vehicle range, power constraints, limited fuel availability, and exaggerated safety concerns may also have been decisive. The relatively substantial incremental on-site costs at CNG fueling stations may also have weighed heavily on the natural gas utilities. Finally, with CNG representing such a small portion of the income of gas utilities, it might very well have been unable to attract the management attention required to make the program work.

Still, CNG has experienced some notable successes, and not surprisingly in the high fuel-consumption end of the market. Six CNG-fueled buses in Hamilton, Ontario, have performed so well in EMR-sponsored demonstrations that the Transportation Commission in nearby Toronto has announced its intention to purchase up to 125 CNG-fueled buses over the next three years to replace part of its fleet of diesel buses and electrically operated street cars. An order for 25 buses will be placed shortly for delivery in the fall of 1989. Should they perform as expected, the remaining 100 will be purchased. In addition, transportation commissions in Hamilton, Ottawa, and Mississauga (Ontario) are also considering the purchase of similar buses if the Toronto test shows promise.[9] This is a clear indication that financial imperatives and environmental pressures are opening doors for CNG (and propane) in these specialized and so-called niche sectors where fuel consumption is high and range is limited (thereby minimizing the need for fueling stations). As these markets sort themselves out, the alternative fuel of choice will be the one that best meets the technical, financial, environmental, and safety imperatives of the moment.

As regards the future, the Canadian Gas Association (CGA) adopted a natural gas vehicle business plan in 1986 that appears to address the major problems encountered earlier. Most important is the decision of the natural gas utilities to take a more active role in promoting conversions. Some in fact have hired special sales forces to solicit new business. Since they are specifically targeting higher fuel-consumption vehicles in urban areas, it augers well for a financially sound evolution of this phase of development of the CNG market.

CONCLUSION

Propane currently fuels about 1 percent (140,000 vehicles) of the small-vehicle fleet in Canada. Note that the market shares cited here are national shares and hence only show the national significance of each fuel. But these fuels are not marketed nationally, nor is national distribution likely in the near future, and thus these shares do not adequately demonstrate the actual penetration of the markets in which they do compete.

If current trends persist, propane could double its national share to about 2 percent (300,000 vehicles) by the turn of the century. This works out to an annual conversion rate of 3 to 4 percent of new vehicles. Since this continued market penetration is a function of the retail fuel price differential, this share could shift markedly with changes in the relative prices of oil and natural gas or with changes in government politics toward preferential tax treatment of gaseous fuels. Finally, we are satisfied that the auto-propane industry is in a healthy financial state in Canada and can lead a life essentially independent of additional government subsidies. Its fueling station network also appears to be adequate to meet the needs of a large increase in the propane fleet, should market forces or national policy ever require it.

CNG, on the other hand, still has a way to go. Its current 0.1 percent (16,000 vehicles) national share of the vehicle fleet is obviously embryonic. However, with the latest plans of the CGA to increase its marketing effort, we are confident that this share can grow to about 50,000 vehicles by 1991 and 100,000 vehicles by the turn of the century. This assumes a yearly conversion rate of 1 to 2 percent of new vehicles. Although this may seem ambitious when compared to the current market share enjoyed by CNG, it could also be viewed as the minimum penetration required to maintain the interest of the natural gas utilities upon whom the success of this program depends. Since the utilities are monopolies and usually dominate the fields in which they operate, it is difficult to visualize them maintaining their interest for very long in a market where profitability may be marginal, market penetration is low, and their monopoly status provides no benefit unless they are permitted to rate-base this portion of their business. So far, this option has not been tested before rate-granting authorities.

The attainment of these strategically important CNG shares depends on several factors, not the least of which is continued government support. Also critical is a retail price differential with gasoline that does not narrow. Also, price spreads notwithstanding, equally important is the absolute level of gasoline prices: even if they remain at current levels, they are low enough to make it very difficult to attract the consumer to CNG. Further, the utilities certainly must preinvest in the fueling station network required to support their target market segments. No matter what they do to promote conversions, they cannot succeed without an established fueling station network. However, they can minimize the potentially huge expenditures required in this area by property targeting their markets. If they plan to tap the entire vehicle fleet they will maximize their investment in infrastructure and hence minimize financial returns. On the other hand, if they exploit markets at the upper end of the fuel consumption scale they can minimize investment and maximize financial benefits.

What About Future EMR Involvement With CNG And Propane?

At a general level, one may legitimately ask why the bother for such a modest result. Less than 1 percent share of market for CNG and under 3 percent in aggregate for both gaseous fuels hardly seems to be of strategic significance. However, it must be noted that the potential is actually much higher: 15 to 30 percent of the small-vehicle fleet at current price differentials if the consumer would only take note. And from a government perspective, the potential is significant and could be tapped in times of an oil supply crisis, at no cost to the treasury, by simply increasing consumer taxes on gasoline and diesel fuel relative to gaseous fuels. Also, in an oil supply crisis, the price of oil-based fuels relative to gaseous fuels is likely to rise and hence improve the competitive position of these alternative fuels. Finally, it is becoming clear to policy planners that the transportation sector in the future will be fueled by a variety of alternative fuels: the days of the single transportation fuel are over. It therefore makes long-term sense to introduce consumers to a range of alternative fuels during this transition phase.

Consequently, in order to demonstrate leadership, EMR has identified candidates for CNG and propane conversions within the federal fleet. It is our intention to open doors to this market for both gaseous fuels. It will then be up to the industry protagonists to carve out their own shares of this potential.

We also intend to survey those who received conversion grants from EMR over the past several years to better understand what went right and wrong in the past in order to guide our future decisions. Since we already know that high conversion costs are a known drag on this program, we shall direct our R&D efforts toward reducing the cost of CNG conversions in particular by one-third by 1991. If successful, this is expected eventually to eliminate the need for federal government conversion grants. This research effort will be carried out in cooperation with the Canadian Gas Association whose NGV (Natural Gas for Vehicles) R&D program receives an annual grant from EMR. We are also exploring similar opportunities with provincial governments.

Since demonstration programs have been so successful in the past, they will continue to receive our preferred attention. In order to optimize future commercialization benefits we shall concentrate our efforts in areas where potential benefits are great: urban buses, heavy-duty trucks, and all locomotives—high-consumption vehicles generally.

In summary, the federal government intends to continue its support of the CNG and propane industries, but from now on in a slightly different fashion. Where market penetration so far has depended substantially on federal government financial intervention, the plan hereafter is to play a more complementary role. The industry, and especially the natural gas utilities, must take the lead.

In summary, propane and CNG, two wings of a single policy, are competing to satisfy the small though important demand for alternative fuels in relatively high fuel-consumption market segments. The success or failure of either or both fuels depends primarily on their respective abilities to exploit the market opportunities before them. Equally important, however, is the continued existence of an adequate retail price advantage over gasoline and diesel fuel. This is a function of international market forces over which the governments of Canada and its provinces have no control, and of taxation policy on fuel consumption over which they do. We shall work cooperatively with industry, other federal departments, and provincial governments to ensure that both gaseous fuels are still with us as we approach the age when as yet promising though unproven future fuels may provide us with more permanent and more environmentally benign transportation energy.

NOTES

1. National Energy Board, *Canadian Energy Supply and Demand 1985–2005*, Ottawa (October 1986): 139, 143.

2. Derived from data provided by the Transportation Energy Branch (TEB) of EMR and the Canadian Automobile Association.

3. TEB estimates.

4. TEB estimates. $1 Canadian = $0.80 U.S. approximately.

5. TEB estimates.

6. Superior Propane Inc., TEB staff.

7. TEB records and estimates.

8. TEB records and estimates.

9. TEB records and estimates.

10

SERGIO C. TRINDADE
and ARNALDO VIEIRA DE CARVALHO, JR.

Transportation Fuels Policy Issues and Options: The Case of Ethanol Fuels in Brazil

Brazil's Proalcool (pronounced approximately *proalcohol)* is the largest alternative transportation fuels program in the world today. About 7.5 million metric tons of oil equivalent (MTOE) of ethanol were consumed as transportation fuels in 1987 (equivalent to 150,000 barrels of crude per day). About 4 million vehicles, some 25 percent of the total auto fleet, run on hydrous ("neat") ethanol (96 percent ethanol and 4 percent water by volume). There is no pure (or "neat") gasoline available at the pump in the country today, but only gasoline C, a 20 to 22 percent (by volume) blend of anhydrous ethanol in gasoline.

The June 1988 retail prices for automotive fuels in Brazil were U.S. $0.61 per liter (U.S. $2.30 per gal.) for gasoline C, U.S. $0.42 per liter (U.S. $1.59 per gal.) for hydrous ethanol, and U.S. $0.29 per liter (U.S. $1.10 per gal.) for diesel fuel.

OVERVIEW

Ethanol fuels have been occasionally used all over the world since the inception of the automobile.[1] That was the pattern for many years in Brazil as well. In 1975, however, a new ethanol fuels policy set the stage for a radical change in the national transportation fuels market. Since then Brazil has become the first market in the world where ethanol fuels are a significant and durable feature.

In addition to supplying the domestic market, Brazil has exported large volumes with a peak of 1 million cubic meters (264 million gal.) in 1984, mostly for fuel use in the United States, but also for potable alcohol in Japan. Objections by American grain ethanol producers resulted in a substantial decrease in Brazilian alcohol exports to the United States after 1984.

Table 10.1
Transportation Fuels Consumption in Brazil, 1973–1987

Year	Diesel Fuel	Gasoline	Ethanol	Total Otto Fuels	Ratio Diesel/Otto Percent
1973	6.1	10.4	0.2	10.6	57.5
1975	7.5	11.0	0.1	11.1	67.6
1979	10.7	10.0	1.7	11.7	91.5
1980	11.3	8.7	2.0	10.7	105.6
1985	12.7	5.9	5.6	11.5	110.4
1987	14.6	5.8	7.5	13.3	109.8

Source: CNE

Ethanol's share of liquid fuels for ground transportation (in both diesel and Otto engines) increased from 0.5 percent in 1975 to 26.9 percent in 1987. The pace of market penetration of ethanol fuels is better illustrated, however, by the displacement of gasoline from the Brazilian domestic market, as both gasoline and ethanol are used primarily in Otto engines. On an energy basis the share of ethanol in this market increased from 0.9 percent to 56.4 percent during 1975–1987. Also significant in this period is the market penetration of diesel fuel. Meanwhile total liquid fuels consumed in Otto engines grew only by 19.8 percent to 13.3 MTOE in 1987. Table 10.1 contains the corresponding data.

This speedy change in the transportation liquid fuels market took place during a period of overall modest economic growth amid two international oil price shocks and increasing foreign debt.

Technology development in response to the rapidly expanding ethanol fuels market helped improve considerably the economics of production, transport, distribution, and utilization of ethanol. However, even when oil prices were still high in December 1985, ethanol fuels in Brazil were barely economical relative to gasoline for Otto engine applications. On the other hand, simple economic comparisons do not take into consideration ethanol's value in reducing oil imports, ethanol's superior environmental characteristics, and higher value as an octane booster. It is reasonable to expect that if demand for ethanol fuels keeps expanding, further technological improvements and therefore cost reductions will materialize. On the other hand, the long-term marginal cost of oil is likely to increase. Hence there may evolve a situation in the future where even in market economic terms, ethanol fuels can become a rational option for some countries.

The learning process that started in 1975 with the inception of Proalcool could benefit other countries as well, regardless of their different resource endowment and needs. This learning process actually started in Brazil in the early 1920s, then was accentuated in the 1930s in the wake of the prevailing international economic crisis. Since the 1930s, ethanol had been occasionally added to gasoline to hedge the important sugar industry against the vagaries of the international sugar market.

Nevertheless, ethanol fuels cannot be recommended without reservations to all countries in the world. In fact, only a limited number of countries, particularly those that generate an agricultural surplus but lack and are concerned about energy self-sufficiency should consider such a program.

Ethanol consequently does not seem a good universal fuel candidate in the sense that hydrocarbon fuels are today. Ethanol fuels however have an important role to play in many countries as a transition fuel into the future as well as an octane-boosting additive and clean-burning alternative to gasoline. In Brazil ethanol is already playing that role.

For various reasons methanol is likely to become a larger-volume transition fuel. Even in Brazil, methanol may play a role in the future, perhaps in conjunction with ethanol. Development of methanol-fueled engines could also benefit from the experience acquired with the use of ethanol fuels in Otto and diesel engines.

There are at any rate definite limits to market penetration of ethanol fuels connected with:

1. the inherent difficulties of penetrating the diesel fuel market;
2. the disruption of the crude slate of oil refineries caused by the market penetration of ethanol fuels at the expense of gasoline alone;
3. improvement in prices and availability of hydrocarbon fuels from international sources;
4. the initially unfavorable economics of ethanol valued strictly as a hydrocarbon fuel substitute;
5. the competition from other alternative fuel sources: natural gas, oil shale, tar sands, coal, and others; and
6. the development of longer-term transportation fuels solutions (e.g., generalized use of electric vehicles, hydrogen as fuel).

BRAZIL'S ENERGY SCENE

At the inception of Proalcool, Brazil's energy problem was characterized by uncertainty about prices and supply of liquid fuels for transportation and industry. Since then, major changes have taken place in energy production and use.

By 1975 most of the crude processed in the country was imported and was costing 4.5 times more in foreign exchange than just 2 years earlier. Furthermore, domestic oil production was lower than 10 years before.

During the period 1973–87 increased exploration for oil resulted in new findings of oil and gas. Currently over 600,000 barrels per day of domestic oil are being produced (mainly from offshore fields), corresponding to about 60 percent of total consumption of liquid fuels. Incremental oil reserves added per year have been larger than annual consumption. It is unknown how long this trend will persist and at what marginal cost future domestic oil will be obtained.

At the same time, a combination of economic recession, substitution for oil by alternative sources, and conservation have curtailed Brazilian industry's consumption of liquid hydrocarbon fuels, especially fuel oil. Also ethanol as hydrous fuel and in gasoline C has displaced over 50 percent of gasoline consumption. Market penetration of ethanol fuels in Brazil was aided by the availability of a distribution infrastructure for hydrous ethanol as premium gasoline was phased out of the market.

Consequently diesel fuels, for which no clear alternative was available, and whose price relative to gasoline decreased substantially during 1973–87, have gained the largest share of hydrocarbon fuels demand (52.3 percent of total transportation liquid fuels consumed in 1987 on an energy basis). Hence, the supply of diesel fuel today constitutes the core of Brazil's concern over self-sufficiency in transportation fuels.

In summary, Brazil's energy picture has gone through a period of fast change (compared with the usually slow pace of movements in energy systems).[3] The country's overwhelming reliance on hydroresources for power generation (39,000 MW out of 46,000 MW total installed generating capacity) has confined the uncertainties about hydrocarbon supply and costs to transportation and industry. Most of Brazil's primary energy is made up of renewable resources: hydroresources, wood, and sugarcane (converted into ethanol and bagasse). Total primary energy consumed by the Brazilian economy was 177.6 MTOE in 1986.

ETHANOL FUELS VECTOR

The concept of ethanol fuels ''vector'' constitutes a powerful and convenient representation of the chain of activities from production to utilization. The ''vector'' represents the ensemble of source–conversion–form–transport–storage–distribution–end-use. Applying this concept to ethanol results in Brazil's case with:

source: sugarcane

conversion: milling, fermentation, distillation

form: ethanol (hydrous, anhydrous)

transport: the infrastructure necessary to move ethanol from the distillery to storage and distribution centers and from there to retailers

storage: the storage infrastructure required to assure a smooth supply throughout the year to retailers

distribution: the infrastructure necessary to prepare ethanol-gasoline blends (gasoline C) and to move both gasoline C and hydrous ethanol to retailers, and sales to the public end-use: the final utilization of ethanol fuels in engines to provide the required transport services

The following discussion attempts to understand the policy issues and options involved in ethanol as alternative transportation fuels in Brazil and particularly the motivations of the many economic protagonists concerned.

THE KEY STAKEHOLDERS

The cast of actors playing significant roles in the market penetration of ethanol fuels in Brazil includes government, industry, and consumers.

The role of government is played by policy-making bodies and financial institutions. The ministries of Industry and Commerce (MIC), Energy and Mines (MME), Agriculture (MINIAGRI), Ministry of the Interior (MININTER), the Planning Secretariat of the President (SEPLAN), and the National Petroleum Council (CNP) were the early policy makers in this area. A National Alcohol Commission (CNA) was established, chaired by MIC and composed of the ministries previously mentioned. CNA was disbanded in 1979 and replaced by a National Alcohol Council (CNAL). That same year the National Energy Commission (CNE) was established and took a leading role in energy policy-making, including ethanol fuels. Within MIC the key player has been the Executive National Alcohol Commission, known as CENAL.

A number of incentives, financial, fiscal, and otherwise, resulting from the above policy-making bodies, supported the steep increase in ethanol production capacity since 1975. The principal actors in this field are the Ministry of Finance (MF), the Central Bank, national and regional development banks, and the Bank of Brazil (BB). The package of incentives has been gradually reduced over the past few years.

Industry is engaged in ethanol fuels through sugarcane growers, sugar millers, and ethanol distillers; distillery makers; liquid fuels distributors (freighters and marketers); oil refiners (Petrobrás); and automakers and auto dealers.

The overwhelming majority of ethanol produced in Brazil derives from sugarcane. Most sugar mills have attached ethanol distilleries. These byproduct distilleries are invariably owned by the sugar mills that supply them with molasses. They produce the cheapest ethanol in Brazil. The independent distilleries produce ethanol directly from sugarcane juice. They also own the bulk of the sugarcane they use. This activity is totally in the hands of the private sector.

Liquid fuels distributors in Brazil are all private (Shell, Esso, Atlantic, Ipiranga, Texaco, etc.) except for the largest one, the state-owned Petrobrás, which plays a pivotal role in the bulk distribution of ethanol fuels in the country.

The oil refineries with few exceptions are owned and operated by Petrobrás.

Automakers in Brazil are all private. The leading ones are Autolatina (a joint venture of Volkswagen and Ford), GM, and Fiat. Vehicles marketed in Brazil are built from scratch there. There is a heavy tariff wall against auto imports.

Fully assembled autos, completely knocked down (CKD) autos, and autoparts are important exports.

The motivations and rewards of the economic protagonists have not necessarily coincided during this 13-year period of fast market penetration of ethanol fuels.

GOVERNMENT MOTIVATIONS

Lingering behind the new ethanol fuels policies, the original Proalcool of 1975, and the neat ethanol program of 1979, were government concerns over security of supply of liquid fuels; prospect of severe imbalances in international trade; and projected slow pace of economic growth.

These concerns were exacerbated in 1979 with the onset of the Iranian crisis. This led to policies supporting a more radical approach to promote market penetration of ethanol fuels. Ethanol moved from its earlier role of hydrocarbon fuel extender to a fuel in its own right. These developments took place during the presidency of General Ernesto Geisel (1974–79), a former president of Pe-trobrás.

By 1979 some 85 percent of Brazil's oil consumption was imported, the bulk of it from Iran and Iraq. Brazil's 1979 oil imports cost more than 10 times the 1973 oil imports; the trend continued strongly upward through 1981.

The cautious Proalcool of 1975 was not providing the expected relief in the volatile environment of oil prices of the 1970s. The decision was thus made to bring a new nonhydrocarbon fuel into the marketplace at the fastest rate possible. This new fuel was hydrous or neat ethanol. Modified Otto engines, adapted or new out of the factory, were required to use this higher octane fuel.

The rush to launch neat ethanol encountered many difficulties relating primarily to the initial poor performance of many adapted vehicles and some new cars, and to the fluctuations of the price ratio of neat ethanol/gasoline.

These initial problems were overcome within a year to a large extent by the introduction or modification of the pertinent government policies. Today three-quarters of the ethanol consumed in Brazil is hydrous ethanol used by some 4 million vehicles. Over 90 percent of the new cars produced in Brazil currently are fueled by neat ethanol. The balance of the fleet runs on gasoline C. A U.S. $250 million loan by the World Bank in the early 1980s gave Proalcool a degree of international credibility.

Meanwhile the international oil market has undergone a radical change of its own. Government concern over security of international supply has waned. Oil prices during 1975–87 have moved up and then down as shown in Table 10.2. But all the time, despite significant technological improvements along the ethanol fuel vector, the economic value of gasoline has been lower than the cost of ethanol.[4] Measuring the value of ethanol for its octane-boosting quality, internalizing the environmental net benefits of ethanol fuels, or attributing a premium on the foreign exchange saving impact of ethanol would all improve the economic performance of ethanol.

Table 10.2
Brazilian Petroleum Imports, 1973–1987

Year	Avg. Oil Price US $/Barrel	Volume 10³bbl/day	Value 10⁶ US $	Share of Total Imports Percent
1973	2.54	652.8	605.2	9.8
1975	10.53	703.5	2,704.1	22.2
1979	16.83	1,019.7	6,263.5	34.6
1980	28.98	886.0	9,372.4	40.8
1985	29.70	545.1	5,749.3	43.7
1987	16.62	676.0	4,100.0	27.3

Sources: SOPRAL, "National Energy Balance," MME.
"Brazil - Programa Economico," Brazilian Central Bank.

Nevertheless, since 1975 Proalcool has by and large met the government's concerns, although not necessarily by its own devices.

Concerns over security of supply abated considerably, also due to the increase in domestic crude production which more than tripled during 1975–87, resulting in a two-thirds petroleum self-sufficiency.

The ultimate decrease in imported oil prices and the overall export drive of Brazil reversed the trade-balance picture. Analyses of the impact of Proalcool on the trade balance of Brazil have been highly positive when imports and exports of alcohol chemicals, ethanol, foreign capital and imported materials, gasoline, crude oil and derivatives, and lost agricultural exports are all considered. An estimate suggests a positive contribution by Proalcool to the balance of trade approaching U.S. $1 billion in 1983.[5]

Unfortunately the foreign debt kept increasing during this time, and its service offset most of the trade surpluses accumulated with the contribution of Proalcool, among others. Table 10.3 illustrates the argument. The growth in oil prices since 1973 is one of the causes of the increase in the outstanding Brazilian debt.

The rate of economic growth fluctuated during the period considered. Nevertheless most analysts agree that agricultural employment and industrial activities connected with ethanol fuels production, distribution, and utilization benefited from Proalcool.

MOTIVATIONS OF SUGARCANE GROWERS, SUGAR MILLERS, ETHANOL DISTILLERS AND DISTILLERY MANUFACTURERS

Sugarcane agriculture began in Brazil with the arrival of the first Portuguese colonists in the early 1500s. Since then it has been a traditional crop. It began

Table 10.3
Brazilian Trade Balance and Foreign Debt, 1973–1987

Year	Exports	Imports	Balance	Net Foreign Debt
1973	6.2	6.2	0	6.2
1975	8.7	12.2	-3.5	17.1
1979	15.2	18.0	-2.7	40.3
1980	20.1	23.0	-2.8	46.9
1985	25.6	12.2	+13.4	84.2
1987	26.2	15.1	+11.1	121.3

Source: FGV Conjuntura Economica, "Brasil - Programa
          Economico," Brazilian Central Bank

first in northeast Brazil, then spread to the rest of the country, particularly to the southeast region. More recently nontraditional areas in the south, central west, and north regions have been added as a result of Proalcool requirements.

Brazil plays a key role in the international sugar and molasses markets, and now in the newly created world ethanol market. Brazil has in all these markets the potential tonnage that could have an impact on prices for these commodities. Consequently Brazilian moves in these markets must be cautious to avoid potential damages to its interests.

Brazil is a country of continental dimensions with an area of 850 million hectares. Total cultivated area in the country is approximately 50 million hectares, of which sugarcane takes about 8 percent. Consequently in the case of Brazil there is no issue of availability of land for sugarcane growing. The issue is rather the competition for actual scarce resources, such as agricultural management skills, extension services, and agricultural credit. The sugarcane, sugar, and ethanol industries constitute a powerful economic group with strong political influence in the country.

There are two sugarcane harvest seasons in Brazil each lasting about 6 months: October through March in the north and northeast regions and May through November in the south, southeast, and central west regions.

Although the northeast region was historically the first sugarcane growing area in the country, today production is concentrated in the southeast. The new areas of north, south and central west have shown the fastest expansion in output, albeit from a very low base. This geographical spread of sugarcane and ethanol production promotes regional development, helps to check urban migration, and cuts down ethanol distribution costs.

Sugar millers and ethanol distillers in Brazil as a rule own the bulk of the sugarcane area required. They responded well to the incentives resulting from government policies to stimulate growth in ethanol output. From a low base, about 600 distilleries are in place today with a total ethanol capacity of 16 million cubic meters (4.2 billion gallons).

Ethanol producers, in both byproduct and independent distilleries, and sugarcane growers have in general been well rewarded by Proalcool. In the words of one of the industry's leaders, Cicero Junqueira Franco (SOPRAL)[6] on the occasion of the tenth anniversary of Proalcool in 1985:

Over the last 10 years of Proalcool, more than 50 million cubic meters (13.2 billion gallons) have been produced, equivalent to a gross income of over U.S. $20 billion shared as follows:

    34 percent to sugarcane growers

    27 percent to ethanol distillers

    12 percent to government as fiscal revenue

     7 percent to freighters, distributors, and retailers

    20 percent as superavit

The overall investment of Proalcool reached U.S. $6.5 billion, with U.S. $2 billion on sugarcane agriculture and U.S. $4.5 billion on ethanol distilleries. About 60 percent of the investment benefited from subsidized financing, and 40 percent came from owners' equity. Proalcool has saved an estimated U.S. $9 billion in foreign exchange by reducing the need for oil imports. Ethanol productivity increased from 3.1 cubic meters per hectare of sugarcane to 4.5 cubic meters per hectare in 10 years. Ethanol fueled engines have, on an energy basis, performances equal to or better than gasoline-fueled engines. Some 580,000 new direct jobs and over 2 million indirect jobs were created as a result of Proalcool, while the Brazilian economy as a whole suffered setbacks.

Other sources indicate that for a 16 million cubic meters per year ethanol capacity about 900,000 direct jobs have been generated, whose average pay is higher than jobs in the rest of the country.[7] Other experts also state that before Proalcool, on a national average basis, agricultural productivity for sugarcane was 46.4 tons per hectare and conversion productivity for alcohol was 57.4 liters per ton. By 1986/1987 these figures were 53.9 tons per hectare and 70.7 liters per ton respectively.

In the state of Sao Paulo, agricultural productivities up to 94 tons per hectare have been reached. Similar gains were obtained in conversion. Technological development due to Proalcool was disseminated to other agricultural fields. Future conversion efficiencies are expected to improve by 30–40 percent (33 percent on extraction, 60 percent on distillation, etc.); that means an additional 10 million cubic meters on the existing sugarcane acreage.

Makers of ethanol distilleries increased in number, introduced new technologies, and besides supplying and expanded domestic market, began to export as well.

The main concern of ethanol producers is with the pricing policy and the handling of purchases by Petrobrás and the cost of ethanol storage. Under the highly inflationary economy of Brazil, ethanol producers can become quickly insolvent if the ethanol prices are not adjusted in a timely manner for inflation. Government pricing policies are therefore crucial to Petrobrás, the key commercialization agent, and can have a marked impact on the cash flow of ethanol producers.

MOTIVATIONS OF LIQUID FUELS DISTRIBUTORS

The distributors of liquid fuels include the freighters, the marketers, and the retailers. The only marketer that has its own source of oil products is Petrobrás. All others buy from Petrobrás. Hence, the rewards of these private distributors focus on the margins they can realize. These margins are established by CNP and are the same to all distributors. Therefore their performance will depend primarily on their financial management, particularly in the inflationary economy of Brazil. Without Proalcool their total income during 1979–85 would have certainly been much smaller as a result of strong pressure on gasoline supply.

MOTIVATIONS OF THE OIL REFINER—PETROBRÁS

Besides being a distributor, Petrobrás plays a major role in the commercialization of ethanol to all other distributors. Petrobrás also provides all of them with their required oil products. As Petrobrás buys ethanol at prices set independently from the prices of fuels, such as gasoline C and hydrous ethanol, it has frequently experienced negative cash flows. This has led on occasion to confrontations between Petrobrás, the ethanol producers, and the government.

Furthermore, the abrupt penetration of ethanol in the fuel market has displaced gasoline, which has since been exported in increasing quantities, particularly to the United States. But the combination of surplus gasoline and increased demand for automotive diesel fuel imposed a severe stress on oil refining operations of Petrobrás. The objectives of the Brazilian oil refining policy has been self-sufficiency in oil products. As a result the oil refining capacity in the country has been kept systematically above oil products demand. To meet the growing demand for diesel fuel entirely out of domestic refining would generate surplus gasoline (in view of the penetration of ethanol into gasoline markets) that would not find an easy market abroad. The importation of diesel fuel on the other hand would weigh heavily in the trade balance. A perfect match between refinery crude slate and the peculiar demand profile for fuels in Brazil could only be met at great expense in refining investment.

This situation is currently under discussion at the CNE.[8] It becomes therefore clear that Petrobrás's motivations and rewards do not necessarily coincide with the government. Tables 10.4, 10.5, and 10.6 illustrate the above points.

Table 10.4
Production and Consumption of Oil Derivatives, 1986–1991

Oil Product	1986		1991 (Projected)	
	Refinery Crude Slate	Demand Slate	Refinery Crude Slate	Demand Slate
Diesel Fuel	32	34	37	40
Gasoline	16	13	10	8
Fuel Oil	18	17	18	18
Liquefied Petroleum Gas - LPG	7	11	8	12
Naphtha	10	11	10	10
Other	17	14	17	12

Source: CNE

Table 10.5
Production and Demand Annual Growth Rates for Oil Products Through 1995

Oil Product	Refinery Crude Slate	Demand Slate
LPG	10.9	12.6
Gasoline	13.3	5.3
Naphtha	10.4	11.2
Diesel Fuel	38.8	45.1
Fuel Oil	16.7	15.0
Other	9.9	10.8

Note: Total demand for oil products in 1995 estimated at 1.56
            million barrels/day.

Source: CNE

MOTIVATIONS OF AUTOMAKERS AND AUTODEALERS

The automobile dealer network in Brazil reached 2,500 in 1985.[9] The auto dealers are in a situation similar to the distributors as they are both intermediaries of products they do not manufacture in a market where most prices are directly or indirectly controlled by government. Hence they are supportive of Proalcool as long as their market benefits from ethanol fuels.

Table 10.6
Ethanol and Gasoline Exports, 1980–87

Thousand Barrels/Day

	1980	1981	1982	1983	1984	1985	1986	1987ᵈ
Gasoline Exports	6	25	27	35	72	81	58	96
Domestic Gasoline Consumption	192	184	180	151	136	128	147	122
Alcohol Consumption	46	14	64	59	112	139	184	184
Combined Domestic Consumption of Gasoline and Alcohol	238	228	244	240	248	287	331	306

ᵃEstimate

Source: CNE

The auto industry in Brazil is a strong economic agent contributing about 10 percent of the industrial GNP. Brazil's car fleet is the ninth largest in the world. The potential for growth is large as the number of vehicles per inhabitant is as low as 0.08 whereas it is over 0.5 in the United States. As an automobile producer Brazil is tenth largest in the world. As a vehicle exporter the country ranks eleventh in the world.

The contribution of the auto industry to the trade balance was negative through 1975. Since 1976 the industry has consistently generated a surplus of foreign exchange. From 1976 to 1985 the cumulative net balance reached U.S. $8.6 billion.

The direct jobs (mostly skilled) provided by the Brazilian auto industry amounted to 146,000 in 1985. An estimate of total employment related to automaking in Brazil in 1985 was 3.7 million people.

This powerful industry, mostly transnational, was naturally reluctant in the late 1970s to enter into the new market of neat ethanol engines, but in the end went along with the government programs to improve the performance of these new engines. The industry invested heavily in tooling and research and development related to ethanol fueled engines. The neat ethanol engine technology ended up benefiting as well the gasoline engines made in Brazil in terms of improved engine efficiency.

The introduction of neat ethanol engines into the market was not at all smooth as illustrated in Figure 10.1. Swinging demand put a severe stress on the industry during 1980–82. Stability has been achieved since 1983 as a result of improved reliability of engines and security of supply of neat ethanol at a price not higher than 65 percent that of gasoline C.

However, the level at which the market stabilized—around 95 percent of Otto

Figure 10.1
Sales of Neat Ethanol Vehicles as Percentage of Total Vehicle Sales (Passenger Vehicles)

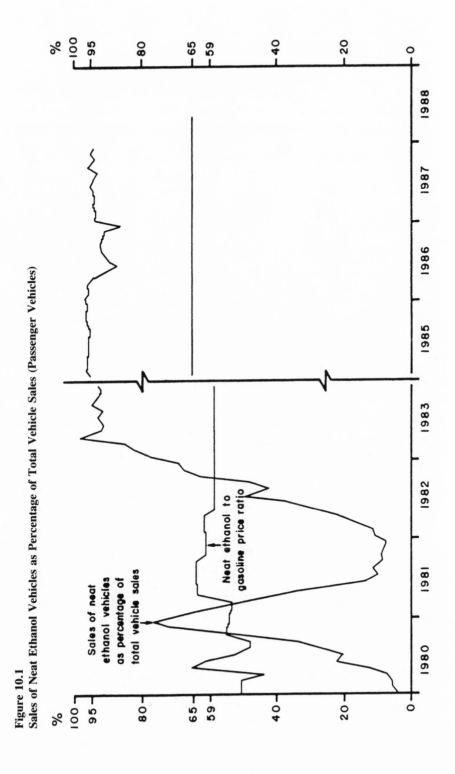

Table 10.7
Brazilian Vehicle Sales in the Domestic Market, 1973–87

Percent by Fuel Type

Year	Cars			Light Commercial			Heavy Commercial		
	Gasoline	Ethanol	Diesel	Gasoline	Ethanol	Diesel	Gasoline	Ethanol	Diesel
1973	100.0	-	-	99.5	-	0.5	36.4	-	63.6
1975	100.0	-	-	99.4	-	0.6	20.9	-	79.1
1979	99.7	0.3	-	82.6	0.9	16.5	1.9	0.0	98.1
1980	71.5	28.5	-	63.8	15.2	21.0	0.6	-	99.4
1981	71.3	28.7	-	37.5	11.1	51.4	0.1	1.6	98.3
1982	61.9	38.1	-	24.4	24.3	51.3	0.2	1.9	97.9
1985	4.0	96.0	-	4.8	68.5	26.7	0.1	3.0	96.9
1987	5.9	94.1	-	6.5	66.3	27.2	0.1	0.8	99.1

Source: ANFAVEA

Table 10.8
Brazilian Vehicle Sales in the Domestic Market, 1973–87

Thousands, by Fuel Type

Year	Cars			Light Commercial			Heavy Commercial		
	Gasoline	Ethanol	Diesel	Gasoline	Ethanol	Diesel	Gasoline	Ethanol	Diesel
1973	558	-	-	106	-	1.0	26.0	-	45.0
1975	661	-	-	118	-	1.0	17.0	-	62.0
1979	826	2.0	-	79	1.0	16.0	1.0	0.0	89.0
1980	567	226.0	-	60	14.0	20.0	1.0	-	93.0
1981	319	129.0	-	25	8.0	35.0	0.0	1.0	64.0
1982	344	212.0	-	21	21.0	44.0	0.0	1.0	48.0
1985	24	578.0	-	5	67.0	26.0	0.0	2.0	61.0
1987	23	387.0	-	8	72.0	24.0	0.0	0.5	66.0

Source: ANFAVEA

car engines—will result in gasoline being wiped out of the market in the near future. Tables 10.7 and 10.8 illustrate the point.

On the other hand, the auto industry was helped in the domestic market by the sales of neat ethanol engines in the early 1980s when sales of gasoline-fueled cars began to falter. Total domestic car sales in 1979 reached a peak in Brazil at 828,733 vehicles (826,462 on gasoline and 2,271 on ethanol). The following year the total was 793,028 (566,676 on gasoline and 226,347 on ethanol). However in 1981 the total was drastically smaller at 447,608 (318,929 on gasoline and 128,679 on ethanol) when the public was concerned with the effects of the Iran-Iraq conflict and was disappointed with the neat ethanol program. By 1985 the total level had partially recovered to 602,069 (23,892 on gasoline and 578,177 on ethanol); by then the neat ethanol engines and the supply of fuel had gained public confidence and the new engines had 96 percent of the incremental domestic car market. The smaller absolute figures for 1987 in Table 10.8 reflect a slowing down of economic growth in Brazil.

It is interesting to note that because automakers have increased exports of gasoline cars since 1985—by 1987 gasoline car production equaled ethanol car production—the Brazilian automakers are not so vulnerable to domestic market shifts.

MOTIVATIONS OF THE CONSUMERS

Public acceptance was crucial to the successful market penetration of ethanol fuels. The segment of the public most concerned are the owners of private cars and taxi drivers in particular. They constitute a minority of the Brazilian population as there is only one car for every 12 Brazilians in the country.

Taxi drivers are an important media to disseminate information about fuel and engine performance. After the initial fiasco of the neat ethanol-fueled cars in 1980–81, luring the taxi drivers into buying new cars at about 50-percent discount (i.e., waiving all taxes) was crucial to the success of the program. The improving quality of the engines was also important in regaining public confidence.

Originally consumers of neat ethanol were motivated by the lower capital and operating costs of the new vehicles relative to gasoline-fueled vehicles. Government made sure that taxes and credits would make neat ethanol cars more attractive to consumers. Consumers have also been attracted by the improving quality of ethanol cars. Moreover, the pump price of neat ethanol has been kept at 65 percent or less of the gasoline price, while ethanol cars consume 15–30 percent more fuel per mile than comparable gasoline cars. Maintenance costs are not appreciably different. Today the main incentive to consumers is the lower operating cost of ethanol-fueled cars mainly due to the lower fuel cost. Any changes in the price ratio of neat ethanol/gasoline elicits an elastic response from the public, as was demonstrated during the early 1980s.

On June 23, 1988, in a policy shift the ratio of retail prices of hydrous ethanol and gasoline C was raised from 0.65 to 0.69. This move narrows the gap between the operating cost of vehicles fueled by hydrous ethanol and those powered by gasoline C. This may encourage increased sales of gasoline-fueled vehicles at the expense of neat ethanol cars. The retail price of diesel fuel remained relatively low so that demand growth for this fuel is likely to remain unabated.

CONCLUSIONS

Government policy makers in Brazil must soon introduce new policies with the consumer in mind to arbitrate conflicts among the key economic stakeholders in Proalcool: Petrobrás, as commercializer of ethanol and petroleum refiner and marketer; ethanol producers who take most of the value added generated by the program but who are exposed to sudden bankruptcy as a result of pricing policies; and the automakers, whose industry depends on the availability of fuels.

Petrobrás faces the most crucial dilemma. As commercializer of ethanol it is at the mercy of the pricing policies set by government which have led on many occasions to negative cash flows. As oil refiner it is confronted, under a refining policy of self-sufficiency, with the need to invest heavily in refining to make the crude slate approximate the demand slate, where diesel fuel is fast approaching 40 percent or more of demand.

Should the refining configuration remain as it is today, the volumes of surplus gasoline generated may become harder to move economically. Furthermore the diesel-fuel volumes that would have to be imported to satisfy the growing domestic demand could push prices up. On the other hand, diesel fuel remains as a strategic fuel for the country since intercity transportation for both goods and passengers relies strongly on diesel-powered trucks and buses.[10]

The following are policy options to cope with a market where hydrocarbon fuels coexist with a large element of ethanol fuels:

1. bring gasoline back into the market; hold ethanol production at current levels; increase the supply of gasoline-fueled cars;
2. move gasoline into diesel engine markets (e.g., heavy Otto engine fueled by gasoline);
3. move ethanol into diesel engine markets (e.g. heavy Otto engine fueled by neat ethanol);
4. change refinery configuration at some cost (to minimize gasoline and maximize diesel fuel output).

Options 1 and 2 would help decrease gasoline surplus and cut a bit on diesel fuel expansion, and would likely be supported by Petrobrás and opposed by ethanol producers.

Options 3 and 4 would be supported by the ethanol producers and would likely be opposed by Petrobrás.

The fuels price policy change of June 23, 1988 is consistent with options 1 and 2 above. Gasoline is likely to make a comeback at the expense of incremental ethanol consumption. The relative retail price of diesel fuel remains very low (48 percent that of gasoline C and 69 percent that of hydrous ethanol). Consequently demand for diesel fuel is likely to expand unabated. If a refining policy of self-sufficiency prevails, then expensive refining capacity will be required.

The issue of the economics of Proalcool is an unsettled matter.[11] The methodologies used to assess social cost benefits of the program are not very useful as they invariably depend on a forecast of the oil price over time. On the basis of unregulated prices there is considerable evidence that ethanol has always been more expensive than gasoline. In the inflationary economy of Brazil the meaning of prices and costs gets a bit blurred. Keeping a price fixed against a backdrop of inflation is a practical way of providing subsidies informally. This happens often in Brazil to the benefit of consumers and at the expense of ethanol producers and of Petrobrás.

Valuing ethanol for its octane quality, internalizing environmental net benefits of ethanol fuels, and giving a premium value to foreign exchange saved by ethanol could change the economic assessment in favor of ethanol.

A 1986 producers' estimate valued anhydrous alcohol production cost at U.S. $0.27 per liter (U.S. $1.03 per gallon). That compares with an estimated value for gasoline of U.S. $0.15 per liter (U.S. $0.56 per gallon) in Brazil for crude at U.S. $18 per barrel. Table 10.9 details alcohol production costs.

Proalcool is a political program where economics can wait for the rebound of oil prices. It resulted from the political will of decision makers at the time. In the process of implementation it found a natural ally in the ethanol producers and an accommodation with automakers and Petrobrás. The consumers have been kept in mind, particularly after the initial debacle of the neat ethanol program.[12]

Other countries could learn from the Brazilian experience that there is a delicate balance among the interests of the alcohol producers, the refiner/marketer and the automaker/auto dealer. Careful attention to the potential consumer (and in the case of Brazil to the taxi drivers) pays handsomely in practice.[13]

In summary the following lessons can be learned from the Brazilian case:

- Decisions on alternative fuels must be seen in a long-term perspective.

- The issue of domestic production of motor fuels cannot be based on an uncertain and unpredictable crude-oil price picture. It may be better and safer to spend two dollars at home than one dollar on imports.

- The decision to start to use alternative fuels may be adequate against the background of most forecasts on future crude oil availability, origin, and price.

- Coordination of the refining industry and production of other motor fuels is a necessity, and preferential taxation of one oil product (diesel oil) should be avoided. The Brazilian refining industry seems not to be optimized concerning product slate or product qualities.

- Consumers react affirmatively to economic incentives relative to the cost of ownership and operating costs of motor vehicles and to consistent farsighted policies.

- Consumers can be very sensitive to the initial reputation of vehicles fueled by alternative fuels.

- Air quality improvements can be achieved.

Table 10.9
Anhydrous Alcohol Production Costs

COST COMPONENT		US $/m^3	US $/gal
DIRECT COSTS			
DIRECT CONVERSION COST			
Production			
Labor		6.06	
Inputs		3.14	
Fuel/lubricant		1.68	
Electricity		1.74	
Transportation		0.79	
Storage charges		1.82	
Other		5.04	
Administrative Expenses			
Labor		9.80	
O&M		9.80	
Other		11.40	
Revenues/tax		5.07	
Financial Charges		1.48	
Depreciation		18.97	
	SUBTOTAL	76.79	0.29
DIRECT AGRICULTURAL COST		121.88	0.46
	SUBTOTAL	198.67	0.75
INDIRECT COSTS			
INDIRECT CONVERSION COST			
Working capital cost		8.41	
Investment cost		26.00	
	SUBTOTAL	34.41	0.13
INDIRECT AGRICULTURAL COST		40.09	0.15
	SUBTOTAL	74.50	0.28
TOTAL ANHYDROUS ALCOHOL PRODUCTION COST		273.17	1.03

Source: CNE, SOPRAL March 1986

• Future supplementary domestic production of methanol from other feedstocks should be considered on economic grounds and is compatible with the ethanol usage.

Very few countries in the world should actually embark on an ethanol fuel program. Likely candidates include those with food surpluses and energy deficits. Landlocked countries without their own oil resources offer a more economic prospect for the penetration of ethanol in the transportation fuels market.

NOTES

The authors benefited from comments on early drafts of this paper generously provided in the form of personal communication by Ake Brandberg (SDAB-Sweden), Alberto Mortara (ANFAVEA-Brazil), Harold L. Walters (Ford-USA), Julio M. M. Borges (COP-ERSUCAR-Brazil), and Lourival C. Monaco (CNE-Brazil). Thanks are also due to Gilson G. Krause (PROMON-Brazil) who researched and checked the statistical data.

1. D. Sperling, "Brazil, Ethanol and the Process of System Change," *Energy* 12 no. 1 (1987): 11–23.

2. S. C. Trindade, *Implementation Issues of Alcohol Fuels: An International Perspective*, Proceedings of the VI ISAF, Ottawa (May 1984).

3. *Annual Bulletin of the Brazilian National Committee of the World Energy Conference*, Rio de Janeiro, 22, no. 32 (1988).

4. R. Seroa da Mota, *Proalcool: A Social Cost-Benefit Study*, Conference on the Brazilian Alcohol Program, University College, London (24 May 1985).

5. S. C. Trindade, *Brazilian Alcohol Fuels: An International Multisponsored Program*, Rio de Janeiro (1984).

6. *VI Encontro Nacional dos Produtores de Alcool ("National Meeting of Alcohol Producers,"* SOPRAL, São Paulo (1986).

7. J. Borges, *The Brazilian Proalcool: Its Development and Prospects*, Alcohol Week Washington Conference on Alcohol, Washington (1985).

8. *Política de Combustíveis Líquidos Automotivos ("Automotive Liquid Fuels Policy")*, CNE, Brasilia (1988).

9. *Brazilian Automotive Industry: Statistical Yearbook 1957/1986*, ANFAVEA, São Paulo (1986).

10. A. Oliveira, J. Lizardo, and R. H. de Araujo, *O PNA no Cenário Energético Nacional ("PNA in Brazil's Energy Scenario")*, IV Brazilian Energy Congress, Rio de Janeiro (August 1987) 313: 938–47.

11. W. Annicchino, *Evoluçao Económica e Social do Proalcool e Perspectivas ("Social and Economic Development and Perspectives of Proalcool")*, National Forum on the Future of Proalcool, São Paulo (28 September 1987).

12. J. Borges and C. Caracciolo, *"Economia e Eficiência na Produçao de Alcool no Brazil."* International Symposium on Genetics for Biological Efficiency in Production, Piracicaba (August 1987).

APPENDIX

THE ETHANOL FUELS POLICIES IN BRAZIL

Proalcool

On November 14, 1975, via Decree no. 76,593 the president of Brazil established the National Alcohol Program—later known as Proalcool—to supply the needs of domestic and foreign markets and the needs of the automotive fuels policy (article no. 1).

Accordingly production of ethanol from sugarcane, cassava, or any other substrate was to be promoted through the expansion of the supply of raw materials, with special emphasis on increasing agricultural productivity, modernization of new distilleries, byproduct, and independent distilleries and storage capacity (article no. 2).

The implementation of the Proalcool was assigned to the MF, MINIAGRI, MIC, MME, MININTER, SEPLAN, and the National Alcohol Commission (CNA) then established, whose members are representatives of the above ministries and whose chairperson was the Deputy Minister of Industry and Commerce (article no. 3). In 1979, via Decree no. 83,500, CNA was disbanded and a National Alcohol Council (CNAL) was established together with an executive secretariat (CENAL) to manage the implementation of the program.

Among the functions of CNA and later CNAL, performed in practice by CENAL, were the definition of the role of the various government bodies engaged in the program, to expand ethanol production and to establish criteria for location of new distilleries bearing in mind the following guidelines:

1. reduction of regional income disparities;

2. availability of production factors for agricultural and industrial activities;

3. ethanol transportation costs;

4. harmonizing raw material supply (e.g., sugarcane to expansion of existing distilleries and to new neighboring distilleries).

CNA was also charged with the responsibility of planning annual requirements of various grades of ethanol specifying end-users and to approve projects for ethanol making (article no. 3).

The National Monetary Council (CMN) was told what the subsidized financing of investment would be for distilleries and sugarcane (or the relevant agricultural substrate) development (article no. 5).

The crucial area of pricing of ethanol was defined in article no. 6. CNP and IAA were involved, and the parity price was such that 44 liters of anhydrous ethanol would have the value of 60 kilograms of crystallized "standard" sugar FOB sugar mill or FOB distillery. Today's "parity" price is set at 37 liters per 60 kilograms of sugar.

While the price to ethanol producers was set in "parity" with sugar, the price to fuel distributors was set independently by CNP. The ethanol consumers in the chemical industry would pay for the liter of 100 percent ethanol (at 20° C) the equivalent of 35 percent of the price of one kilogram of ethylene, as set by the appropriate government bodies (article no. 7).

Other provisions of the Proalcool decree covered the commercialization of molasses, the mechanics of the handling by CNP of cash flows generated in the purchase and sales of ethanol from producers to fuel distributors, and the role of IAA. CNP's role in financing ethanol commercialization of fuel purposes was later taken over by Petrobrás, a significant development at the root of the current issues facing Proalcool.

CNE—National Energy Commission

On July 4, 1979, via Decree no. 83,681, the president of Brazil established the National Energy Commission (CNE) to provide guidelines and criteria aiming at rationalizing consumption and the increase in the national oil production and its substitution by other energy sources.

On April 2, 1980, Decree no. 87,079 aimed at rationalizing energy utilization to save on energy inputs and gradually to replace oil products by domestically produced alternatives.

Pricing as a policy tool was the subject of CNE resolutions no. 3 of July 2, 1980 and CNE no. 4 of August 26, 1980 which set the following guidelines:

1. energy pricing policy should be flexible to stimulate production and consumption of domestic energy inputs;
2. gradual and continuous transfer of international oil prices (then expected to rise forever) into the domestic economy in harmony with the policy to combat inflation;
3. domestic energy inputs should be priced to ensure a reasonable return to private investment in the energy sector, resorting to subsidies if necessary;
4. make sure that conjunctural subsidies do not become sources of inflation.

The application of this policy to ethanol, gasoline, and diesel fuel would result in having gasoline and diesel prices reflect the move in international prices, and ethanol price to cover costs and profits for the producers.

Furthermore, resolution CNE no. 4 contained the following suggestions:

1. pricing of domestic energy inputs should encourage investment by the domestic private sector;
2. such pricing should consider:

    • price to producer to cover cost plus profit;
    • price to consumer to make domestic energy inputs cheaper than imported energy inputs;
    • gradual and systematic removal of subsidies;

3. price of unit of energy contained in domestic energy input should always be lower than that of the imported energy input;
4. hydrous ethanol price at the pump should be at the most 65 percent of the price of gasoline C.

Despite all efforts Brazil's dependence on imported crude increased from 80 to 85 percent during 1975–79. A radical departure from the original Proalcool policy took place. Ethanol fuel use would move from being a fuel extension (e.g., 20-percent blend with gasoline) to being a totally distinct fuel, hydrous ethanol, to penetrate the gasoline market at a fast pace.

CNE resolutions nos. 10 (May 14, 1981), 12 (January 5, 1982) and 14 (March 24, 1982) decided to expand ethanol output, to encourage automakers to produce neat ethanol vehicles, to change oil refining configuration to increase output of diesel fuel and liquified petroleum gas, and to motivate consumers via reduced excise tax on new neat ethanol vehicles and a lower ratio of neat ethanol to gasoline price at the pump (59 percent of gasoline C).

Further measures included provision at gas stations to ensure easy checking by drivers of the quality of hydrous ethanol such as direct reading densimeters in each pump. Ethanol is miscible with water in all proportions, but the fuel must not contain more than 4 percent water by volume.

The MIC was requested to work with the automakers to improve the performance of neat ethanol-fueled cars. The National Association of Automakers (ANFAVEA) on behalf of its members committed itself, under the Automobile Fuel Conservation Program (PECO), to substantial improvements in fuel efficiency of both gasoline and hydrous ethanol-fueled engines. To cut down the rate of increase in diesel-fueled light commercial vehicles, the excise tax on them was substantially raised whereas the reverse applied for hydrous ethanol-fueled light commercial vehicles.

The cost of storing ethanol was to be shared among producers, distributors, and government. Expansion of ethanol output to 14.3 million cubic meters (3.8 billion gal.) in 1983 and later to 20 million cubic meters (5.3 billion gal.) was authorized.

GLOSSARY OF ACRONYMS AND TERMS

ANFAVEA—Brazilian National Association of Automakers

BB—Bank of Brazil

CENAL—Brazilian National Executive Alcohol Commission (since 1979)

CKD—Completely Knocked Down (vehicle)

CNA—Brazilian National Alcohol Commission (1975–79)

CNAL—Brazilian National Alcohol Council (since 1979)

CNE—Brazilian National Energy Commission

CNP—Brazilian National Petroleum Council

DIESEL ENGINE—Compression-ignition internal combustion engine

ETHANOL—Ethyl alcohol or simply alcohol

FGV—Getulio Vargas Foundation

GNP—Gross National Product

MF—Brazilian Ministry of Finance

MIC—Brazilian Ministry of Industry and Commerce

MINIAGRI—Brazilian Ministry of Agriculture

MININTER—Brazilian Ministry of the Interior

MME—Brazilian Ministry of Mines and Energy

MTOE—Million metric tons of oil equivalent

MW—Megawatt (10^6 watts)

NEAT ETHANOL—A solution of 96 percent ethanol and 4 percent water by volume

NEAT GASOLINE—Gasoline without alcohol

OIL—Petroleum

OTTO ENGINE—Spark-ignition internal combustion engine

Petrobrás—Brazil's state oil company

PROALCOOL—National Alcohol Program of Brazil

SEPLAN—Brazilian Planning Secretariat of the President

SOPRAL—Brazilian Sugar and Alcohol Producers Society

11 JAYANT SATHAYE, BARBARA ATKINSON, and STEPHEN MEYERS

Promoting Alternative Transportation Fuels: The Role of Government in New Zealand, Brazil, and Canada

The use of non-oil-based fuels for motor vehicle propulsion has a long history, but it is only in the past 10 to 15 years that they have received serious attention as important fuels for the future of transportation. Interest in alternative fuels such as liquefied petroleum gas (LPG), compressed natural gas (CNG), and ethanol increased greatly after the 1973 oil price shock. The strong alternative-fuels policies pursued by some countries were shaped by world oil market conditions prevailing in 1979 and the early 1980s. It was expected that oil prices would continue to rise or at least remain at their high levels. Several countries with abundant natural gas or biomass resources chose to use these to produce substitutes for imported oil. The goals of increasing energy self-sufficiency and reducing the oil import bill generated the political support necessary for major programs.

Many countries began programs in the 1970s to do research and testing of alternative transport fuels. Historic use of alternative fuels such as LPG in Holland and CNG in Italy continued, ethanol blends were marketed, and a few countries began to substitute LPG in urban vehicles. Only in New Zealand, Brazil, and to a lesser extent, Canada, did the national government implement programs designed to bring about large-scale use of alternative fuels. Government efforts to develop a market for alternative fuels met with varying degrees of success. In this chapter we describe the programs that were implemented in New Zealand, Brazil, and Canada, the market adoption of alternative fuels, and the interaction between the government, industry, and consumers.

NEW ZEALAND

In the 1970s, New Zealand was largely dependent on imports for oil supply, but possessed substantial natural gas reserves.[1] The alternative fuels effort began in 1979 and was prompted by a large increase in the oil import bill, which by 1980 had risen to about 21 percent of export earnings. The government made a strong commitment to increase energy self-sufficiency, and this included substitution of petroleum used in transportation.

The Liquid Fuels Trust Board was formed to manage efforts to reduce the use of oil in transportation. By 1986, gasoline demand was displaced 35 percent by synthetic gasoline (produced from natural gas), 10 percent by CNG, and 3 percent by LPG. Synthetic gasoline was emphasized because its use required little infrastructure change and no adjustment of consumer behavior. The large government expenditure on synthetic gasoline is a questionable investment in light of current oil prices. We do not discuss synthetic gasoline here. Substitution of methanol produced from natural gas was rejected mainly because the country did not have its own integrated auto industry or extensive expertise in automotive engineering. Its domestic automotive market was small (about 90,000 new cars per year), and thus it had limited leverage in convincing overseas automobile manufacturing companies to produce dedicated methanol cars. Low-level methanol blends would have had too small an impact on overall gasoline consumption.

The CNG Program

The CNG program, launched in July 1979, was a joint effort by the government and the private sector. The target was the conversion of 150,000 vehicles to CNG by the end of 1985. This included all suitable government vehicles with access to CNG fueling stations as well as post office vehicles with their own fueling stations. The goal was later modified to 200,000 vehicles by the end of 1990. CNG use was restricted to the North Island for supply reasons. By 1986, 110,000 vehicles, or 11 percent of all cars and light trucks, had been converted, and 400 fueling stations were in operation.

To overcome institutional barriers, the government established a series of standards covering both vehicle conversion and refueling stations, and set up a CNG Coordinating Committee to address technical and regulatory issues confronting the industry. With the incentive of government grants, conversions began as planned in 1979, but in 1980 the rate began dropping off. In November 1980, the government increased the amount of the grant for conversion kits from N.Z. $150.[2] The cost of vehicle conversion then was about N.Z. $1,500: about U.S. $1,400 (U.S. $750 after the 1984 devaluation of the N.Z. dollar). The new grant offset the import tax levied on imported CNG cylinders and conversion components as well as the sales tax. The 25 percent grant for fueling stations now included related expenditures as well as equipment. Tax write-offs were extended to new vehicles converted in the factory. The road user-tax for CNG vehicles was replaced by a lower fuel tax.

Active Government Role. The government actively promoted the program with

Table 11.1
New Zealand: Government CNG Incentives

Grant Programs:

1979	N.Z. $150 grant for conversion kits.
	Grants to station owners to cover 25 percent of the wholesale cost of CNG compressors and storage equipment.
	Balance eligible for tax write-offs.
1980	N.Z. $200 grant for conversion kits.
	Fuel station grant extended to include related expenditures.
1982	N.Z. $200 grant extended to cover compressed biogas.
	Conversions encouraged on assembly lines.
1983-85	N.Z. $150 grant for conversion kits.

Loan Program:

1983	Low-interest loan program; N.Z. $500-5,000, 6 to 36 month loan.

Reduced Incentives:

June 1985	Limited loans to $1,600 a month.
	Raised interest rate from 10 to 17 percent.
	Required cash deposits -- 25 percent of conversion cost.

---

information and publicity campaigns. It encouraged dealers to maintain the price of CNG at half that of gasoline and publicly stated that any fuel tax changes would not disadvantage CNG. The retail price of CNG has varied between 65 percent (in 1979) and 42 percent (1984–85) of the price of gasoline. The average payback period for conversion, with government grants, was about two years. By the end of 1980, the conversion rate was 1,000 per month. 32,000 vehicles had been converted by the end of 1982 (Figure 11.1), and 83 public and 33 private fuel stations had been constructed.

Despite the incentives the program moved more slowly than expected. Private motorists were reluctant to convert to a new technology and had concerns about

Figure 11.1
New Zealand Program CNG and LPG Conversions

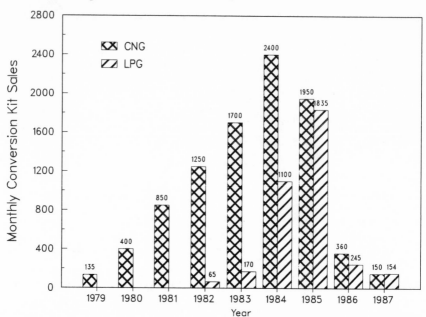

fuel availability. Fuel cylinder approvals were delayed, causing the most popular make to be withdrawn from the market. The quality of conversions performed by some installers was poor. Because of the slow rate of conversions, no large company moved into the refueling business. The private sector saw promotion as the responsibility of the government.

In 1983 the government strengthened support of the program and amended the target to 200,000 vehicles by 1990. A low-interest loan program was introduced to overcome motorists' reluctance to incur a high first cost for conversion kits. The loan program was run by the Ministry of Energy through the trading banks. The loans ranged from N.Z. $500 (U.S. $320) to N.Z. $5,000 (U.S. $3,200), and ran from 6 months to 3 years. No down payment was initially required, and interest rates were around 10 percent. Market research indicated that the availability of the loan was a deciding factor for over 60 percent of those converting their vehicles. Over 70,000 loans had been issued by the end of 1986.

A second important development was the financial commitment of Caltex to the establishment of a CNG network involving online stations and two mother/daughter systems. (Mother/daughter systems feature a main station connected with the natural-gas pipeline that serves remote satellite stations by means of trucks carrying compressed natural gas.) Industry formed the Alternative Fuels Association and the CNG Federation, and launched the CNG voucher scheme, under which motorists could receive up to N.Z. $300 (U.S. $200) worth of free

CNG in the two years following conversion. The need to improve the quality of conversion was recognized.

In 1984 the Ministry of Energy concluded that the level of public confidence remained fragile, the program was viewed as "belonging" to the government, the spread of refueling facilities was uneven, and certain features of the programs (quality of conversions) remained vulnerable to abuse.[3] To bolster the program, the government enlisted greater support of the private sector for marketing and publicity. The fuel station grant program focused on priority areas, using mother/daughter stations to fill gaps and to extend the coverage of CNG stations to the entire North Island. An installer registration program was tied to loan eligibility, and consumer complaints dropped sharply.

An announcement that financial incentives would terminate on December 31, 1987, coupled with rising gasoline prices brought an increase in conversions in 1984. By the end of 1985, about 100,000 vehicles had been converted, and about 300 refueling stations were supplying CNG. Industry was optimistic as the government was considering a large aid package for the following year.

Cutback in Government Support. The dependence of the program on government incentives was demonstrated in 1985. A newly elected administration believed that the rapid growth in conversions indicated that government support could be reduced without greatly impairing the program. The number of CNG conversion loans was limited to 1,600 per month, interest rates increased from 10 percent to 17 percent, and cash deposits equal to 25 percent of the conversion cost were required (the amount of deposit was later reduced to 10 percent). These changes were introduced overnight in June 1985. The only remaining government incentives were for fueling stations established in new areas. Conversions fell off dramatically, from 4,200 in June to 1,000 in July to 240 in December 1985, as the public no longer perceived sustained support for CNG. The conversion industry felt abandoned with their new equipment stocks no longer moving. The oil price drop in early 1986 decreased activity even more.

In July 1986, the government attempted to revive the program by enlisting private industry to take over the leadership. Gas wholesalers are now promoting lower-priced retrofits (often below cost). Conversion prices have fallen by as much as 50 percent as the industry struggles to remain competitive. Gas Association no-deposit conversion loans at market interest rates replaced government loans in late 1986.

Analysis. Since CNG was sold at market prices, the program cost less than it would have had fuel price supports been used. Three-fourths of the government expenditure was in repayable loans, and a savings on foreign exchange of N.Z. $83 million was realized between 1980 and 1985. CNG displaced 875,000 barrels of gasoline in 1985, which amounted to about N.Z. $49 million (U.S. $25 million).[4] The savings thereafter have been less because of the lower price of gasoline. Our cost-benefit analysis of the program indicates that with 1985 prices of fuels the CNG program was clearly beneficial to the economy, but at the lower prices of fuels in 1986 the conversion to CNG was barely economical.

The CNG industry is in a period of retrenching, with sales of new conversion kits much lower in 1986 and 1987. The industry is trying to export CNG technology to countries in Southeast Asia, Latin America, Canada, and Australia.

The LPG Program

The government promoted LPG less vigorously than CNG. LPG supply and distribution facilities were limited, and a greater national benefit was perceived from the use of CNG. LPG distribution facilities were slower in being constructed, and the rate of vehicle conversions was much more gradual. There was a definite bias towards CNG in the government-backed publicity programs, which offset some of the disadvantages of that fuel compared to LPG. At first LPG supply was constrained by production capacity; later the bottleneck became transport capacity. A new bulk LPG distribution company, Liqui-Gas, was formed comprising national and international petroleum companies. The national bulk distribution system came into operation in 1985, and LPG supply finally surpassed demand. LPG was promoted more vigorously on the South Island because it was difficult to transport natural gas from the North Island.

The government promoted LPG by providing grants and soft loans to the distribution system. LPG installations received grants, and the government paid a N.Z. $0.19 per gallon subsidy for LPG shipped to the South Island. A single national price for LPG from the bulk depots was established (which is then varied at distribution points downstream). The 25 percent fueling station grant program for CNG was extended to LPG once supply was readily available. Conversion incentives applied to LPG as well as CNG. Grants of N.Z. $150 (U.S. $110) for conversion kits were available for LPG on the South Island. The energy conservation loan program, new car grants, and tax write-offs were available for LPG facilities and conversions. Promotional schemes were similar to those for CNG. In 1988, the only remaining government incentive for LPG is the grant for fueling stations, which applies mainly to the South Island.

LPG conversions climbed strongly in 1984 and 1985, but dropped sharply in 1986 and declined further in 1987. Analysis done by the Ministry of Energy shows payback periods for LPG conversion ranging from 2.4 to more than 5 years, depending on annual distance traveled.

BRAZIL

Brazil has had the most ambitious program for alternative transportation fuels in the world.[5] Domestically produced ethanol supplied 22 percent of automotive transport demand in 1985, and ethanol use was nearly as great as gasoline use (Figure 11.2). In 1986, 90 percent of new car sales were dedicated ethanol vehicles. In addition, a program to use CNG as a diesel substitute in transport began in 1986, with a goal of eventually converting one-third of the truck and bus fleet to CNG.

Figure 11.2
Brazil: Fuel Use in Road Transport

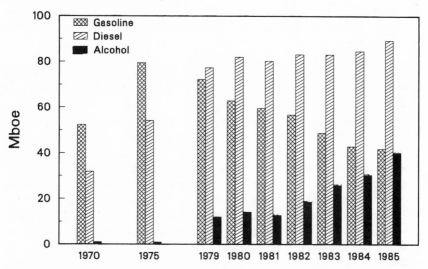

Source; *LBL Data Base, May 1987*

The Ethanol Program

The ethanol fuel program began in 1975 as a response to a combination of world and domestic economic conditions. In part due to higher oil prices, the annual growth rate of the economy fell and the foreign debt became a large burden. At the same time, the large sugar industry, which had modernized and increased its output by the 1970s, suffered from a sharp drop in world sugar prices in 1975. Against this backdrop, Proalcool, the National Alcohol Program, was born and implemented in two phases. The first phase (1975–79) focused on increasing the ethanol percentage in gasohol to 20 percent nationwide. (Alcohol in the form of ethanol produced from sugarcane had already been used in some gasoline as a low-level blend since 1931, mainly to aid the sugar industry.) The second phase (1979–85) was a major shift to produce and supply dedicated ethanol vehicles. A third phase, proposed in 1985 to expand the use of dedicated ethanol vehicles and increase ethanol production, is now on hold due to the world oil price drop. However, the popularity of ethanol in the automobile sector achieved in the second phase continues.

Responsibility for the Proalcool program was distributed among various agencies that managed distillery construction, ethanol production and prices, gasoline blending, and credit. As a producer incentive, ethanol was purchased (originally by the National Petroleum Council and then by Petrobrás, the national petroleum company) on a sugar-equivalent basis: prices and quotas were fixed by the

National Institute for Sugar and Alcohol on a cost basis. Additional incentives included producer credit subsidies, which paid for up to 75 percent of investment. The government assured that sugar mill/distilleries received a 6 percent return on their investment as long as they promised to produce ethanol and not export the sugar instead. They were thus able to diversify their investment and sell ethanol to the government while sugar prices were low.[6]

Despite early difficulties, Proalcool's first production goal was reached by 1979. Alcohol fuel use in transport rose from 2,600 barrels per day oil equivalent (BDOE) in 1970 to 33,000 BDOE in 1979. This first phase was successful because it required little technological or institutional change. Since ethanol was already being blended with gasoline and the new blend was only 20 percent, vehicles and fueling stations did not require modification. Nor was supply a problem. Distillery equipment was already manufactured domestically. Many distilleries at existing sugar mills were underutilized and were easily brought into production. New distilleries at mills could be built within one year.

The automobile industry played a supportive but cautious role regarding future expansion of the program. It did research demonstrating that there were no technical barriers to dedicated ethanol use in automobiles but was reluctant to produce ethanol vehicles in large quantities without a guaranteed market.[7]

Promotion of Dedicated Ethanol Vehicles. The 1979 oil price rise combined with a mounting foreign debt prompted the government to launch a new phase of the ethanol program. This called for a massive switch to dedicated ethanol vehicles, aiming to reach 50 percent of vehicle sales by 1985. The ethanol production target was 2.8 billion gallons by 1985, which required capacity to increase by 150 percent. The investment goal was $5 billion in fuel production and distribution facilities. Most of the funds came from an Energy Mobilization Fund, which was to generate $1.25 billion annually from fuel taxes, vehicle licensing fees, and other sources. In addition, $1.2 billion was borrowed from a consortium of 51 foreign banks, and the World Bank made a $250 million loan.

Consumer incentives to buy the new vehicles were lower purchase taxes, lower registration fees, smaller down payments, greatly extended repayment periods, and lower fuel costs. The pump price of ethanol was now guaranteed to be no more than 65 percent of gasohol, giving ethanol about a 20 percent advantage in cost per mile. Improvement in the efficiency of alcohol cars has increased this advantage. The actual ratio of ethanol to gasoline price varied over time. Automobile manufacturers were encouraged by the strong government commitment, consumer incentives, and assurance of sufficient fuel supplies.

Initial sales of ethanol cars and retrofits were so high that some Brazilian vehicle manufacturers implemented only the minimum modifications required for all-ethanol vehicles, resulting in less-than-optimum efficiency and consumer dissatisfaction. (This was mainly due to lack of knowledge.) Unauthorized mechanics began to offer inferior conversions at lower prices. Mounting complaints caused a fall of sales, but ethanol supplies were still insufficient to meet the demand as it became more profitable for some producers to export the alcohol.

It appeared that the economic consequences of the program were not as beneficial as had been hoped. The generous credit required to begin the program had contributed to inflation. World sugar prices were predicted to rise again, increasing interest in exporting sugar rather than producing alcohol. In response, in late 1980 the government began to increase ethanol prices from as low as 40 percent of gasoline toward the 65 percent limit. In June 1981 credit subsidies for distilleries were suspended. With the sudden shift in the price ratio, sales of ethanol vehicles plummeted, revealing the dependence of the program on strong government support as well as the speed of consumer response to a change in the commitment.

Meanwhile, the massive foreign debt continued to rise, petroleum import supplies were uncertain and prices were unpredictable, and world sugar prices remained low after all. The government reassessed its position and moved to bolster the ethanol program in late 1981 and 1982. To restore consumer confidence, ethanol prices were held at 59 percent of gasoline prices for two years. This gave ethanol about a 29 percent cost advantage over gasohol. Previously removed incentives such as reduced ethanol automobile sales taxes, favorable credit conditions, and ethanol availability on Saturday were restored. Manufacturers were encouraged to improve vehicle quality and increase consumer incentives, measures that had a strong effect on sales.

Alcohol producers received more attractive credit and were exempt from an agricultural production tax imposed in 1982. Even though the level of investment subsidy was reduced from Phase 1 levels, the government still paid an estimated two-thirds of capital costs for the portion it financed. Production of ethanol began to increase, and a substantial reserve had built up during the bust period. Consumer incentives caused ethanol vehicle sales to increase rapidly in 1983, and demand remained strong in 1984 and 1985 (Figure 11.3).

A third phase was proposed in 1984 calling for increasing ethanol production to 3.7 billion gallons by 1987. This was to be met by enlarging the cultivated area of sugarcane and by improved production efficiency and management. At a fairly advanced stage in the negotiations, a proposed World Bank loan was frozen due to the 1986 oil price drop.

Analysis. While the technical and implementation success of Proalcool is certain, many critics argue that with today's oil prices, the program's costs exceed its benefits. Government investment for Proalcool amounted to $3.7 billion between 1976 and 1985, and industry invested $2.7 billion. Estimated foreign exchange savings were $8.9 billion. Some analysts placed the real cost of ethanol in 1983 at $40 to $65 per barrel of gasoline displaced for southeast production, and $100 for northeast production.[8] Weiss argues that replacement of gasoline imports by producing ethanol and dedicated ethanol cars is not justified at an oil price less than $30 per barrel, assuming operation of existing

Figure 11.3
Brazil: Vehicle Sales by Fuel Type

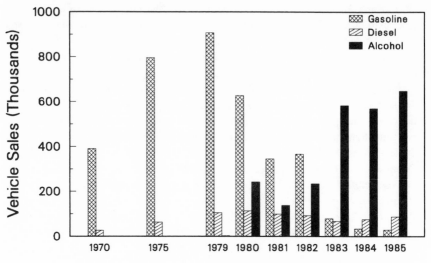

Source: *Industria Automobilistic Brasiliera, pp 84-86, 1986*

ethanol plants with current technology.[9] With improved production technology, the threshold drops to $23 per barrel.

Proalcool has social benefits and costs not directly treated by economic analysis. One widely debated issue is that of "food-vs.-fuel," whether sugar production decreases the amount of land available for food crops or whether sugar is grown on lands that would be otherwise marginal or pastureland. Job creation has occurred through the huge infrastructure of the program, though the nature and distribution of these jobs is controversial. The effect on concentration of land ownership is another issue of debate. The main environmental impacts have been from distillery emissions in rural areas and improved air quality in the cities from burning alcohol rather than gasoline.

Despite low oil prices, the government has remained committed to the ethanol program. While construction of new distilleries was prohibited after the oil price drop, fuel price subsidies continued. In 1985 ethanol prices were decreased to below the cost of production to maintain the relative advantage compared with gasoline. While IMF austerity measures have been implemented throughout most of the economy, alcohol production programs have been partially exempt from subsidy cuts. The large expenditures for the ethanol program (17 percent of the 1983 budget) have been justified by the belief that the balance of payments deficit would benefit. This perception has been encouraged because distillery equipment as well as ethanol vehicles are produced domestically.

Proalcool was successfully implemented because the government played a critical role in moving past the blending stage to use of neat ethanol. By putting a ceiling on ethanol prices it circumvented market signals that would otherwise have delayed transition to alternative fuels. The strategy of concentration in one region also benefited the program: initially more than half the alcohol was produced in the state of São Paulo, and most of the fuel was consumed there. This allowed economies of scale in production and distribution, and increased the density of fuel outlets and support services.

Despite the criticism from some quarters, Brazil's ethanol program has become entrenched within the economy. Powerful interests have been created in the sugarcane and alcohol production industries, and over one million jobs depend on the survival of Proalcool. Expansion of land devoted to sugarcane is being limited, however, and Petrobrás is resisting purchasing all of the alcohol that is produced.

CNG for Diesel Fleet Vehicles

In 1982, extensive natural gas fields were discovered in Brazil. A pipeline from the new fields to São Paulo should be completed in 1988. As the cost of developing a distribution network for households is high, the national energy plan calls for increased use of natural gas in transportation. Currently, the focus is on substitution of natural gas for diesel fuel in urban transit. The first target is diesel buses, with trucks and light vehicles expected to form an additional market. Eventually, 40,000 city buses are slated for conversion. A dozen cities now have CNG-powered buses, and the municipal bus company of São Paulo has plans for using 500. Mercedes Benz of Brazil has developed an advanced diesel engine with a full gas mode option, and has started production of dedicated CNG buses.

CANADA

The National Energy Policy formulated in 1980 called for oil self-sufficiency by 1990, with a combined strategy of conservation and oil substitution. Use of alternative fuels in transportation was part of the strategy.[10] Government efforts focused first on LPG, for which there was surplus supply, and then on CNG. Alcohol fuels have received less emphasis, although some provinces have incentives for alcohol/gasoline blends. The success of Canada's alternative fuels programs has been influenced by the policies of both federal and provincial governments.

The LPG program distributed Can. $28 million (U.S. $20 million) in grants, and lasted from 1980 through March 1985.[11] The goal of 130,000 vehicles converted was met with the help of incentives for vehicle conversion. The propane industry has assumed responsibility for continuing the program's momentum. The total of conversions and new vehicles using LPG is now about 20,000 per year, slightly lower than the level in 1985–86. The CNG program has cost the government Can. $7 million (U.S. $5 million) for grants, and is still in place.

The fuel station target has been met, but the conversion goal of 35,000 vehicles has not. Provincial government and gas industry incentives supplement the federal CNG programs. The federal government has also spent about Can. $4 million (U.S. $2.8 million) per year on research, development, and demonstration for alternative transport fuels.

The LPG Program

In 1980, Canada was producing surplus LPG, mostly from natural gas. Prompted by the surplus, expectations of higher gasoline prices, proven LPG use in vehicles, and ready availability of conversion equipment, the federal government began to promote use of LPG. It believed that a threshold level of market penetration was necessary for sustained use of the fuel, as the propane supply industry was neither large nor organized. The objective of the program was to create an LPG market of 1–2 percent of road gasoline demand, eventually to reach a level of 5 percent, or about 400,000 vehicles.

Federal and provincial governments launched a coordinated program in 1980 to remove regulatory, supply, capital, and information barriers to expanded use of LPG. The primary federal program was the Propane Vehicle Grant Program, which gave a taxable grant of Can. $400 (U.S. $280) for converting a commercial vehicle to LPG or for purchasing a new LPG vehicle. In 1984, this program was extended to cover private vehicles as well. The federal government also reduced the tax on LPG.

Provincial government incentives also promoted use of LPG. The main incentive was removal of taxes on LPG.[12] LPG pump prices have been 66–78 percent of those of gasoline on an energy equivalent basis. In British Columbia and Ontario, the sales tax was removed for dedicated LPG vehicles and conversion kits.

Most of the 130,000 vehicles converted between 1980 and 1985 were commercial vehicles. At the end of 1984, LPG vehicles had captured 3.3 percent of the commercial market. The largest market was the manufacturing/wholesale industry, followed by the service industries, especially taxis, and then private vehicles, mostly light trucks.

Sales of tanks and LPG vehicles declined after the program was discontinued in March 1985 (Figure 11.4). The price of gasoline did not decrease until after March 1986 and thus was not a factor in the decline in LPG tank sales, but may have influenced calendar 1986 sales of vehicles. The decline in sales occurred despite the relatively short payback period, which in a typical case was three years with the government grant (and about four years without it).

Although LPG is favorably priced compared with gasoline, there are deterrents to its use. One (not insignificant in the cold Canadian climate) is that LPG

Figure 11.4
Canada: Monthly LPG Tank and Vehicle Sales

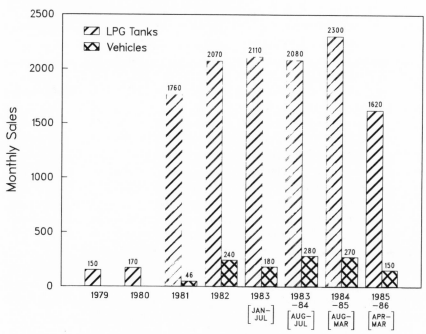

vehicles are not allowed to park indoors in public garages as a safety precaution to avoid explosions from ignition of spills of LPG.

The government discontinued the grant programs because the vehicle conversion target was successfully met. It continues R & D support to improve LPG conversion technology to keep pace with engine development. Further federal subsidies are not planned, although some provincial incentives continue. The government expects industry to continue the momentum and the Propane Gas Association to increase its marketing effort.

Promotion of CNG

Industry and government interest in CNG began to grow in the late 1970s prompted by favorable experience with the fuel in other countries and the growing price spread between natural gas and gasoline. Also, LPG supply was recognized as limited. In 1980, CNG as a transportation fuel was well behind LPG in terms of technology, convenience, acceptance, awareness, and market penetration. The Natural Gas Vehicle Research and Development Program was initiated by the British Columbian government in 1980 and supported by the federal government in 1981. The purpose was to evaluate and improve existing CNG technology. This work led to the establishment of safety regulations, design of carburetion equipment suited to the Canadian environment, and a knowledge base concerning natural gas vehicle combustion and performance.

The conversion costs for CNG vehicles of about Can. $2,500 (U.S. $1,750) were much higher than for LPG. Establishment of CNG fueling stations initially cost Can. $200,000–250,000 (U.S. $140,000–175,000), and the current cost is about Can. $300,000 (U.S. $210,000). In July 1983 the federal government began two grant programs to overcome these obstacles: the Natural Gas Vehicle Program (NGVP) and the Refueling Station Program. The station goal was set at 125 by 1986. The target for converted vehicles was 35,000. Natural gas utilities were expected to supplement government incentives.

The NGVP gave grants of Can. $500 (U.S. $350) to vehicle owners converting to CNG. It was complemented by removal of the provincial sales tax on conversion kits and/or vehicles in some provinces. A grant of Can. $50,000 (U.S. $35,000) was available for establishment of CNG refueling stations. While the program's fuel station goal was met and exceeded, the number of vehicles converted was less than planned. By December 1986, 11,000–12,000 natural gas vehicles were in operation.[13] Of the vehicles converted, 45 percent were automobiles, 23 percent were light trucks, and 21 percent were vans. These are all retrofits since auto manufacturers produce no dedicated CNG vehicles. Two-thirds are in fleets.

Many gas utility companies in the country offer grants and financial packages to encourage vehicle conversions and fuel station construction. Some gas utilities have adopted low profit-margins for natural gas destined for CNG use. Other utilities have adopted the strategy of taking a somewhat higher margin on gas and giving a substantial rebate on conversion costs. Several companies are also developing CNG cylinder leasing programs. A second approach is to offer easy financing for conversion. B.C. Hydro offers 36-month, 9 percent financing for conversion. A customer using 1,300 gallons of motor fuel per year saves more each month than the monthly finance payment. Some utilities in Ontario and British Columbia have been able to negotiate rate basing of new refueling stations with the natural gas regulatory boards.

The price of natural gas was controlled by agreement between the federal and provincial governments at 85 percent (on an energy basis) of that of crude oil from 1975 to 1981 and at 65 percent from 1981 to 1985. Since then the price has been gradually decontrolled, with retail prices staying well below 65 percent of oil. With the plunge in the oil price in 1986, the spread between gasoline and CNG prices decreased dramatically. It is widening again as natural gas prices continue to drop, but will not go back to the 1985 levels until gasoline prices rise.

Despite the incentives, the conversion rate has been low, even in Vancouver where the majority of the fueling stations are located. For a vehicle using 1,300 gallons per year of CNG, the payback period for conversion (with financial incentives) was estimated by Flynn[14] to be about 12 months in Toronto and about 20 months in Vancouver in 1985. In June 1986, with lower gasoline prices, the payback periods were about 18 months and 32 months respectively.

Outlook. Market surveys show the cost of conversion, fuel availability, and driving range limitation to be the largest barriers to public willingness to use CNG. Reasons given by those in favor of conversion are lower fuel price, lower emissions, natural gas availability in Canada, reasonable cost, better mileage, and longer engine life.

There is competition between LPG and CNG in some provinces. However, they may also be viewed as complementary, since LPG broke the ground as an innovative fuel. LPG can serve areas where the natural gas network does not reach, and its fueling stations are less expensive.

Increased use of CNG will depend on the price relative to gasoline and the degree of government and industry support. Deregulation has tended to lower natural gas prices and therefore favor CNG over gasoline. The federal government will likely extend the NGVP to 1991. Funds come from a program set up by the government of Alberta based on contributions from gas producers seeking to expand natural gas sales east of Alberta. The plan is to place more emphasis on high-volume users in key target areas to form a base of consumers who continue to use CNG vehicles. Natural gas utilities and provincial governments are expected to continue to support CNG usage. The Canadian Gas Association has prepared a strategy that calls for a network of visible fast-fill fueling stations, possibly as joint ventures between the utility and the retailer, selling several fuels. The industry recommends extensive promotion and publicity, cooperation with federal and provincial governments, and better relations with the automobile industry on technical matters.

SUMMARY AND CONCLUSIONS

The experience in the three study countries—particularly Brazil—demonstrates that it is possible to develop a large market for alternative fuels within a reasonable time period if there is a favorable financial environment and efforts are undertaken to overcome uncertainty on the part of industry and consumers. Doing this requires a strong commitment of effort and resources by the government, particularly in the early stages of market development. Fuel suppliers, the vehicle conversion industry, and automakers have been wary of investing in new technology for which the market is uncertain. Consumers have been reluctant to spend sizable sums for vehicle conversions, even when the return on investment is quite favorable. Assurance of fuel availability and technical quality of vehicles have proven to be important factors.

Government and Private-Sector Roles. Federal governments played a strong role in the development and market adoption of alternative fuels. Government programs included grants and low-interest loans for conversion of vehicles and fuel stations, reduced taxes and guarantees of competitive prices for alternative fuels, regulation of the conversion and equipment industries, and promotion and marketing campaigns. In each country, gaining the confidence and cooperation of the public, automobile manufacturers, conversion industries, and fuel companies was extremely important.

Government incentives and financial involvement overcame some uncertainty on the part of the private sector, and considerable market penetration was achieved, especially in Brazil and New Zealand. But they also created institutional momentum and consumer expectations that made change of course difficult when oil market conditions shifted and rendered uncertain the economics of the selected alternative fuels. As oil prices and political attitudes changed, the degree of financial and other support waxed and waned. Lessening of government support caused substantial reduction in vehicle conversions and/or sales of new alternative fuel vehicles. The international experience demonstrates the importance of consistency of government policy but also suggests that it is not easy to achieve.

The willingness and capability of the private sector to provide services for vehicle conversion, to produce dedicated alternative fuel vehicles, and to renovate or construct fuel stations have varied among countries. Capability to convert vehicles to alternative fuels developed relatively rapidly in Brazil and New Zealand, though problems of quality control were encountered. Generally, government regulation was required to ameliorate these problems and to restore the reputation of the programs. In Brazil, the only country to produce large numbers of dedicated vehicles, car manufacturers were encouraged by strong government commitment, consumer incentives, and assurance of sufficient ethanol supplies. Fuel station owners were reluctant to provide a new fuel for which the demand was uncertain in New Zealand. Recent government policies in Canada and New Zealand have encouraged the private sector to partially replace public sector support now that market acceptance has been demonstrated. In Canada there is now support for CNG usage from the natural gas companies, which see motor vehicles as an emerging market with potential to augment their normal heating markets.

Program Emphases. Brazil and New Zealand desired widespread substitution by alternative fuels, and programs were aimed at both fleets and personal vehicles. Canada's programs were much less ambitious, though still important. Fleet vehicle operators have been more likely to use alternative fuels with a lesser degree of incentives than individual owners. Payback periods for fleet vehicles tend to be shorter because they travel more miles per year and they travel within a limited range, facilitating refueling.

Only Brazil has emphasized dedicated alternative-fuel vehicles. In New Zealand and Canada, the main target of programs has been conversion of existing vehicles. In New Zealand, the automobile market was not large enough to warrant production of dedicated vehicles. The main focus has been on gasoline substitution. There was considerably more experience with conversion of spark-ignition engines to fuels such as LPG and CNG than there was with conversion of compression (diesel) engines. In recent years, as diesel engine conversion technology has been further developed, and urban air pollution has generated more concern, countries are turning their attention to diesel substitution. Diesel vehicles tend to have high mileage and thus shorter payback time and are often fueled from a central location. Both Brazil and Argentina are embarking on programs of urban diesel fleet conversion to CNG.

Consumer Attitudes. Reducing the initial cost of conversion through grants and low-interest loans produced considerable consumer response, but many private motorists have been reluctant to convert to alternative fuels. In Brazil, consumers readily bought dedicated ethanol vehicles when offered a clear economic advantage, and supply and cost-advantage of the new fuel were assured by government commitment. National pride was also a factor in the countries with major programs designed to enhance national self-reliance. When governments reduced financial incentives for vehicle conversions or price subsidies, however, conversions and sales of vehicles dropped sharply. Revitalizing the programs required restoration of government incentives.

Market surveys have shown that concerns about alternative-fuel availability are a major factor in consumer decision-making. Thus it was important for the government to take the initiative in promoting development of a fueling infrastructure. Market penetration of neat ethanol in Brazil was helped by the utilization of the distribution infrastructure for premium gasoline. In New Zealand and Brazil, poor quality of vehicle conversions reduced consumer confidence, and government standards were required to ensure satisfactory performance. In Canada, the cost of vehicle conversion and low gasoline prices have also been cited as barriers to public willingness to use CNG. Fuel cost savings, less harmful emissions, and reliability of Canadian natural gas supplies were factors given in favor of conversion. The shorter driving range was the major problem cited by those who had converted to CNG.

The experience in New Zealand, Brazil, and Canada demonstrates that addressing consumer concerns about adopting unfamiliar fuels must be a major focus of programs that seek to bring about widespread use of alternative transport fuels.

NOTES

The authors wish to acknowledge the assistance of Roy Sage (Ministry of Energy, Mines, and Resources, Canada), Peter Graham (Ministry of Energy, New Zealand), Pierre Moulin (World Bank), and Dan Sperling (University of California-Davis).

This work was supported by the Office of Policy, Planning, and Analysis, of the U.S. Department of Energy under contract no. DE-ACO3–76SF000098.

1. New Zealand has a population of approximately 3 million people and about 1.8 million motor vehicles. Three-fourths of the vehicles are on the North Island.
2. The exchange value of the New Zealand dollar has fluctuated greatly: from U.S.

$0.97 in 1980 to U.S. $0.67 in 1983 to U.S. $0.52 in 1986. The figures expressed in U.S. dollars were calculated using the exchange rate in the relevant year.

3. C. J. Ryder, *The New Zealand CNG Programme: A Partnership Between the Government and the Private Sector* (Wellington, New Zealand: New Zealand Ministry of Energy, 1986).

4. CNG Federation. *CNG Information Brief.* Wellington, New Zealand, 1985.

5. Brazil, a country with 138 million people, has 12.5 million vehicles, of which 10 million are cars. The country has some domestic oil production, and nonassociated gas fields were discovered in 1982. In 1974, oil supplied 46 percent of domestic energy demand, and 76 percent of this was imported. By 1985, oil supplied only 31 percent of domestic energy demand, with 44 percent imported.

6. D. Sperling, "Brazil, Ethanol, and the Process of System Change." *Energy* 12 (1987): 11–23.

7. D. Sperling, *New Transportation Fuels: A Strategic Approach to Technological Change* (Berkeley: University of California Press, 1988).

8. J. Lizardo, and A. Ghirardi, "Substitution of Petroleum Products in Brazil." *Energy Policy* 15.1, February 1987.

9. C. Weiss, "Fuel Ethanol in Brazil: Technology and Economics," unpublished manuscript, 1986.

10. Canada, with a population of 26 million, has 11 million cars and 3 million trucks and buses. Canadian supplies of conventional crude oil are dwindling, but the country is well endowed with natural gas. In recent years Canada has deemphasized the goal of energy independence, and has pursued diversification of its energy mix.

11. The exchange value of the Canadian dollar has fluctuated from U.S. $0.66 in 1980 to U.S. $0.77 in 1984 to U.S. $0.60 in 1986. The figures, given in U.S. dollars, were calculated using a rate of Can. $.1–U.S. $0.70.

12. P. Delmas, "The Supply, Demand and Pricing of Natural Gas for Motor Vehicle Use in Canada," unpublished manuscript, 1986.

13. K. Liko, and K. Deeg, *Natural Gas for Vehicles, Industry Survey 1986* (Ottawa: Energy, Mines and Resources, Canada Report TE87–3, 1987).

14. P. Flynn, "CNG as a Vehicle Fuel: North American Economics and Markets," unpublished manuscript, 1986.

12
ALBERT J. SOBEY

A Global Fuels Strategy: An Automotive Industry Perspective

This chapter will present some of the problems and processes in a transition to alternative fuels and how this may occur in both the developed and developing nations; it will also emphasize the benefits of alternative fuel for the automotive industry, the energy companies, and the oil importing nations, and describe the incentives for the participants.

This chapter will address the benefits that could be provided to the automotive companies and major energy companies if they work together to accelerate a process which I believe is inevitable: the commercialization of alternative fuels. If this process can be started before the next rise in the price of world oil, it could reduce the traumatic effects on the world economy that would otherwise occur. The transition should start in the nations with the most to gain, those without adequate supplies of petroleum.

The organizations that must cooperate in managing a transition include the governments (in particular of those nations lacking petroleum resources), the energy companies, the engineering companies, vehicle and other equipment manufacturers, customers, and investors. They all must understand the benefits of a cooperative strategy.

The strategy is based on several assumptions:

- That alternative fuels will be needed sometime within the next century, beginning with "oil poor" nations.
- That the fuels of choice for ground vehicles will be methanol and compressed natural gas.
- That the objectives of each nation differ depending on its energy resources and economic and political situation.

- That the greatest source of growth of demand for petroleum products will be in the developing nations, and that a world strategy for easing the transition must start with those nations.

- That cooperation between the public and private sectors is required to provide the transition with the least economic or social cost.

The four alternative fuels of primary interest include compressed natural gas, alcohols (ethanol and methanol), synthetic gasolines and diesels, and solid fuels, primarily coal.

Since we do not have a complete understanding of the technology and economics, the possibility of other fuels cannot be ignored. When natural gas and the alcohols, ethanol and methanol, will be competitive in price with petroleum is debatable and will depend on the rate of economic growth and the future cost of production, and it will be different in individual nations.

THE NEED FOR ALTERNATIVE FUELS

The primary reasons for alternative fuels are to replace exhausted economic petroleum reserves, to reduce dependence on limited sources of supply (Persian Gulf and OPEC), to reduce international trade deficits, and to improve the environment.

Alternative fuels for the world's transportation systems will probably be needed sometime within the next half-century as the economic resources of petroleum outside of OPEC are exhausted. They will first be economically attractive in the oil-importing nations. It would be to the advantage of all oil-consuming nations to have more than one energy source for their transportation systems, thus minimizing the risk of serious economic harm that could come from another world oil price spike.

In the more economically advanced nations, the first interest in the use of alternative fuels is to provide some independence from the world oil monopoly pricers, OPEC.

A second (and at the present more popular) reason for interest in methanol in the developed nations is the desire to improve the environment. While this can be significant, the benefits can be overrated and should not be the sole or even primary reason for an interest in alternative fuels. The automobile industry is concerned about attempts to mandate or accelerate the use of methanol. It is not yet confident that it can build methanol engines with the life and reliability customers expect. There is no good solution, yet, to the cold-starting problems. Attempts to accelerate the use of methanol, even for socially beneficial purposes, could lead to reliability and operating problems and set the process back by years if not decades.

The countries that can benefit most from alternative fuels are probably the developing or newly industrialized nations without significant petroleum resources. Many of these have other indigenous energy resources, primarily natural gas or coal and biomass.[1]

In the "oil poor" nations, the need for economic energy resources is critical. The technical problems are similar in the developed and developing nations, although some developing nations lack the technical infrastructure and the skills available in developed nations. The implementation opportunities, problems, and strategies, however, are quite different.

In the last energy price shock these nations spent a disproportionate share of their international earnings on imported oil (in at least one case more than the total earnings from exports). The purchase of imported oil still takes a large portion of their international earnings.

Many of these countries are in a state of transition from dependence on traditional energy supplies, such as firewood or cow dung, to more modern forms of energy. They are divided into a rural (traditional or small-town) sector that has changed little over the centuries, and a growing modern sector trying to emulate the developed nations. The energy needs of both sectors are critical.[2]

The traditional sector is using an increasing share of the available firewood, in many places more than is being replaced. The gathering of wood for the family to cook with and provide heat and protection from animals often takes the majority of the time of one family member. Imported oil may be used to run a diesel electric generator, to pump water, or to cook. The quantity required is relatively small, and the villages and farms are usually remote. All of these functions can be served by alternative fuels if they are liquids and safe to handle in cans and drums. The sale of automotive products (primarily trucks) is small and unlikely to grow significantly in decades. The most pressing equipment need may be for small electrical generators and pumps.[2]

The growth of the modern sector has been a major cause of the increased demand for crude oil seen in the last decade. The oil consumption of the developing nations is expected to increase from 14 million barrels per day in 1985 to between 20 and 22 million barrels per day by 2000, while consumption in the rest of the world (outside of the communist nations) is expected to be relatively constant.[3]

The modern sectors tend to be concentrated in the capital cities, usually ports. Their energy uses are similar to those of the developed nations a few years ago. Electrical power still uses a large portion of the oil (about 30 percent) because useful substitutes are not readily available. These nations are usually in warm climates, so relatively little oil is used for heating. Transportation and industry each use about one-third of the oil.[2]

In the rush to find oil after OPEC I, many of the "oil poor" nations were disappointed to find natural gas. It has little local market and is difficult to transport. Natural gas resources may be remote, and collection pipelines may not be justified at today's oil prices. Most of the nations that have not found natural gas have geological formations that promise significant natural gas supplies. Some of the nations that produce crude oil for export would like to find other energy supplies for their domestic consumption. Most of these countries have substantial natural gas supplies.[4]

Other nations have coal or land that can be used to grow biomass fuel crops. There are only a few countries, primarily island nations, that lack any significant fossil energy resources or are using their existing resources to the fullest extent.

Since many of the developing nations consume much of their oil under boilers and for electrical power generation, use of natural gas is a relatively easy technical step. To develop the natural gas supplies, they will have to find the capital to drill the needed production wells and to build the gas pipelines. Once the infrastructure is in place, the use of compressed natural gas and methanol in cars and trucks will become logical.[4,5]

Because it will require several years for most of these nations to develop an adequate supply of natural gas, the process should be initiated soon. The best approach for each nation will depend on its specific needs, available resources, and economy.

AVAILABILITY OF PETROLEUM-BASED FUELS

Before committing resources to alternatives, we should understand the potential for the effective use of the petroleum resources we have. The availability and cost of the alternatives should be tested against the probable scenarios for the availability and cost of petroleum-derived fuels. The price path will depend on economic and technical developments as well as the availability of oil resources outside of the Persian Gulf. There are two families of scenarios on the availability and cost of foreign oil supplies: "shock" scenarios in which the oil supplies from the Middle East are interrupted as a result of military or political action; and economic-driven or "no shock" scenarios with oil prices rising or decreasing with world demand and supplies.[6]

Short-term oil prices are very responsive to supply and demand (and politics), as is true of any commodity. The world demand for oil will change with price, in particular the price relative to other substitutable energy resources such as domestic natural gas and coal.

Prices could exceed those of the 1980s for a short time in an oil price shock but only if the supply interruption were more severe than that of OPEC II. We do not expect such an interruption to last more than six months to a year, as the oil-producing nations are dependent on oil revenues for their internal programs.

In the economic or "no shock" scenarios, oil would always be available at some price, increasingly from the Persian Gulf. While the short-run prices will respond to supply and demand like other commodities, the long-run average prices tend to be bounded by two factors: the "low" long-run price would be bounded by the cost to discover and produce the marginal barrel of oil from non-OPEC or noncartel sources, and the "high" long-run price will be bounded by the reduction in total earnings of the cartel nations when consumption decreases and prices rise.

Oil prices could exceed or drop below these lower and upper bounds for a short time, but the average annual oil price should range between the two bounds.

The low long-run price has been relatively constant for many years. Oil industry analysts suggest that the price of the marginal barrel outside of OPEC is in the $7 to $10 per barrel range. It is expected to remain near that level for about a decade, then increase exponentially as it becomes necessary to tap increasingly costly resources.[7] It may be possible to construct alternative fuel facilities for less than it costs to invest in further oil exploration and development.

Many energy analysts expect the high long-run price of oil to rise significantly toward the end of the century when the low-cost resources of oil outside of the Persian Gulf have been exhausted. Their forecasts of the real long-run price show oil prices increasing to about $30 to $35 per barrel by 1995, then increasing about 1 to 2 percent per year with no end in sight. We think that this need not occur if alternatives are introduced in a timely manner.

The high long-run price should be bounded by the use of alternative fuels, conservation, new technology, and other measures. These can act like a cap on world oil prices. There appears to be a long-run price above which OPEC's earnings will not increase enough to offset the decrease in demand, and thus the revenues of the cartel nations will peak. This will be due in part to conservation and reductions in the use of oil products for industry and transportation, but more significantly to the substitution of alternative energy, in particular coal and natural gas, which can be produced profitably at lower prices. The price at which this will occur is debatable, but several studies have shown it to be between $25 and $45 per barrel.

U.S. production by the year 2000 may drop to less than one-third of the domestic consumption of oil products or 5 to 6 million barrels per day, against a demand of 16 to 18 million barrels per day.[7] If world oil prices rise to only $25 per barrel, this would represent an expenditure on oil imports of $90 billion per year, a figure for concern given our negative international balance of trade, which in 1986 amounted to a deficit of about $124 billion for everything but energy. This problem is more serious in other oil-importing nations.[8,9]

The much-expanded efforts to locate oil in the 13 years since OPEC I have lead to the discovery of large quantities of natural gas. The current estimates are that the world has about twice as much energy in natural gas as in petroleum. Some people believe that there may be ten times as much. Equally important, the resources of natural gas are widely distributed across the world. World Bank studies indicate that most nations have adequate natural gas (confirmed or probable) to operate their economies for decades. Only a few nations—in Europe, Japan and some islands—do not have adequate resources.[2]

If the economic projection of the cost of production of these resources is confirmed, it is plausible to expect that known resources of difficult-to-recover oil will not be produced. Other options, in particular the use of alternative fuels, will be more economic.

DEVELOPMENT OF ALTERNATIVE FUELS

The four alternative fuels of interest include compressed natural gas, alcohols (ethanol and methanol), synthetic gasolines and diesels, and solid fuels, primarily coal.

Compressed natural gas, where available, is and should remain the lowest-cost fuel suitable for transportation. The technology for its use has been demonstrated, although both performance and cost can be improved.

The primary problems with natural gas include locating and developing the fields, transporting the gas to the users, the limited range of the vehicles (usually under 100 miles), and the cost of converting vehicles from gasoline to natural gas. Despite the apparent benefits there are only about 30,000 cars and trucks adapted to natural gas in the United States. Most CNG vehicles are in California and in a few utility fleets. CNG is successful in Italy where there are nearly 300,000 vehicles operating on compressed natural gas (the number has been nearly constant for several decades). In total, there are more than 400,000 natural-gas-compatible vehicles worldwide.[6]

From an economic viewpoint natural gas is justified for use in truck and automobile fleets, where the range of about 100 miles between refueling is often acceptable.

Major oil companies have studied retail sales of compressed natural gas in the United States and developed nations but have found the cost in the United States for compressing the natural gas prohibitive if industrial-grade pneumatic compressors and controls are used. There are several ways to reduce this cost. First, if the market is large enough to equip about 10 percent of the stations to supply compressed natural gas, the compressors could be designed for this service. With volume production the equipment cost should come down significantly. Second (in a few limited locations), commercial stations could be connected to the intercity pipelines where the higher pressures would reduce the compression requirements and consequently the costs.

Most of the developing nations will require construction of gas pipelines from the fields to the centers of demand. These pipelines may be long and may have to cross jungles and large bodies of water. Some small fields, relatively close to points of demand, can be tapped by compressed natural gas trucks. Where larger supplies are available, the gas could be liquefied and moved to its destination in insulated trucks. While the gas at the wellhead is essentially without value if it cannot be moved to the market, the cost of these collector systems must be included in evaluating the economics of the use of gas. In many cases the resulting cost per Btu will be comparable with the present price of oil, but there is a security premium, a value to the country, of being able to moderate another oil-price spike.[2,4]

For vehicles where a longer range is required, liquid fuels are preferred. Despite large expenditures for research on gasoline-like synthetic fuels, methanol made from natural gas appears to be the probable future fuel of choice for ground vehicles. Methanol can be made from many resources: corn stalks, peat, natural gas, and possibly even as a byproduct of future nuclear reactors.

There is little incentive for investment in synthetic gasoline facilities unless there is a high probability that the fuel produced will be significantly less costly than methanol. Chemical engineers indicate that this is unlikely unless made from heavy oil or other petroleum-like resources. Coal-derived synthetic gasoline fuels are about twice as expensive as methanol made from natural gas (perhaps more as the technology is not as well understood). The U.S. synthetic-fuel programs assumed that the automobile engine designs would not change, which was a misunderstanding of both the technology and the economics.

The alcohols are better fuels than gasoline for automobile engines. They have higher octane ratings and are relatively clean burning. In an engine optimized for their use, 1.7 gallons of methanol will replace 1 gallon of gasoline. In the long run, alcohol's primary benefits may come from the fact that they are simple chemicals, not a mixture, and as a result the engine can be more effectively designed. They are also compatible with new propulsion systems like fuel cells.[10]

The alcohols are felt to be environmentally more acceptable than gasoline. They biodegrade easily, can reduce emissions (depending on the engine design), and appear to be less toxic than gasoline. As with other engineering-environmental problems, there are tradeoffs. If higher energy efficiency is to be provided, some of the emission benefits may have to be given up. There would be sufficient time before the alcohols are directly competitive with gasoline to complete the engineering required, to address the material and cold-starting problems, and to replace the fleets.

Methanol technology is well understood. The United States has nearly 2 billion gallons per year of methanol capacity. Most of it is derived from natural gas. In order to understand better the economics of large-scale alcohol fuel programs, we in the auto industry went to the chemical engineering companies that designed most of the world's methanol facilities. We wanted to know the cost of facilities if they were built in serial production.[10] The data indicated that an investment in a methanol plant in the United States would provide a 15 percent return on investment (above inflation) when the retail price of gasoline rose to the range of $1.30 to $1.40 per gallon (in 1985 dollars). This is based on the cost of moving a vehicle (with an optimized engine) the same distance per dollar as with gasoline. The methanol costs were based on new U.S. plants with 100 percent equity financing. The profits could be larger if the construction costs were leveraged.[10]

These methanol costs assume the use of U.S. natural gas at a wellhead price of $2.50 per thousand cubic feet (tcf) as the feedstock. In 1986 the average wellhead price of natural gas in the United States was less than $1.20 per thousand cubic feet. Home and industrial natural gas prices were in the $3.00-to-$5.00

range and varied nearly two to one between different regions of the country. This was during the period when the average price of crude oil was about $16.00 per barrel. The estimates of wellhead gas prices for 1995 range from less than $2.50 to more than $6.00 per tcf, depending on the price of competitive fuels, in particular crude oil.[9] If the cost of oil increases to $50 per barrel, which would lead to retail prices of gasoline in the $1.40 per gallon range, the price of natural gas at the wellhead should increase to $2.50 or slightly higher.

It is unlikely that any significant additional U.S. methanol capacity will be built (except for military purposes or where gas has a negative value, as on the north slope of Alaska). However, in many nations natural gas is a wasted or underutilized asset. If the natural gas feedstock is valued at $1.00 per thousand cubic feet or less, the gasoline price at which a 15 percent return on a methanol investment can be realized decreases by $0.20.

Small methanol plants are attractive in remote natural gas fields to convert the gas into an easily transported form of energy. The plants would be simpler and less efficient than the normal full-scale plants but much lower in construction cost and could be transportable.

New Zealand has demonstrated what can be done. It has built a large "natural gas-to-methanol" and a "methanol-to-gasoline" plant. The country would not have had to build the second stage if the automobile industry had been prepared to produce methanol-compatible cars.

Several studies of the comparative cost of alternative fuels reached the conclusion that methanol would cost about half as much per Btu as ethanol, but any conclusions that methanol will always be less costly than ethanol should be viewed with some caution as chemical engineers are seeking less-costly processes and may be successful. Most fuel ethanol is made from feedstocks (such as corn) that could be used for food, while methanol can be made from the cornstalks and thus could be an attractive byproduct for farmers and a potential source of fuel to operate their tractors, pumps, and other machinery.

The primary interest in ethanol will be in nations such as those in the Caribbean dependent on sugarcane for their international trade and local economies. Ethanol is a major fuel in Brazil, which developed a sugarcane-to-ethanol fuel industry to reduce its dependence on OPEC. Subsequent oil and gas discoveries coupled with the low prices of world oil have lead some people to suggest that Brazil should reevaluate its commitment to ethanol. The processing equipment for ethanol is relatively simple, but as the Brazilian experience demonstrated, it requires careful quality control, something that is difficult if there are many small stills.[11,12]

Some Pacific nations are also interested in developing a blend of coconut oil and diesel fuel as a diesel-fuel extender. While the economics are not fully evaluated, they appear attractive.

Interest in coal liquefaction is continuing because because of the very large coal reserves in this country. Engineering studies in the early 1980s led to estimates of prices of $90 to $100 per barrel. However, during the last few years improved processes have been developed, for example a process that can produce two fuels, oil and char (which can be burned in boilers). The most optimistic projection I know of to date is that the costs might be reduced to $25 per barrel including an allowance for profits. We doubt that making synthetic gasoline from coal will be economically justified. The coal can be made into methanol at less cost and with less energy loss.[8]

Solid coal can be used as a vehicle fuel. It was used by the railroads when steam locomotives were the most advanced form of power. Steam locomotives are still being built by the Chinese, but the engines are less efficient than modern diesel or electric locomotives. Advanced technologies for use of coal in transportation such as fluidized bed combustors and micronite coal (very fine cleaned coal) are being developed. Research by Consolidated Coal has concluded that the cost of cleaning coal so that it could be burned directly, in gas turbines for example, is probably greater than the cost to convert it to methanol. Some of the new coal-combustion technologies, fluidized beds for example, may be used for ship propulsion.[6]

Phillips (Holland) developed the stirling external combustion engine for the express purposes of using traditional fuels (primarily solids) to provide small amounts of power for lights, radios and appliances. This use may still be of interest to developing nations, and new stirling designs in the power range of light diesels could be attractive for small villages and for remote industrial sites.[6]

ALTERNATIVE-ENERGY TRANSITION STRATEGIES

There are two basic transition strategies for introducing alternative fuels to the United States and other nations: to mandate their use for security, environmental, or financial reasons; or to allow the market to drive the transition when the economic benefits become large enough.[6,12]

The first strategy inevitably leads to misuse of resources. The use of more costly (alternative) fuels will increase more rapidly than is economically justifiable, thus leading to excessive national costs. Equally important, it may lead to premature introduction of technology or exploitation of incompletely developed resources, causing technical problems and diverting resources from more beneficial purposes. A serious concern is the risk of a public failure that could set the use of alternative fuels back for years if not decades.

The second strategy, which depends on private-sector investment and public acceptance, may delay introduction of the fuels beyond the time at which they are economically beneficial. This could lead to social problems and economic losses as the nations expedite the development of these fuels. The delays may be caused by time lags required to construct facilities, by regulatory problems, or by lack of public support.

The ideal transition strategy will be difficult to define, much less institute. The process can be made less costly if the problems are well defined and the public and private sectors cooperate to identify and implement the solutions—

the private sector to continue research on the production and consumption of alternatives, and the public sector to address the problems of the regulatory and policy environments. Both must address the need for and availability of capital to produce the needed facilities.[8]

I am confident that the transition to methanol or similar fuels could occur in the United States without government intervention. There will be a point at which the investment in natural gas or methanol facilities will require less capital than further exploration for oil. However, if investors must wait until they are confident that the oil price will not drop below their costs, the transition may not start until the 2000s, causing unnecessary social costs due to the excess cost of oil during the decision-making and facility-construction periods.

An extended transition of perhaps 20 years is likely during which both gasoline and methanol fuels would be available in the developed nations (while gasoline-using cars and truck wear out). This transition may be similar to the phase-out of leaded gasoline. Methanol would first replace high-octane or super fuels because of its high performance and image. The first cars to use methanol may be sports cars or large cars like station wagons that cannot meet corporate average fleet economy (CAFE) or gas-guzzler tax criteria.

The successful introduction of alternative fuels in developing or oil-poor nations will require that the automobile companies, the sponsoring energy companies, and perhaps chemical companies develop a coordinated strategy in which the domestic natural gas resources are more thoroughly explored, wells drilled, and pipeline built as necessary.[13]

There will be no market for automobiles that use alternative fuels unless there is an available supply of economic fuel and vice versa. While we tend to emphasize the introduction of fuels for cars and trucks, the logical first products to use domestic natural gas in these countries may be in boilers and small stationary electrical-power generators.

The transition strategy for developing countries will involve a number of steps:

1. developing a plan in cooperation with the energy producers, local business people, and the government;

2. locating financing for the construction of the pipelines and other facilities;

3. approval of the process by local authorities;

4. construction of facilities to produce the fuel, upgrade it, and start importing the vehicles to consume the product; and

5. introduction to the market.

The first step will be to develop a plan to implement the system in cooperation with energy producers, local business people, and the local governments. Each nation will be different because of different resources, requirements, and regulations. It would be appropriate to select first those that have the most to benefit, that have adequate resources, and that are willing to experiment, and then to work with the nations where the benefits are more difficult to establish. This will enable them to be prepared to respond in a timely manner when the price of world oil rises. Ultimately, methanol facilities should provide a source of international earnings for the sponsoring countries.

The second and perhaps most difficult step will be to locate the financing for the plan. This will include obtaining funds for additional exploration and drilling when needed and the construction of natural gas pipelines and the facilities to upgrade the fuel. Few of the countries of interest have surplus capital; in fact, most are in debt. Local savings rates are low when compared to the cost of these programs. Private capital may be interested in the oil exploration and processing equipment if the market can be assured and investments protected.

Cooperation between the public and private sectors is required to accomplish the transition with the least economic or social cost. Energy producers will need to provide the raw materials, processors will need to convert the energy resource into a usable product, and distributors will have to make it available to the industrial and private users. Equipment manufacturers will need to provide the products that will use the energy. Local and national governments must create the regulatory and business environments required to make the process possible and to encourage investors (local and international) to provide the capital to create the supply system and support the purchase of the energy-using equipment.

INCENTIVES TO THE MAJOR PARTICIPANTS

The major participants in introducing new fuels are national and state governments, energy supply companies, civil engineering firms, chemical processing companies, equipment manufacturers, customers, and investors. These participants have sufficient common interests to bring them together to develop a common strategy.

Since the greatest increase in demand for petroleum products will be in the developing nations, a world strategy for easing the transition should start with those nations. The incentives or objectives for those nations fall into four areas:

- to provide energy security and to minimize the economic impact of another major supply interruption;
- to minimize negative balance-of-trade problems and to use international earnings to purchase other valued goods, for example electrical power stations, trucks, cars, and consumer goods;
- to encourage economic growth by providing low-cost energy for their industry and consumers; and
- to improve the environment.

The ranking of these benefits will depend on the economic status of the nations. In the United States the environment may be the primary reason for near-term interest; for the undeveloped nations, where survival is a major problem, the environment may be of little concern.

The incentives for the first group of participants, the governments of each interested nation, will differ with its resources and economic status. They can be classified in many ways, but for this discussion they may be divided into four major categories. The first includes the oil exporters, in particular the OPEC nations. They may want to discourage this transition, at least until their low-cost oil resources have been used up. The second category includes the industrialized oil-consuming nations with domestic supplies, such as the United States. They view alternative fuels as a benefit for security and environmental reasons. The third is the oil-consuming nations with little domestic petroleum resources that would like to use their other domestic energy resources such as natural gas and coal as a source of economic vehicle fuels. The last category is the energy-poor nations with few fossil energy resources who may have to depend on biomass, solar, or geothermal energy as sources for fuels.

Studies of the oil-poor nations indicate that the most important need may be for electrical power. Investments in electrical power can justify concurrent investments in natural gas production and distribution systems. When the natural gas infrastructure is in place, then the use of natural gas for transportation becomes a logical step, with methanol following behind.

The second group of participants, energy companies, whether they are large international majors, local private companies, or national companies, are critical to the introduction of alternative fuels. While I do not have personal experience in this industry, discussions with oil companies lead me to identify four major incentives for their participation.

The first is profits. The level of interest in a specific market depends on its relative size and the cost of the competitive energy supplies. At some time in the future it will become less expensive to produce alternatives rather than try to continue the search for and production of petroleum. For example, the president of Shell USA gave a speech in 1983 in which he estimated that it would cost over $2 trillion to continue U.S. oil production at its present level through 2010. The cost of building methanol-from-natural gas facilities to provide a similar energy supply would be much less than that.

The second and closely related incentive is to create a hedge against long-range depletion of petroleum. Ultimately the oil companies will exhaust their economic oil reserves (and those that they can purchase at a reasonable price), at which time the oil companies must be prepared to provide other equivalent products or enter new markets. The development and marketing of alternatives will have the least "shock" as they are very similar products.

The third incentive for the energy companies could be the use of otherwise unproductive assets. As was pointed out earlier, much of the world exploration for oil has lead to the discovery of natural gas. One of the best ways to market natural gas would be to encourage its use in electrical power generation and industrial processes, then to develop methanol facilities to provide an easily transported fuel for sale in other markets.

The fourth incentive is growth: entering new markets that would not otherwise be available to them. This would include providing new services and encouraging the economic growth of the nations they serve.

The third major group of participants needed for the creation of an alternative fuel supply system in developing nation is the civil engineering and chemical processing companies. I found them interested and potentially willing partners. If the market develops and the energy resources can be made available, they will find a way to design and build the needed conversion facilities. Some have even expressed an interest in investing in natural gas-to-methanol plants.

The fourth group or industry that must be involved is the equipment manufacturers, primarily (but not only) those companies that can produce trucks, buses, and cars for the interested nations. Access by the major U.S. automobile manufacturers to the developing-nation markets is restricted, either by the economic size of the market, or by local content rules and import duties. Since the development of alternative fuels requires a cooperative strategy to make available the products to be used (natural gas and methanol) and the products to consume them (automobiles, trucks, etc.), it is necessary that some relief from restrictive provisions be provided to the manufacturers sponsoring such developments.

An advantage to large automotive companies is the relatively small investment required to penetrate initially small markets. The warm climate of many oil-poor developing nations is also an advantage to manufacturers in that the cold-start problems are less severe. These markets are definable and can be effectively monitored to provide information which will be of value when the methanol products are introduced into the United States and other industrialized nations.

The interest of the U.S. automobile industry in these markets is just beginning. Today the U.S. automobile companies are concentrating on meeting the competition from overseas and meeting increasingly stringent regulatory standards. The markets in the nations where alternative fuels will first be used are small. The Pacific Rim, where many Japanese, Korean, and national companies are competing, had in 1985 a market of 7.5 million trucks and cars that was growing at 2 to 3 percent per year, similar to that of the United States. In South America the market is growing more rapidly at 6 to 7 percent per year, but is expected to reach only 2.5 million vehicles per year by 1995.

The next and sometimes overlooked group is the customers. They must have an incentive to purchase and use the products. They must be confident that the product is and will continue to be available, that the operating cost will be low (significantly less than the product they are used to), and that the cars and trucks can be maintained easily with available resources and skills. If they are unwilling or unable to purchase the products, no strategy will work.

Finally, the investors must be confident that the technology is mature, that the market exists, and that they can obtain a return on investment commensurate with the risk. Several international financial organizations have expressed an interest in sponsoring these kinds of developments, but the final decision on investment will depend on whether the needed capital can be generated internally, how much time it will take to recover the investment, and other investment opportunities for the money.

CONCLUSIONS

U.S. industry should take the leadership in encouraging the use of alternative fuels in developing (or developed) nations for the following reasons:

1. It would improve the ability of the United States to resist a third oil-price shock, thus also helping to maintain the world economy and the market for automobiles and trucks.
2. It could increase the market for cars and trucks by improving the economies of the developing oil-dependent nations.
3. It would provide energy companies with additional sources of income and with experience in introducing new fuels.
4. It could provide an opportunity for the United States to demonstrate leadership in marketing alternative fuels and compatible products.
5. It could lead to the creation of a world alternative energy supply, independent of OPEC.
6. It could help to place a long-run cap on the price of world oil.

Of course a single industry cannot introduce this fuel alone. The development of an alternative fuel infrastructure will require the cooperation of other companies as well as governments. It will require the energy companies to provide the feedstocks, chemical engineering companies to devise the most effective process for cleaning and upgrading the feedstocks, local distributors to sell the fuel to industrial and retail customers, automobile manufacturers to produce the products to use the fuel, national and international financial institutions to provide the capital, and national governments to create the business climate that will make it possible to produce these products profitably.

The various interested parties need to meet and discuss problems and opportunities, leading to the formation of a strategy mutually beneficial to all parties. The timing is critical because a long lead time is required to obtain agreement and funding and to build capacity, and time may be running out.

REFERENCES

1. Joy Ramsey Dunkerley, *Energy Strategies for Developing Nations*. Johns Hopkins Press, Resources for the Future, 1981.

2. World Bank, *Energy in the Developing Countries*. Washington, D.C., 1980.

3. Alan S. Manne, *Energy, International Trade and Economic Growth*. Washington, D.C.: World Bank Working Paper 474, 1981.

4. Gunter Schramm, *The Changing World of Natural Gas Utilization*. Washington, D.C.: World Bank, 1983.

5. D. G. Fallen-Bailey, *Energy Options and Policy Issues in Developing Countries*. Washington, D.C.: World Bank Working Paper 350, 1979.

6. A. J. Sobey, *GM Viewpoint on Alternative Fuels*. Joint U.S.-Canadian Meeting on Alternative Fuels, June 25, 1985.

7. U.S. Department of Energy, *Energy Projections to the Year 2010*. Washington, D.C., 1983.

8. U.S. Department of Energy, *Energy Security, Report to the President of the U.S.* Washington, D.C., 1983.

9. National Petroleum Council, *Factors Affecting U.S. Oil and Gas Outlook*. 1987.

10. A. J. Sobey, *Technology and the Future of Transportation: An Industrial View*.Transportation Research Board, 2020 Conference, June 24, 1988.

11. World Bank, *The Energy Transition in Developing Countries*. Washington, D.C., 1987.

12. Jayant Sathaye, Barbara Atkinson, and Stephen Meyers, *Alternative Fuels Assessment: The International Concert*. International Energy Studies Group, Applied Science Division, Lawrence Berkeley Laboratory, January, 1988.

13. OECD/DAC, *Investing in Developing Countries*. November, 1982.

13

NELSON E. HAY
and PAUL F. McARDLE

Regulated Utilities in the Vehicular Fuel Market: Toward a Level Playing Field

Fueling vehicles with natural gas, as opposed to gasoline or diesel fuel, is preferable from economic, environmental, energy security, and safety perspectives.[1] Some 30,000 natural gas vehicles (NGV) are on the road in the United States, yet vocal proponents of NGVs in this country have been relatively few, even among natural gas distributors and pipelines. Today, this situation is changing dramatically.

In August 1988 The Natural Gas Vehicle Coalition opened its doors in Washington, D.C. This coalition, which will facilitate NGV demonstrations and technology development, is a concrete manifestation of the growth in interest in NGVs. Through the coalition, natural gas distributors and pipelines will seek to direct a new order of magnitude of resources to NGV promotion with a particular emphasis upon factory-built dedicated NGVs. And increasingly their efforts are being joined by gas and oil producers, original vehicular equipment manufacturers, the environmental community, and state and municipal governments.

This chapter will examine the factors that have retarded NGV development in the past, the changes that have now stimulated a new level of support for NGVs, and the impediments that remain to be corrected.

PAST IMPEDIMENTS TO NGVs

In the 1960s and again in the late 1970s, significant interest in NGVs developed in the United States gas utility industry. Yet, in both cases, the fledging development of NGVs was quickly terminated. Why?

Three major factors were responsible for limiting the development of NGVs:

Regulatory/Gas Supply Environment. The first impediment to NGVs was the prevailing regulatory environment of the time. Natural gas was viewed as a precious resource that should be husbanded for particular "high priority" uses. It was believed that natural gas would be in short supply and that prices would rise significantly. Acting on these beliefs, federal regulators in the early 1970s and again in the early 1980s took actions that sharply increased gas prices and effectively killed the developing NGV market.

Insufficient Economic Advantage. Even before regulatory action increased natural gas prices, the economic advantage of compressed natural gas (CNG) was insufficient by itself to stimulate significant market penetration. Consequently, gas utility industry projections of NGV market potential have been quite small.

Poor Technology. Early efforts to build the market were dependent upon retrofits of existing vehicles. The industry had no mechanism for improving NGV technology, not only to overcome existing limitations (e.g., to optimize performance on CNG, and reduce fuel tank weight and size), but also to keep pace with changing gasoline-vehicle technology. All three of these major impediments to NGVs are now diminishing rapidly.

U.S. Gas Supply

The history of U.S. gas supply provides important lessons regarding the availability of gas resources and is recounted here as an introduction to the subject of resource cost and availability. The natural gas industry is just entering the fourth phase in its modern development (post–World War II). The first era (1945–72) was characterized by growth and optimism. While this period was dominated by extensive governmental interference in all aspects and segments of the gas industry, it nevertheless saw rapidly increasing demand and market share for natural gas. Overall, it was an extremely optimistic time for natural gas with most forecasters regarding gas as a fuel of the future. This was the "golden age" of gas.

The end of the first era witnessed decreasing supplies—a specter of the dark days ahead. The combination of field price controls, rising gas demands, and declining supplies resulted in the second era (shortages and pessimism). The second era (1973–78), like the first, was characterized by tight regulatory controls over all aspects of the gas industry. During this period natural gas picked up a negative image and was discussed as a "premium fuel" or "finite resource." These catch phrases were disastrous for natural gas development.

The third era began in 1979. It was the result of new rules decontrolling natural gas (phased in through January 1, 1985) in the Natural Gas Policy Act of 1978. Analyses of the prospects for gas in this era dealt with rules and regulations left over from the second era. Resolution of these problems—high take-or-pay requirements in nonmarket-responsive contracts, a regulatory morass ill suited for a competitive market, and pessimism over future supplies and markets—were the focus of attention throughout the period. However, these were problems left over from the regulations of the past. They are not the issues of the future. This was an era of transition.

The fourth era, which we have just entered, will be most like the first period and least like the second and third eras which were both reactive in nature.

The first, second, and third eras have all proven that both gas supply and gas demand are very price elastic. On the demand side, when prices were controlled, consumption soared; but when prices later soared, demand collapsed. Demand proved to be very price elastic within each sector (i.e., 25 percent conservation in the residential sector, 45 percent in the industrial) as well as between fuels. On the supply side the evidence is nearly as dramatic. The higher prices brought about a dramatic shift in gas availability and reserve additions. On the deliverability side, a period of shortages and curtailment was replaced by a period of surplus. In the 1970s, new additions to natural gas reserves equalled only 46 percent of proven reserves; in the 1980s, this increased to over 90 percent. These price-elasticity lessons are important because they presage a future free of the abrupt swings of the past.

Natural Gas Resources

Whenever natural gas is discussed in relation to energy policy, the question of the adequacy of gas supplies is critical. The availability of gas supply is dependent on the wellhead gas price. While this appears to be an overly simple and indisputable statement, until recently it has been heretical to many large energy users and energy modelers. The dissenters—and there are still some— believe in a "finite resource" theory that has as its basic tenet a steeply sloping gas supply curve in the future. The wide array of gas supply sources available to new markets rebuts this hypothesis.

In May 1988 the U.S. Department of Energy (DOE) published a comprehensive report on domestic natural gas supplies that details the resources known to be technically recoverable today—1,059 Tcf or 60 years' supply at the current consumption rate—and concludes that over half this supply can be made available at $3 per Mcf or less (1987 dollars).[2] U.S. natural gas production in 1987 was 16.3 Tcf. The nation has discovered and has available to be produced about 160 Tcf of proven reserves in the lower 48 states. The Potential Gas Committee (PGC), the industry group that evaluates undiscovered gas resources, has estimated that there are an additional 620 Tcf of conventional gas resources that can be produced with foreseeable technology and economics. Therefore, conventional lower 48 states gas resources total nearly 800 Tcf: 160 Tcf of proved reserves and 620 Tcf of potential. This is equivalent to nearly 50 years' supplies at the current usage level. Analyses indicate that most of these conventional supplies could be made available at very competitive prices, e.g., a report prepared for A.G.A. estimates that 75 percent of the 620 Tcf could be produced at $3 per Mcf or less (1986 dollars).

The perception that the nation is using up the conventional natural gas resource base (the finite resource theory) is the rationale for projections of much higher gas prices beyond 2000. This theory is wrong. In addition to conventional gas resources, unconventional gas resources add about 200 years of supplies. These unconventional sources include Eastern Devonian shales, Western tight sands, coal seam methane, and enhanced gas recovery. Technology development will result in these vast unconventional supplies becoming conventional.

A.G.A.'s assessment of gas supply capability in the lower 48 states is supported by the fact that the replacement of production with new reserves (i.e., reserve additions) has far exceeded the expectations of just a few years ago. In the period 1981–86, aggregate reserve additions in the lower 48 states were 93 percent of production, with 1986 at 96 percent and 1987 estimated at 75 percent. What is interesting is that even when demand and prices are poor, as in 1986 and 1987, gas supplies held up better than expected. These numbers prove that if gas demand is there and the price is right, supplies will be discovered and available. The strong showing for gas must be contrasted with oil, which saw a sharp decline in domestic production and reserve replacement.

Until the last decade, the development of the natural gas resource base was a stepchild of oil exploration and development activities. This has changed somewhat, and will change dramatically over the next decade. The relative size of the undiscovered resource base as well as the economics of exploration and production will increasingly favor natural gas. In the early 1990s, natural gas production in the lower 48 states should be at least 50 percent greater than crude oil production. A survey of annual reports by major and independent oil companies confirms the view that the exploration and production industry is increasingly viewing gas as the fuel of the future.

Gas exploration success over the last five years further confirms our resource assessment. Recent gas discoveries have opened up new frontiers; they reinforce the resource estimates made by the Potential Gas Committee. For example, major new gas finds in the eastern and western Gulf of Mexico and deep offshore in the gulf will likely stabilize production capability in this high-productivity area in the foreseeable future. Exploration successes in frontier on-shore areas have the promise of increasing future national production capability.

Is There A Gas Bubble?

A.G.A. has forecast that excess domestic production capability ("gas bubble") will be gone by 1990. There has been all too much emphasis on the gas bubble, and for the wrong reasons. Some people think the end of the bubble will mean shortages and much higher prices. This is wrong. Others have overlooked the negative impact of the bubble on gas supply activity, including reserve additions. There will be a healthy market in equilibrium after the gas bubble disappears, with only slightly higher prices.

Supplemental Gas Sources

In addition to lower–48 gas, a number of gas supply sources will help satisfy domestic requirements as we move up the gas supply curve, thus slowing price increases as production is maintained or expanded.

Supplemental sources of gas (i.e., sources other than lower–48 production) currently provide less than 5 percent of the total U.S. natural gas supply. This percentage is likely to increase over the next 25 years. By the year 2010, sources other than the lower–48 supplies are expected to contribute from 17 to 22 percent of the total gas supply.

Pipeline deliveries from Canada are likely to be the dominant supplemental supply source over the next decade. Canada has 97 Tcf of proven reserves and 320 Tcf of potential resources, with only about 2 Tcf of domestic Canadian requirements. Thus, access to U.S. markets is essential to the development of these extensive gas resources. Other imports, such as gas from Mexico or LNG from Algeria, the Caribbean, or other overseas sources, could resume or be initiated in this period but are expected to contribute less than 1 Tcf to total U.S. supply before the late 1990s. Domestic supplemental gas resources could be of significance after the turn of the century. Recent announcements by the sponsors of the Alaskan Natural Gas Transportation System (ANGTS) and a pipeline down the Mackenzie Delta by Imperial and Shell Canada indicate that the cost of bringing North slope gas by pipeline to the lower 48 states has been significantly reduced, thus greatly enhancing the likelihood that these gas supplies will be available by 2000. Synthetic gas production processes, particularly those based on coal, could provide increased volumes of gas energy after the year 2000.

It is expected that all of these supplemental sources could be contributing to the nation's gas supply at some point. However, changes in gas markets (prices and volumes), national energy priorities and policies, and technological advancements will influence the rate at which each of these sources is integrated into lower–48 state supplies.

In summary, continued and expanded natural gas production can continue into the foreseeable future with only small increase in cost and price.

The Negative Impact of Regulation

Gas utility regulation has had both positive and negative attributes. There is no denying, however, that the effect of many regulations has been to inhibit creativity and innovation, reduce gas production, and make the industry less efficient.

Historically, the focus of natural-gas regulation was largely the protection of "captive" customers in terms of price and availability. The process was highly restrictive and slow, generally characterized by a full cost-of-service approach to pricing, the subsidizing of smaller customers by larger customers, and full proceedings to establish changes in price or expanded service. Outright restrictions on new hookups and most types of advertising and promotion also became common in the 1970s.

As is now well established, this regulatory regime became acutely counterproductive in the 1970s and 1980s as energy markets became more volatile. The 25-percent decline in annual natural gas sales and emergence of the take-or-pay problem are among the effects that have been well documented. In response, progress has been made in the past decade toward increasing the role of market forces in the natural gas industry and in reducing the regulatory impediments to responding to market forces. At the federal level we have seen *de jure* or *de facto* deregulation of most wellhead gas prices, institution of a highly flexible purchased gas adjustment process, repeal of the Powerplant and Industrial Fuel Use Act (FUA) and incremental pricing provisions of the National Gas Policy Act (NGPA), development of a meaningful spot market, and substantial opening of the system to carriage and gas-to-gas competition.

A market-based philosophy underpins ongoing regulatory changes at FERC and some state regulatory commissions. This philosophy involves the unbundling of utility rates and services, rates that reflect the market value of the level of services received, and accountability on the part of the purchaser for its gas purchase decisions. These are three straightforward concepts providing sellers (both pipelines and distributors) the opportunity to offer a wide array of services as well as the traditional one-stop shopping, and to respond to new market opportunities such as natural gas vehicles. It will increasingly result in those services being priced to respond to market conditions and opportunities.

Changes in NGV Economics

NGV economics have been dealt with extensively elsewhere and will not be reviewed here.[3] It has been shown than NGVs, especially dedicated NGVs, are economically attractive for a broad range of energy price conditions.

Technology Development

One of the most important factors in the development of the NGV market is improved vehicle technology. After several decades of stagnation, NGV technology is now entering a period of revolutionary change. These improvements are needed to catch up with the gasoline engine as well as to improve performance and emission in diesel engines. For over 80 years automotive engineers were devoted to improving and optimizing motor vehicles for gasoline; now a similar effort is needed to redesign engines and vehicles for natural gas.

A key to the success of NGVs, therefore, is investment in technology R&D. The natural gas industry must play an important role. One step was the establishment of the NGV coalition as described earlier. A second step is increased vehicular R&D activity by the Gas Research Institute (GRI), the research arm the gas industry in the United States (administered by the Federal Energy Regulatory Commission). A third step is creation of initial market through expanded purchases of NGVs by vehicle fleets owned by governments and businesses.

Gas Research Institute Vehicular R&D Activity

The GRI is currently exploring research and development avenues that could greatly increase the marketability of NGVs. Dedicated natural-gas engines, retrofit kit technology, cylinder certification procedures, compressor station media and metering devices, and baseline emission data for duty vehicles are some of the areas that GRI is presently focusing on.[5]

Specifically, the GRI programs include the development of dedicated natural-gas engines for four-cycle transit buses to the 1991 EPA standards and industrial vehicles to meet the 1994 EPA industrial and light-duty vehicles now powered by spark-ignited gasoline engines (this application would also include the development of a cost-competitive fuel metering injection system). In the retrofit area, GRI is developing a cost-effective retrofit kit to permit dual fuel or spark ignition natural gas-only operation of existing two-stroke diesel transit buses. The establishment of a scientifically based time interval for CNG cylinder recertification and the development of onboard CNG recertification methods are GRI programs aimed at reducing the cost and inconvenience of CNG cylinder inspection. CNG refilling-station initiatives include the development of an improved high-pressure meter to make possible the accurate filling of CNG cylinders to 3,000 per square inch (psi) in five minutes and improved natural gas storage media to reduce storage pressure to 500 psi or less. GRI is also attempting to establish baseline data from a dedicated CNG heavy-duty engine under the federal transient emissions test procedure (GRI is currently testing a Caterpillar 3400 fast, lean-burn natural-gas engine in this application).

Conversion and Demonstration Programs

Emblematic of the heightened interest and investment in new NGV technology is the wide array of entities getting involved in NGV conversions and demonstration programs. Gas companies, state and local governments, colleges and universities, mass transit authorities, and school districts are all investing substantial sums in NGV development.[6]

This investment has manifested itself in the CNG conversion of diesel and gasoline vehicles. Gas companies recently converting vehicles include Washington Gas (2 diesel buses), Southwest Gas (300 vehicles recently converted to add to existing fleet of 400 NGVs), Brooklyn Union Gas (2 gasoline transit buses converted), Columbia Gas (18 diesel trucks converted), Atlanta Gas Light (13 diesel vehicles converted) and Minnegasco (10 buses converted). State and local governments launching NGV programs include the state of Arizona (CNG is recognized as a viable clean-air fuel by the state, and 90 NGVs are currently being tested), Phoenix (1 diesel bus converted), Tucson (6 vehicles converted, with plans to convert its entire fleet by 1993), Scottsdale, Arizona (23 gasoline vehicles converted to add to their current fleet of 84 NGVs), and Glendale, Arizona (40 vehicles converted). Pima College of Tucson (25 gasoline vehicles converted) and Arizona State University (100 vehicles converted) are among the universities and colleges embracing NGVs. The Los Angeles County Transportation Commission (10 dedicated CNG buses and 2 diesel/gas dual-fuel buses), Sun City Transit of Sun City, Arizona (4 gasoline minivans converted with 8 more conversions planned), and the Garland (Texas) School District (81 buses and 10 maintenance vehicles converted) are among the transit and school districts investing in NGVs.

LEVELING THE PLAYING FIELD FOR NGVs

In spite of the improvements in the gas supply outlook, regulatory environment, and technology area described above, significant impediments to gas marketing remain. In a 1982 American Gas Association survey, 78 percent of respondents said there were restrictions on natural gas advertising imposed by their state public utility commission.[7] More recent information indicates that such restrictions are only modestly less prevalent today.[8] A common restriction, for example, is that only advertising with an energy conservation message can be included in the utility's cost of service. Promotional rates for development of new markets and applications remain a regulated matter for utilities, though not for their competition, and are also far less commonly approved for natural gas utilities than for electric utilities.

The ability of regulated natural gas utilities to respond quickly and competitively to the marketplace remains well below that of oil, coal, and sometimes electric utilities. In fact, the "recent regulatory progress" has to some extent left local gas distribution companies (LDCs) and pipelines at a distinct disadvantage compared to direct sale/purchase players in the gas marketplace. Similarly, utility taxes on sales by LDCs (but not under transportation agreements) have placed utilities at a competitive disadvantage. Resolution of the take-or-pay and contract demand reduction issues would constitute enormous progress, but fundamental questions will remain for LDC marketing.

Simply stated, the issues are whether or not it is desirable that natural gas utilities be able to market more effectively, and how regulated utilities can more effectively compete with less-regulated and unregulated entities. While these questions may have general national answers—and importance to national policy-making—they must ultimately be resolved on a system-by-system basis.[9]

As stated earlier, regulators have historically taken a narrow view of the desirability of gas marketing. From this traditional perspective, the question is whether increased gas sales will lower or at least not raise prices to "captive" or "core" customers. The case in favor of more aggressive development of natural gas markets, viewed from the narrow perspective of traditional regulatory logic, is as follows:

1. Total U.S. natural gas sales have declined by one-quarter since the early 1970s, while the transportation and distribution system has been substantially enhanced. There is clearly a surplus delivery capability on an annual national basis. Consequently, unit costs to customers could, in many service areas, be reduced by achieving fuller system utilization.

2. Increased natural gas sales could alleviate the take-or-pay problem, again reducing unit costs to customers. Further, reduction of the take-or-pay problem would stimulate gas drilling, increase securities values and ratings, and eliminate a major barrier to open transportation access and increased competition. All of these would contribute to lower costs.

3. A surplus of natural gas deliverability in the United States and Canada has persisted for eight years. While the surplus exists, sizable demand increases could likely be accommodated with little wellhead price impact. More fundamentally, events of the past decade have greatly changed perceptions regarding natural resources and the cost of bringing those resources to market. To the extent that these perceptions are analytically captured, it is possible to demonstrate in many instances that even after the "bubble" is gone, wellhead price increases associated with reasonably expected levels of gas demand growth would be more than offset by the savings enumerated in items 1 and 2.

Whatever the results of the traditional LDC regulatory analysis just outlined, such analysis is no longer adequate in an increasingly competitive environment. Regulators and policy makers must avoid provisions that discourage the marketing of gas when it is economically justified. They must not prevent the full utilization of the nation's cleanest fuel. The consequence of ill-conceived regulatory barriers to natural gas marketing is economic inefficiency, including:

- higher consumer costs for gas, vehicle fuels, electricity, and manufactured goods;
- reduced domestic industrial output and employment, and greater reliance upon imports of energy and energy-intensive industrial goods;
- decreased gas exploration activity and negative economic consequences in oil-and gas-producing regions;
- distorted capital formation decisions; and,
- unnecessary and costly environmental pollution.

Increased gas marketing would encourage introduction of new gas utilization technologies that can be of key importance in meeting the energy and environmental needs of the 1990s. Natural gas vehicles are a case in point.

As stated above, the U.S. natural gas industry has experienced a 25-percent decline in sales since the early 1970s, primarily as a result of various regulatory initiatives. As a consequence, the industry has a surplus today of both production and delivery capability. For the long-term as well, the industry has a strong supply outlook, but " . . . has an urgent need to develop new applications as it reaches a saturation point in most of its traditional end-uses. The opportunity for natural gas to enter and capture a new market is at hand . . . as an alternate fuel for the transportation industry.''[10]

Because of the natural gas industry's history as one of the most regulated industries in the nation, and because neither the natural gas industry nor regulators are accustomed to thinking of vehicular fuel as a major natural gas market, many regulatory and institutional factors place the gas industry at a disadvantage in developing this new market. Fortunately, the industry is in the midst of major changes in its regulatory framework that are opening the system to greater competitive responsiveness. Recognizing both the need for the NGV market and the changing regulatory environment, the gas industry increasingly will be seeking relief from these impediments. In particular, in order to stimulate the market and achieve scale economies, utilities will be seeking to expand the number of utility-owned, rate-based, and public refueling stations.

While there are approximately 300 utility NGV refueling stations in the United States, there are only 19 public NGV refueling stations. In comparison, Canada has more than six times this number, most of which are located at existing gasoline stations. Governmental and utility-provided incentives have encouraged corporate investment in NGV stations in Canada.

The U.S. stations are typically located on gas utility property, and they service the utility's fleet as well as outside NGV fleets. The NGV rate at the public station, which must be approved by the utility's regulatory commission, is comprised of several items besides the cost of gas. Operation and maintenance costs are almost always included in the NGV rate. State road taxes, where applicable, can be collected at the pump or paid directly to the state by the fleet owner, depending on state law. Return on the investment in the station can also be included in the rate, or the utility may decide to include the entire cost of the station in its rate base (the rate base is the value of the utility's assets on which the regulatory agency allows the utility to earn a return, collected through its gas rates). In the latter case, the outside fleet would not pay any additional price for the recovery of the station investment.

Other forms of utility ownership of NGV stations are being considered. One method involves the utility owning and operating the refueling facility on another fleet owner's property, with the facility being on the utility's side of the meter. Alternatively, the utility could participate in a joint-ownership arrangement with an NGV refueling system equipment manufacturer. The utility would provide some financial support and would sell gas to the station while the manufacturer would provide the equipment (at cost) and operate the facility, with both splitting the profits. Regulatory permission would be required if the utility intended to ''rate base'' such investment.

Each of these measures can contribute significantly to development of the NGV market by improving the economics and lowering the price of gas for the transportation sector. For example, gas utilities that have finely segmented their commercial sector and instituted special NGV rates are currently selling natural gas for compression at roughly 20 percent below their general commercial natural gas rate. Based on this differential, the cost of compressed natural gas at fleet-owned and operated refueling stations with special NGV rates can often be reduced to 75 percent of the cost of gasoline.

Utility-owned and operated refueling stations further enhance the economic advantage of NGV fleets due to economies of scale and the possibility of rate basing station costs. The equivalent cost per gallon of gas from a utility that fully recovers its investment in the station through the compressed natural gas can typically be closer to 70 percent of the gasoline cost. The lower rate relative to a fleet-owned and operated station is due to economies of scale and the lower return on investment required by utilities. A utility-owned and operated station whose investment cost is recovered through the general rate base (not just through the sales of compressed natural gas to NGVs) can provide compressed natural gas at 65 percent of the cost of gasoline.

Opening a utility-owned and operated NGV refueling station to the public, also, offers numerous advantages to both fuel purchasers and sellers. By eliminating the need of a private fleet to install a refueling station, the original capital outlay and operating cost of that fleet, and thus its risks, are significantly reduced. As a result, the payback period for the fleet's conversion will be reduced. Public refueling stations increase natural gas sales and system utilization since more fleets, particularly small ones, find it profitable to convert their vehicles to dual-fuel operation. For an incremental per-vehicle investment of less than half the investment required for typical residential home hookup, a converted vehicle consumes about 1.25 to 2.5 times the amount of gas that the average residence consumes annually. The utilities' investment in a refueling station may be viewed in a light similar to investing in mains in areas of new residential, commercial, or industrial demand. Opening facilities to outside fleets increases the use of the utility's refueling station, thereby reducing the per-vehicle capital cost of the facility, a benefit to both the utility and outside fleets.

In addition to these measures to stimulate the market, the industry will be seeking to remove other common regulatory impediments and disadvantages, including the following.

- *Prohibition by many state regulatory authorities of the "sale-for-resale" of natural gas.* Sale-for-resale prohibitions harken back to the 1930s when the gas industry was afraid of landlords in multifamily buildings submetering natural gas to residents. These same prohibitions now are roadblocks to utilities opening up third party–owned public refueling stations.
- *Lack of federal government funding for NGV research.* Money is currently being spent on research for electric vehicles, methanol vehicles, and the Stirling engine. DOE historically maintained that NGVs are commercialized, and that they have no mandate to do NGV research and development. They do, however, acknowledge that natural gas diesel technologies could be a viable R&D topic, and more recently they have acknowledged the potential benefits of a new generation of CNG technology.
- *Corporate Average Fleet Economy Standards including natural gas vehicles.* Natural gas vehicles should be included with other alternative fuels in legislation pertaining to

CAFE standards to allow auto manufacturers to average-in alternative-fuel vehicles in calculations of fleet efficiencies.

- *Developing a methane hydrocarbon standard.* Before manufacturers will be willing to commit to building dedicated natural-gas vehicles on a commercial scale, an NGV hydrocarbon emission standard must be promulgated by EPA. Such a standard already exists for methanol vehicles.

- *Vehicle Warranties.* There is some uncertainty as to the degree to which manufacturers will extend vehicle warranties in the event of a conversion to natural gas. Most companies will warrant their gasoline equipment as long as any engine or mechanical problem cannot be attributed to the natural gas system. Clearly, warranty restrictions would present additional barriers to the increased utilization of NGVs.

- *Local Codes and Ordinances.* In some locations NGVs are prevented from traveling through tunnels and parking in indoor and underground garages. A continuation of the safety tests that have demonstrated that NGVs are safer than gasoline vehicles will help allay the fears of local authorities.

SYSTEM UTILIZATION BENEFITS OF NGVs

Replacing gasoline vehicles with NGVs will result not only in a more efficient allocation of U.S. energy-producing capacity, but also in a more efficient utilization of natural gas and gasoline transmission and distribution capacity. Both natural gas and gasoline systems exhibit seasonal peaks, natural gas in the winter and gasoline during the summer.

Since vehicular energy consumption peaks in the summer, increased natural gas vehicular consumption at the expense of gasoline consumption will increase off-peak seasonal natural gas demand (summer) by more than the increase in peak natural gas demand (winter). The converse will be true for gasoline. Therefore, both the natural gas and gasoline pipeline systems will experience a reduction in the differential between peak and off-peak utilization rates. Smaller differentials between peak and off-peak capacity utilization in both systems will reduce the need for redundant idle capacity during off-peak seasons.

Irregularities in daily natural gas demand can also be exploited. Natural gas demand normally experiences major peaks in the early morning and minor peaks in the early evening. The utilization of overnight slow-fill refilling facilities for fleet vehicles will tend to equilibrate peak and off-peak demand, and therefore result in a more efficient utilization of natural gas distribution networks.

CONCLUSIONS

While NGVs offer a number of economic, environmental, energy security, and safety advantages to gas consumers and the nation, their introduction has been slow because of historical regulatory and institutional disadvantages. These disadvantages are now quickly diminishing. The new national focus on alternative fuels combined with the transition to a more competitive regulatory environment in the gas industry provides an opportunity to eliminate further these impediments. Without those impediments, NGVs could compete on a level playing field, allowing the market to choose among vehicular fuel alternatives.

NOTES

1. J. Winston Porter, *Preliminary Analysis of the Safety History of Natural Gas-Fueled Transportation Vehicles* (Arlington, Va.: American Gas Association, 1979). American Gas Association, *An Economic and Efficiency Comparison of Alternative Vehicular Fuels: 1987 Update* (Arlington, Va., September 22, 1987). J. Winston Porter, *Preliminary Analysis.* American Gas Association, "Natural Gas Vehicle Safety Survey—An Update," (Arlington, Va.: American Gas Association, 1987). Charles K. Ebinger et al., *Natural Gas Vehicles: A National Security Perspective* (Washington, D.C.: Center for Strategic and International Studies, 1985). U.S. Department of Energy, *Energy Security: A Report to the President of the United States,* Washington, D.C., March 1987, 114, 116, 120. Milton R. Copulos, *Natural Gas: The Vital Energy Security Link* (Alexandria, Va.: The National Defense Council Foundation, 1985). Chem Systems, Inc., *A Briefing Paper on Methanol Supply/Demand for the United States and the Impact of the Use of Methanol as a Transportation Fuel* (Arlington, Va.: American Gas Association, September 1987). American Gas Association, *Fact Book: Energy, the Environment and Natural Gas* (Arlington, Va.: October 1983). Aerospace Corporation, *Assessment of Methane-Related Fuels for Automotive Fleet Vehicles* (Washington, D.C.: U.S. Department of Energy, 1982). Wisconsin Gas Company, *Review of Compressed Natural Gas Program, City of Milwaukee,* Milwaukee, Wisc., 1986. Alcohol Fuels Systems, Inc., *Compressed Natural Gas Vehicle Performance and Emissions Study* (Sunnyvale, Calif.: Pacific Gas and Electric Co., May 12, 1987). Nelson E. Hay, *Vehicular Emissions* (memorandum, Arlington, Va.: American Gas Association, February 11, 1988). B. C. Research, Ltd., *Exhaust Emissions Requirements of Natural Gas Fueled Vehicles* (Vancouver, B. C., Energy, Mines and Resources [undated, circa 1986]. *Guidance on Estimating Motor Vehicle Emission Reductions From the Use of Alternate Fuels and Fuel Blends* (Ann Arbor, Mich.: Emission Control Technology Division, U.S. Environmental Protection Agency, January 1988). Thomas M. Baines, U.S. Environmental Protection Agency, *Economic and Environmental Studies on Compressed Natural Gas Vehicles* (presentation before "Natural Gas and Clean Air: An Alliance for America's Future," Washington, D.C.: April 19, 1988).

2. Argonne National Laboratory, *An Assessment of the Natural Gas Resource Base of the United States* (Washington, D.C.: U.S. Department of Energy, May 1988).

3. See, for instance, AGA, *An Economic and Efficiency Comparison of Alternative Vehicle Fuels: 1987 Update* (Arlington, Va.: 1987); Mark A. DeLuchi, R. A. Johnston, and Daniel Sperling, "NGV vs. Methanol: A Comparison of Resource Supply, Performance, Fuel Storage, Emissions, Cost, Safety, and Transition Issues" (*SAE,* Paper 881656, 1988).

4. American Gas Association, *Natural Gas Vehicles: The International Experience* (Arlington, Va.: May 13, 1988).

5. Ibid.

6. Ibid.

7. American Gas Association, *Gas Utility Advertising: Analysis of Recent Trends,* Arlington, Va., June 11, 1982.

8. Jeffrey P. Sprowls, "Gas Utility Advertising: An Overview of Recent Commission Decisions," *The Natural Gas Lawyer's Journal* 1, no. 1 (July 1985); *A.G.A. Member Company Comparative Advertising Practices,* (Arlington, Va.: American Gas Association, April 1988, unpublished).

9. Portions of this material are excerpted from Nelson E. Hay, *LDC Marketing/ Growth Policy* (working paper prepared for the Stanford University Energy Modeling Forum, 1988).

10. American Gas Association, *Natural Gas Vehicle Action Plan*, Arlington, Va., A.G.A. Advanced Technology Task Group, December 9, 1987, pp. 1–2.

14

GERALD H. MADER
and ORESTE M. BEVILACQUA

Electric Vehicle
Commercialization

In the last two years, the Electric Power Research Institute (EPRI) and automobile manufacturers have developed an electric van that can compete with petroleum-powered vans in the service-fleet market. The successful introduction of this van—and future electric vehicles (EVs)—depends on the careful development of EV markets and the EV service infrastructure. Through its work to build these markets and the EV infrastructure, the Electric Vehicle Development Corporation (EVDC) will ensure that today's EVs become a permanent part of the national transportation picture. National interest in EVs is increasing as transportation-related air quality problems continue and dependence on foreign oil rises. This chapter reviews the coordinated activities of EPRI, EVDC, the government, individual electric utilities, and automobile manufacturers working together to commercialize EVs.

BACKGROUND

EV commercialization requires the parallel development and evolution of technology, market, and infrastructure. Since the mid–1970s, significant effort and resources have been committed to EV technology research and development. Until 1984, however, little attention was directed to defining and building an initial market for EVs and establishing the support systems required to keep EVs operating and productive in the field.

Recognizing this situation as well as the long-term potential for EVs, a group of electric utility companies formed the Electric Vehicle Development Corpo-ration in 1984. EVDC's sole purpose is to commercialize EVs in North America by consolidating the interest and activities of key stakeholders. EVDC is sup-ported by its membership, which includes electric utility companies that collectively account for some 40 percent of the electricity generated in the United States, as well as major automotive and component manufacturers such as General Motors Corporation (GMC), Chrysler Corporation, Ford Motor Company, Chloride EV Systems (CEVS), and Powerplex Technologies, Inc.

EV technology development is currently being performed by the U.S. Department of Energy (DOE), the Electric Power Research Institute, and private industry. EVDC works in partnership with these organizations to review market needs and technology R&D priorities. It also plays a significant role in organizing and monitoring joint development programs. Much of EVDC's success can be traced to building good working relationships with DOE, EPRI, manufacturers, user groups, and EVDC member organizations. These relationships are essential to achieving steady and progressive commercialization.

Because the EV represents a substitute for the petroleum-powered vehicle, the automotive industry is not motivated to take the lead in EV commercialization. This situation gave rise to the need for an organization such as EVDC to shepherd EV technology. EVDC's efforts to bring together the various elements necessary for EV introduction represent a model for commercializing a new technology. This chapter provides an overview of EVDC's ongoing programs and plans for ensuring the timely development and introduction of market-compatible EVs.

EVDC'S EV COMMERCIALIZATION STRATEGY

EVDC has designed and adopted a near-term EV commercialization strategy to advance the market debut of cost-effective electric vans for commercial fleet applications. On the basis of personal-use-vehicle and fleet-vehicle market assessments, vans were selected as the most suitable near-term vehicle technology. These assessments also targeted the fleet market as an ideal starting point for EV introduction for several reasons:

1. The performance capabilities of near-term EVs are more compatible with the requirements of fleet vehicles than those of personal-use vehicles.
2. Fleet vehicles in general, and vans in particular, are garaged at central locations during the night, which facilitates overnight battery recharging.
3. Most larger fleets perform their own vehicle maintenance, which simplifies providing EV service and support.
4. Fleet buyers are more likely to evaluate alternative vehicles based on total life-cycle cost rather than initial cost, an important factor when comparing the economics of EVs to conventional vehicles.
5. Fleet managers have flexibility to assign EVs to missions that are compatible with their performance capability.

As the technology and infrastructure systems mature, the market for EVs will broaden. Whereas the first electric vans will be marketed as replacements for conventional vans in fleet use, the next generation of electric vans, with enhanced

Figure 14.1
Introduction of Prototype Electric Vans

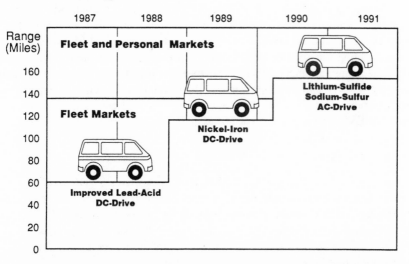

Evolution of the EV market is directly dependent on breakthroughs in EV technology. Two key improvements, a high-performance battery and an AC powertrain, should give EVs the range and performance capabilities to compete in a broad mix of transportation markets, including the vast personal car market.

performance capabilities, will be suitable substitutes in fleet use for conventional passenger vehicles. As the vehicles prove themselves and are accepted into the market, these improved vans could begin to penetrate the personal-use market. However, the large-scale use of EVs as personal vehicles is tied to the availability of advanced batteries and drivetrains that will make long-range, high-performance EVs practical. Figure 14.1 illustrates EVDC's commercialization strategy. To implement this strategy, EVDC is focusing its efforts in four areas: vehicle demonstrations, infrastructure development, market development, and industry/government relations. The following sections briefly describe these activities.

GRIFFON DEMONSTRATIONS

In 1985, EVDC launched a major project to introduce the Griffon, an electric van manufactured by GM in England, into North American service fleets. The Griffon was the first modern-day van to be produced on an assembly line; the same line was used to produce gasoline-and diesel-powered vans as well. The Griffon uses a propulsion system made by CEVS that has accumulated more than 7 million miles in field operations worldwide. EVDC chose the Griffon for the first commercial demonstration of EVs because of its performance capability, its proven reliability as a fleet vehicle in the United Kingdom, and the fact that it was backed by GM.

Over a period of almost three years, the 32 Griffons employed in the demonstration have logged over 400,000 miles of service at 11 electric utilities across the United States and Canada. The vans have been used in applications ranging from mail delivery, employee shuttles, van pools, and plant maintenance, to security and public relations. This project has provided a real-life laboratory for demonstrating the suitability of electric vans in fleet applications, identifying and evaluating van improvements, creating the critical vehicle service and parts support systems, and developing marketing methods and incentives. The Griffon demonstration has successfully broadened interest and confidence in the feasibility of EVs. The key elements of the demonstration are described below.

User Feedback

User reactions to the Griffon have been very positive. EVDC has initiated a reporting system to monitor van operations and to solicit user feedback on the suitability and acceptability of the van and its support systems. The EVDC-implemented data system collects, analyzes, and reports on the vans' service requirements and operations. In addition, a users' group serves as a direct communication link between EVDC and Griffon operators.

The demonstration program has helped to identify product and support-system improvements that will increase EV fleet integration and market acceptance. Recommended vehicle design changes include modification of heating and ventilation controls, addition of a parking pawl, improvement of the suspension system and of handling, and the availability of air conditioning as an option. Many of these features are now being incorporated into the design of new electric vans. For example, interest in improved handling prompted investigations into "electric" power steering for future electric vans.

Infrastructure

The establishment of a service and support system—an initial EV infrastructure—was one of the most important factors in conducting the Griffon demonstration. The system put in place for the demonstration consisted of:

1. Manuals and training: Vehicle operations, maintenance, and parts manuals were provided to all van operators. Comprehensive maintenance and repair training programs were also available.
2. Service assistance: GM and CEVS service technicians provided remote and on-site vehicle service assistance. A maintenance program option was also offered through which a trained CEVS technician would conduct all regularly scheduled service on the vans' electric propulsion systems.
3. Spare parts: standard procedures and arrangements were established for ordering, distributing, and billing spare parts. A Griffon parts center was set up in Detroit, Michigan, to handle the stocking and shipping of mechanical, electric/electronic, and frame parts.

4. Warranties: the Griffon was sold with a standard 12-month warranty guaranteeing all vehicle components except for the propulsion battery, which was covered by a 48-month warranty. Warranty claim and labor reimbursement procedures were also established.

This service and support system will be the foundation for the comprehensive EV infrastructure required to establish EVs as a viable transportation option.

Marketing

Demonstration marketing activities have focused on increasing fleet operator awareness of and exposure to electric vans. Through the Griffon promotional activities, EVDC and utilities have identified potential new electric van users. They have also obtained valuable insights into how to design an effective EV marketing program.

"Ride and Drive" events have been a particularly effective marketing approach. They have given utility and nonutility fleet operators the opportunity to get a first-hand look at the Griffon and to experience the drive and feel of a state-of-the-art EV. Perhaps most importantly, Ride and Drive participants have left with a better perception of EVs. Utilities continue to use Ride and Drive programs to educate fleet operators and identify potential customers for new electric vans. For example, more than 100 fleet operators attended a Ride and Drive event sponsored by Detroit Edison conducted at GM's Milford Proving Grounds near Detroit.

In addition, individual utilities are experimenting with other approaches for introducing electric vans to fleet operators and providing incentives for their use. For example, Florida Power Corporation has exhibited a specially modified demonstrator van at several car shows and technology fairs. Detroit Edison Company set up a program through which Griffons were loaned out to be used in selected fleets, including an auto parts distributor, a university, a florist, and a telephone service company. These loans, which lasted from four to six weeks each, were made in order to interest the fleet owners in a subsidized Griffon lease program.

G-VAN DEMONSTRATION

The current phase of EVDC's EV commercialization strategy involves the introduction of the G-Van, the first modern EV to be produced in North America. The product of a cooperative program between GMC Truck and CEVS, this full-size van will look and handle very much like its gasoline-powered counterparts, the GMC Vandura and Rally. The main difference lies in the electric propulsion system, which is based on the technology used in the Griffon. However, an improved battery and upgraded controller and motor allow the G-Van to provide driving performance comparable to that of the Griffon while offering a substantially larger cargo space and many new features.

Based on recommendations from Griffon users, the G-Van is equipped with features and options that should broaden the marketability of the vehicle. For example, the G-Van will be available as a cargo van or eight-passenger wagon. It will come equipped with all the standard features found in conventional GM vans plus electrically-driven power steering and power brakes. Air conditioning and battery heating (for use in cold climates) will also be offered as options. All these additions help make the highly functional G-Van suitable for a wide range of fleet applications.

In 1988, the G-Van development project maintained steady progress. The initial proof-of-concept van was road tested at EPRI's Electric Vehicle Test Facility, and plans were laid for building 20–30 prototypes. These prototypes are scheduled to be field-tested in selected utility fleets, where their design and performance will be evaluated and driver reactions to their capabilities and features will be assessed.

Following the prototype field test, Southern California Edison plans to direct a G-Van market demonstration in Southern California. Perhaps 300–500 production G-Vans will be used in the demonstration, which should begin in mid–1990. Like the Griffon demonstration, this G-Van demonstration will

1. evaluate and exhibit the G-Van's market suitability and acceptability,
2. refine the systems and procedures for EV service and support,
3. test alternative marketing programs and user incentives,
4. verify the G-Van's cost effectiveness and reliability, and
5. establish the air quality effects of EVs.

As part of the demonstration, regional infrastructure systems and facilities for EV distribution and service will be examined and established.

EVDC is also beginning plans for the introduction of the Chrysler TEVan, an electric van based on the popular Chrysler Voyager and Caravan minivans. This van incorporates a high-performance propulsion system that will substantially expand the market for EVs (see Table 14.1 for a comparison of Griffon, G-Van, and TEVan capabilities). TEVan prototypes are expected to be available for field testing in 1990.

INFRASTRUCTURE DEVELOPMENT

The establishment of a comprehensive and accessible infrastructure is a prerequisite for EV commercialization. Recognizing this, EVDC, EPRI, and several member utilities are working together to develop an EV distribution operation dedicated to selling and supporting EV products. This operation will establish the infrastructure that EVs will need to compete in the nationwide automotive marketplace.

Table 14.1
Griffon, G-Van, and TEVan

	Griffon	G-Van	Chrysler TEVan[*]
Top Speed	53 mph	53 mph	65 mph
Range	60 mi	60 mi	110+ mi
Acceleration (0-30 mph)	11 sec	12 sec	7 sec
Payload Capacity	1900 lbs	1800 lbs	1200 lbs
Cargo Space	208 cu ft	256 cu ft	120 cu ft

[*] Projected

Note: Near-term EVs will be suitable for a variety of applications. With its
        large payload capacity, the electric G-Van will be a service fleet
        "work horse," perfect for hauling cargo and for local shuttles. The
        smaller Chrysler TEVan, offering increased range and higher top
        speed, will blend well with highway traffic and help EVs enter the
        passenger fleet-vehicle market.

Plans are being made for a new business that will establish and direct the distribution of EVs. The first product to be distributed by this business will be the electric G-Van. Business functions will probably include marketing EVs on a national scale, coordinating vehicle and component production, and establishing sales policies and procedures. Sales, warranties, financing packages, inventories, and customer service will be handled at the regional or local level. Electric utilities will be encouraged to play a direct role in creating regional distribution operations.

MARKET DEVELOPMENT

EVDC market development activities are pursued at two levels. The first, discussed previously, encompasses the work of EVDC and individual utility companies to identify and test alternative marketing approaches. Ride and Drives, loan-lease programs, demonstrator vans, preferential electricity rates, and special promotions are some of the techniques that have stimulated interest in and awareness of modern-day EVs. Gaining user feedback on the capabilities, performance, and design of prototype electric vans is also a critical part of these marketing activities.

The second level of market development pertains to EVDC's work to define and characterize the market potential for near-term EVs. This information is gathered through comprehensive market surveys.

In 1987, EVDC prepared for the introduction of the G-Van and TEVan by conducting an extensive survey of commercial fleets in 30 of the largest urban markets in the United States. As the most comprehensive study to date of EV sales potential, the survey provides a clear picture of the future demand for electric vans.

The results indicate that the 60-mile-range G-Van could replace 161,000 vans within these 30 metropolitan areas. Analyses show that current-technology G-Vans will have the greatest sales potential in large van fleets (those with more than 11 vans). The survey points up the advantages of targeting this fleet market segment by showing the following:

1. Over 60 percent of commercial vans are used in only 3 percent of fleets; this 3 percent is made up of fleets with more than ten vans.

2. Large fleets have indicated a significantly higher willingness to buy EVs than smaller fleets.

3. Large fleets, which have a higher public profile, are in a better position to profit from EVs' indirect benefits such as improved urban air quality and fuel substitution.

4. Larger fleets have more flexibility to blend limited-range electric vans into their operations.

5. Managers of large fleets use life-cycle costs in making purchasing decisions and recognize the benefits of lower operations and maintenance costs.

6. The in-house service capabilities of large fleets will minimize the need for establishing extensive electric van service and training systems.

7. Selling to large fleets will be more cost effective because of the higher sales potential per customer and per marketing dollar.

The introduction of electric vans with extended-range capability will significantly increase market potential (see Figure 14.2). For example, an electric van with a 90-mile range (50 percent more than the current G-van) will increase the potential demand for EVs to 283,000 vans, nearly 80 percent of the vans in large fleets.

These results indicate that achieving an annual sales volume of 20,000–25,000 electric vans within the top 30 markets is a practical objective. By expanding the market reach of electric vans to include all vehicle market areas as well as small and public-sector fleets, this sales potential could nearly triple.

Although modest by traditional automotive industry standards, this market size is sufficient to obtain an acceptable return on investment for manufacturers, distributors, and providers of sales and service.

As noted previously, EV markets will be developed on a regional basis to ensure a responsiveness to local transportation needs. In addition, regionally

Figure 14.2
EV Market Survey Results

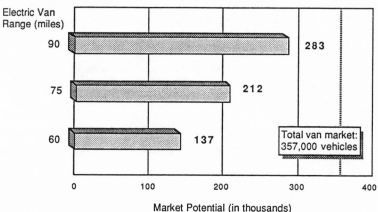

Source: Survey conducted by EVDC in 1987 of commercial vehicle fleets in 30 of the largest U.S. metropolitan areas.

Market potential for electric vans is defined here principally in terms of driving range between battery recharges. This chart shows the result of a 1987 survey of commercial fleets in 30 of the largest U.S. metropolitan areas. The total van market in these areas was estimated to be 357,000 vehicles.

centered marketing will maximize the interest and involvement of local utility companies. The regional surveys conducted as part of the national fleet survey effort support this market strategy by providing region-specific information including the following:

1. Definitions of the fleet van market according to fleet size, industry type, driving requirements, and van configurations.

2. A profile of attitudes among van users toward the benefits and limitations of current-technology EVs.

3. Projections of the effect that improved EV performance will have on potential market share.

The recent market surveys substantiate the large potential for electric vans. As advanced technologies mature, EVs could meet the demands of the vast personal-vehicle market and thereby dramatically expand the sales potential of EVs.

INDUSTRY/GOVERNMENT RELATIONS

Industry/government relations activities are another important part of EVDC's commercialization efforts. Because the development of the EV market is tied to continued progress in EV technology, it is important for EVDC to monitor the status of the extensive R&D being sponsored by DOE and EPRI. To facilitate the exchange of information, EVDC, DOE, and EPRI have established the National EV R&D Steering Committee. This committee provides the three organizations with a forum in which to define R&D priorities and cooperative programs on the basis of market requirements.

By participating in the activities of this steering committee and an EPRI advisory committee called the Electric Transportation Working Group, EVDC obtains information and insight with which to form long-range market and business plans.

To stimulate the market for early EVs, EVDC has recently instituted a Government Relations Program. Through this program, the utility industry will interact with federal, state, and local agencies to develop incentives and regulatory mechanisms to help promote the use of EVs. Justification of government support for such incentives lies in the fact that EVs offer two societal benefits not directly valued by individual users: fuel diversity and reduced vehicle emissions.

The implementation of government economic incentives, such as air quality credits and vehicle tax and registration waivers, could have a significant effect on getting EVs into the market and making them cost-competitive with conventional vehicles. Present economic analyses show that electric vans produced in large quantities would be roughly equivalent to conventional vans on a total life-cycle cost basis. Produced in small quantities, EVs could be more costly to own and operate than conventional vans. Government programs to encourage the acquisition and use of EVs in commercial fleets, even for an interim period, would help build an initial demand for EVs. This demand would reduce EV cost because components and vehicles produced in larger volumes would allow production economies.

EVDC's Government Relations Committee will take the following steps to promote its goals:

1. Monitor legislation and public-sector initiatives that would have an impact on the EV industry. Emphasis will be placed on ensuring that EVs are included in any clean-air or alternative-fuel legislation.

2. Publish a government relations newsletter that highlights these initiatives and promotes additional action at the federal, state, and local levels.

3. Publicize the beneficial effects of EVs on air quality and the role EV technology can play in providing transportation that uses alternative fuel sources.

In addition, EVDC participates in the California EV Task Force. This group includes representatives from several state and regional government agencies. It recommends R&D funding, demonstration programs, and legislative and regulatory actions that will facilitate EV commercialization.

Success in building public-sector support for and involvement in EV commercialization is critical to achieving EVDC's near-and long-term market objectives.

CLOSING

EVDC is putting in place all of the elements needed to foster the timely and systematic commercialization of EVs—demonstrations, infrastructure, marketing, and industry and government support. At the same time, EVDC has worked to involve all key EV stakeholders—the utility industry, the automotive industry, component manufacturers, users, and the government—in its efforts to introduce the first modern-day production EV and to build a new industry.

New technical, organizational, and institutional challenges lie ahead. But EVDC is well positioned and prepared to deal with these uncertainties and looks forward to the successful commercialization of this promising new technology—electric vehicles.

15

KENNETH KOYAMA
and CARL B. MOYER

The Transition to Alternative
Transportation Fuels in California

California, like the rest of the United States, faces a lingering and pervasive problem in its transportation sector: dependence on petroleum for over 99 percent of its transportation energy needs.[1] This dependence on one energy source leaves the state vulnerable to deliberate manipulations of world oil markets. While sources for oil have diversified since the energy crises in the 1970s, the actions of a few key producer-countries could still cause rapid increases in oil prices. Despite our experiences in the 1970s, energy security is still a problem.

Since California's transportation sector relies primarily on petroleum fuels, oil supply disruptions will greatly impact this sector first. Despite advancements in telecommunication networks, the state's economy still depends on the transportation of goods, services, and people. Oil price increases will immediately raise the cost of transportation, causing a multiplier effect on other sectors.

Burning petroleum-based fuels in the transportation sector also causes air pollution. While emissions control systems have steadily improved and could reduce tailpipe emissions from gasoline engines by nearly 50 percent for both hydrocarbon and oxides of nitrogen, even these improvements would not be enough for several cities, including the Los Angeles area, to attain the air quality goals as specified in the 1977 Clean Air Act.[2]

Recognizing these problems with petroleum-based fuels, the California Energy Commission (CEC) began demonstration programs with clean-burning alternative fuels for transportation. These programs have shown the technical feasibility of operating vehicles with alternative fuels. However, bridging the gap between technical feasibility and commercial availability requires a transition period during which fuel-supply infrastructure is put in place.

The CEC proposed a transition strategy for California for moving to clean alternative fuels. The first stage of the transition has already occurred. The CEC implemented an extensive demonstration program to evaluate and test the technology. Based on the results of this program, we have identified several important barriers that must be overcome to continue the transition. These include the lack of a retail fueling-station network and limited range of the vehicle. A unique solution to overcoming these barriers is the fuel-flexible vehicle (FFV) that can operate on any range of methanol-gasoline mixtures. That is the basis of the strategy proposed in this chapter. We will analyze the economic competitiveness of methanol and the role of government in supporting the transition.

This chapter focuses on methanol because it is the fuel the CEC has most experience with and because our working knowledge of compressed natural gas and electric vehicles does not go beyond the available literature and discussions with the industry. However, the CEC also supports the use of other nonpetroleum alternatives. The CEC will sponsor both compressed natural gas and electric vehicle demonstration programs in the 1988–89 fiscal year.

CALIFORNIA'S ALTERNATIVE TRANSPORTATION FUELS PROGRAM

In 1979, California began programs with alternative transportation fuels to address energy security and air quality concerns. The California Legislature passed SB 620 (Mills) to investigate the practicality and cost effectiveness of alternative motor vehicle fuels. The goal of the program was to research and develop an alternative fuel to protect California against gasoline shortages and drastic price increases and that would also improve, or at least not degrade, air quality.

Following the evaluation of potential alternative fuel candidates, the CEC launched an aggressive program of demonstrating and testing alcohol-powered vehicles. These demonstration fleets included converted Ford Pintos, factory built Volkswagen Rabbits, 1981 Ford Escorts, and 500 1983 Ford Escorts. The Pinto and Volkswagen test fleets involved both methanol and ethanol fuels as well as gasoline control vehicles.[3] In 1987, the CEC acquired several prototype fuel-flexible vehicles that can operate on any mixture of gasoline and methanol.

The CEC has also sponsored a methanol demonstration program for heavy-duty vehicles since 1984. The CEC funded two methanol buses, operated by the Golden Gate Transit District located just north of San Francisco, and a multifuel tractor capable of operating on either methanol or ethanol fuels. Based on the success of the bus demonstration program, other transit districts have bought or placed orders for their own methanol buses.

These demonstration programs are based on a strategy established in 1982. At that time the CEC envisioned a three-stage process leading to a market takeoff point when commercial markets could perpetuate methanol growth. The first stage began with design and prototype development of the technology. The second stage used the prototype testing results to build larger-scale demonstration programs. The second stage should have provided enough information and stimulus to lay the foundation for market takeoff. However, this final stage has yet to be realized.

While the goal remains the same, the strategy needs updating. We learned that more steps are required between demonstration and commercial acceptance of methanol.

On the positive side, we found that the methanol vehicles currently operating have provided adequate service for most fleets' needs. In addition the emission data from these fleets showed that methanol vehicles consistently met or fell below California emission standards.[4] Despite the lower emissions there remains disagreement over how large the actual air-quality benefits of methanol will be.[5] Air-quality benefits are the major force currently pushing clean alternative fuels.

California's demonstration program revealed the most significant drawback to alternative transportation fuels: insufficient refueling stations. The 18 methanol fueling stations deprived drivers of the convenience and availability of a neighborhood methanol-dispensing facility. Knowing that the range of methanol vehicles was less than a gasoline vehicle, CEC staff examined trip data from the fleets that were to absorb the methanol vehicles to determine the percentage of trips that methanol vehicles could not take. Over 90 percent of all trips fell within the range of a methanol vehicle, indicating that fleet operators would need a substitute gasoline vehicle for less than 10 percent of their dispatches.

Unfortunately, data notwithstanding, the methanol vehicle users evidently needed the comfort of a broader fueling-station network. While no statistically reliable survey was performed on the drivers of methanol vehicles, the overall impressions from fleet operators left little doubt that methanol vehicles were less desirable than gasoline vehicles principally because of the lack of available fueling stations. Both the operators and drivers needed to take extra precautions before embarking on trips with their methanol vehicles.[6] In some respects these extra precautions added to the drivers' uneasy feelings since they were alerted to the limited fuel availability.

The problem of limited fuel availability can be resolved in several ways. The most obvious method is to increase the number of available methanol fueling stations. The CEC did this by negotiating an agreement with California's major gasoline retailers, ARCO and Chevron, U.S.A., Inc., to establish 50 additional retail methanol stations across California. Still, adequate coverage for the state would probably require hundreds of retail stations rather than a few dozen. Unfortunately, establishing a methanol dispenser and associated equipment costs about $40,000 at an existing station. The cumulative costs of developing a considerably larger fueling station network would easily run into the millions of dollars, beyond the resources of the CEC.

A second method of easing driver concern over methanol fuel availability is to increase vehicle range. Since methanol contains half the energy content of gasoline, a methanol vehicle will not travel as far as a gasoline vehicle on a tank of fuel. Increasing fuel-tank capacity could add vehicle range, but this also raises manufacturers' concerns about vehicle safety and attainment of federal fuel economy standards. Improving the efficiency of methanol vehicles will marginally increase range to perhaps two-thirds the range of gasoline vehicles. While range improvement is important and can serve to ease some driver concerns, there still remains the basic problem of insufficient fuel stations which reduces the utility of methanol vehicles just as it does for compressed natural gas vehicles and electric vehicles.

Recognizing the range and limited refueling availability problem early in the alternative fuels program, the CEC and the major automobile manufacturers shifted their focus to the fuel-flexible vehicles (FFVs) concept. A vehicle capable of operating on either gasoline or methanol or any mixture of the two fuels would nearly eliminate driver concerns over fuel availability. While the FFVs are still in a demonstration and evaluation phase of development, these vehicles may hold the key to a successful transition strategy.

THE ADVANTAGES AND RISKS OF A FUEL-FLEXIBLE VEHICLE TRANSITION STRATEGY

The CEC's dedicated (single-fuel) methanol vehicles made important inroads to the transportation fuels market. It demonstrated technological viability and provided data of on-road performance during actual fleet operations. Apart from the technical achievements, the demonstration fleets attained a high degree of visibility introducing the lay public to alternative fuels. Numerous press conferences, newspaper and magazine articles, and prominent methanol signs on the vehicles have provided opportunities for greater awareness of methanol. The technical success and high visibility were important factors on the path toward methanol fuel's commercial success.

However, as described in the previous section, limited fuel availability is a large barrier to single-fuel vehicles. This barrier can only be overcome by deploying enough underground methanol-compatible tanks and establishing sufficient retail outlets to satisfy the needs of methanol drivers. Anything short of full coverage will lower a methanol vehicle's desirability to consumers.

An "FFV strategy" partially offsets the infrastructure barrier since FFVs can operate on either gasoline or methanol, reducing the initial need for methanol fuel stations.[7] In addition the FFV strategy may reduce ozone and carbon monoxide and thus could be a part of regional clean air programs.[8] Areas violating ambient air-quality standards such as the South Coast (Los Angeles area) Air Basin (SCAB) could implement an FFV strategy without requiring after-market vehicle conversions or requiring automobile manufacturers to develop a specialized methanol car model. Manufacturers are reluctant to invest in vehicles specifically for limited markets because of their desire for large economies of scale in manufacturing. Vehicle conversions are also less desirable due to the high costs and questionable reliability; vehicle conversions are also less desirable than FFVs.

While an FFV may cost slightly more than a comparable gasoline car, the differences are minimal. Ford Motor Company has stated to the CEC that the production costs will increase by only $100 to $300 per FFV.

Another advantage of an FFV strategy suggested by McNutt and Ecklund is the suppression effect on oil prices.[9] The U.S. Department of Energy suggests that the mere existence of FFVs will stabilize oil prices since FFV consumers will always have the option of switching to methanol whenever gasoline prices increase.[10]

Finally, FFVs offer the advantage that owners would not be penalized if methanol proved less desirable for unforeseen reasons, for instance, if methanol production were inadequate or the cost remained significantly higher than competing fuels. Under these scenarios, which we consider unlikely, the FFV can act as insurance until the time when methanol economics become more favorable.

Unfortunately, the FFV strategy also contains substantial risks. The biggest risk is that the strategy offers no assurance that buyers will use methanol. In governmental fleet applications, the fleet operator directly controls the refueling of vehicles. While cost and convenience play an important role in an operator's decision of which fuel to use, public agencies are governed by adopted policies such as to improve air quality. Government-operated fleets are a relatively small market, however, constituting only a modest percentage of vehicles in California.

Other vehicle fleets and general consumers are more problematic; on the one hand their behavior is less predictable, and on the other they are more price sensitive. Many consumer-acceptance analyses suggest that fuel price is the most significant indicator of consumer choice.[11] While we believe other factors can also influence a consumer's choice, we tend to accept the premise that fuel price is the single most important decision variable. We discuss costs further in a later section.

A second risk is that an FFV strategy alone may not create sufficient demand to capture economies of scale in fuel distribution during the early transition period. Transportation costs for shipping methanol would be lowered if tanker ships were used to bring methanol to Los Angeles.[12] However, the smallest tanker ship (35,000 dead-weight tons) for methanol shipments would require a minimum of 8,000 methanol buses or 140,000 FFVs operating exclusively on methanol. Unfortunately, such massive numbers of FFVs and buses will not appear instantaneously. Moreover, the 140,000 FFVs would certainly not operate exclusively on methanol.

A third risk involves the air-quality benefits of FFVs. The greatest benefits will occur when methanol is used in the vehicles, as indicated in Table 15.1. However, since users of FFVs can refuel with gasoline or methanol, air-quality benefits become questionable if only a limited portion of FFV owners use methanol. The key to the FFVs' success, then, is to convince a substantial majority to use methanol.

Table 15.1
FFV Emission Data

	HC (GM/MI)	NOx (GM/MI)
ESCORT:		
Gasoline	0.250	0.630
M85	0.190	0.520
CROWN VICTORIA:		
Gasoline	0.196	0.550
M85	0.131	0.455

Source: See Nichols, "Update on Ford's Methanol Vehicle Experience" in U.S. General Accounting Office (1983), *Removing Barriers to the Market Penetration of Methanol Fuels*, GAO/RCED–84–36.

ECONOMIC COMPETITIVENESS OF METHANOL

In order for methanol or any other alternative fuel to penetrate the market, several layers of risk to the consumer must be reduced or eliminated. The consumer is being asked to purchase a special vehicle that is unfamiliar and has an uncertain support network. In the early stages of a transition period, only a few fueling stations and repair facilities could handle the new vehicles.

Once the vehicle (FFV) is chosen, the consumer then must weigh the risks of using the new alternative fuel, methanol, or sticking with a well-known commodity, gasoline. While the use of methanol may provide the greatest potential benefits from a public-policy perspective, this would not necessarily be the case from the consumer's perspective. While the consumer will assume some risk to obtain added benefits, the consumer is generally risk-averse. As a result, the added benefits of using methanol such as improved performance and air quality must exceed in the consumer's mind added costs of the fuel and the lower range.

Since costs play an extremely large role in determining consumer choices, we will compare the cost of gasoline with the estimated production costs of methanol.[13] We assume that unless methanol can be cost competitive, few FFV users will purchase methanol. In comparing methanol and gasoline costs we assume that consumers will understand that a gallon of methanol will not contain the same amount of energy potential as a gallon of gasoline.

To evaluate the energy-content difference between the fuels, we developed a gasoline equivalent price for methanol. The basis for the equivalent price was the ratio of the energy content of a gallon of gasoline to the energy content of a gallon of methanol. The California methanol demonstration program uses a fuel formulation of 85 percent methanol and 15 percent unleaded gasoline (M85). The Btu content of M85 is 65,275, and gasoline's Btu content is 115,000. The ratio of gasoline to M85 is therefore 1:1.76. The use of the 1:1.76 ratio ignores any efficiency gains theoretically possible with M85. Although the demonstration vehicles in California have shown some efficiency improvements, it is uncertain whether these efficiency improvements will be fully realized in general usage. In addition, using a single equivalency ratio to account for efficiency improvements may not provide accurate results. While we believe that a significant improvement in energy-equivalent fuel economy is achievable, perhaps as much as 25 percent greater, we will assume a lesser improvement and use the 1:1.76 ratio for discussion here.[14]

Using the method presented in Table 15.2 of calculating equivalent prices, we tracked the spot methanol prices with the average retail price of regular unleaded and premium unleaded gasoline. Figure 15.1 shows the rough trends for the past five years. Historically, methanol spot prices were similar to premium unleaded gasoline prices on an energy-equivalency basis. However, as is evident in the graph, crude oil prices and methanol spot prices began to diverge in 1986, suggesting that methanol could not compete with gasoline fuels today. A caveat to these prices is that the methanol prices reflect methanol's value as a chemical commodity rather than as a fuel.

In any case, the current prices of methanol and gasoline are not indicative of cost competitiveness during a transition period especially if, as expected, air quality rules and controls become more strigent in the 1990s. Next, we examine more closely the relationship between methanol and gasoline prices. These price relationships are based on CEC forecasts that premium unleaded gasoline prices will increase up to $1.40 to $1.60 per gallon (1987 dollars) in the 1990s (see Figure 15.2). The methanol prices used in the analysis are the landed prices in California; they do not include local distribution. The emphasis of this analysis is between methanol and premium unleaded gasoline because methanol will be too expensive during a transition period to compete with premium unleaded gasoline and because it has a higher octane rating than premium gasoline.[15]

Of course the CEC price forecasts may not accurately predict future price trends. If the rate of increase for crude oil prices was more rapid than projected, we would expect higher allowable landed prices for methanol without jeopardizing market shares. A slower rate would lead to more lengthy transition period or may even threaten the commercial success of alternative fuels.

Assuming the CEC price forecast for premium unleaded gasoline is accurate, it is shown below that methanol would have to land in Los Angeles at a cost of between $0.35 and $0.43 per gallon. These target prices, while on the lower end, are plausible for the 1990s (see chapters 3 and 5).

Table 15.2
Components of Retail Methanol Price

	Dollars Per Gallon
Landed Cost	0.340
Bulk Storage	0.010
Subtotal	0.350
M85 Methanol Portion: 0.35 x 85 =	0.298
Gasoline Portion: 0.68[a] x .25[b] =	0.102
Subtotal	0.400
Handling Cost[a]	0.003
Truck Delivery[a]	0.030
Excise Tax[c]	0.090
Wholesale Margin[d]	0.030
Retail Margin[e]	0.090
Subtotal	0.643
Sales Tax at 6 Percent	0.039
TOTAL	0.682
GASOLINE EQUIVALENT 1.76 X 0.682[f]	1.20

[a]Data derived from existing methanol fueling station program sponsored by the CEC.

[b]Assumes $18 per barrel crude oil prices and premium unleaded gasoline wholesale price of 0.68/gal.

[c]Excise taxes are assumed to be on an energy equivalent basis with gasoline taxes:

	Gasoline (cents/gal)	M85
California Excise Tax	9.0	4.5
Federal Excise Tax	9.0	4.5
Total Excise Tax	18.0	9.0

[d]Assumed to match independent jobber rate on per gallon basis.

[e]Assumed to match typical retail margin for premium fuel on an energy or turnover basis; also matches existing fueling station program practice in CEC programs.

[f]Energy equivalent ratio of 1.76 gallons of M85 to 1 gallon of gasoline.

Figure 15.1
Retail Prices for Methanol and Gasoline, 1987 Dollars

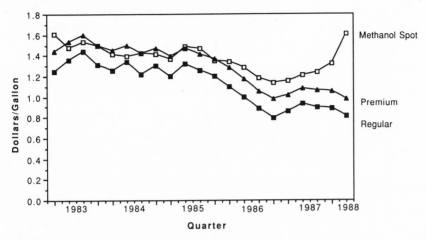

Source: CEC, Fuels Report, 1987

Figure 15.2
Projected Prices, Regular Unleaded and Premium Unleaded, Dollars per Gallon

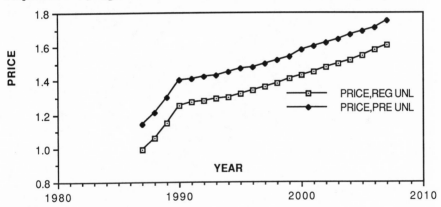

SOURCE: CEC ,1987 FUELS REPORT

A determination of the relationship between methanol and gasoline prices must take into account the fact that methanol production and transport are energy-intensive activities. Energy costs for oil and electricity may account for 20 percent of the total methanol production costs, thereby implying that methanol production costs are a function in part of oil prices. The influence of crude oil prices is reflected in a rather complicated graph in Figure 15.3.

The thin diagonal lines indicate various equivalent retail methanol prices as

Figure 15.3
**Ranges of Methanol Competitiveness Based on Landed Methanol Costs and
Crude Oil Prices**

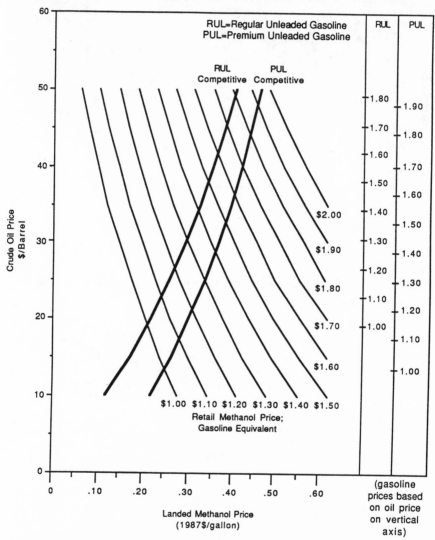

a function of crude oil prices. These diagonal lines were developed assuming the 20-percent energy influence factor. If the target retail price for methanol is $1.20 methanol (gasoline equivalent) and crude oil is $20 per barrel, one follows the $1.20 line to the $20 per barrel level. At that point, reading from the horizontal axis, the landed methanol price cannot exceed $0.33 per gallon.

The two bolder diagonal lines represent the points where gasoline prices and equivalent methanol prices are equal. The left bold line represents unleaded

Table 15.3
Equivalent Retail Prices for Various Landed Methanol Prices

	CENTS PER GALLON					
Landed Price	26.0	30.0	34.0	38.0	42.0	46.0
Retail Methanol Price	60.9	64.5	68.2	71.7	75.3	78.9
Equivalent Gasoline Price	107.2	113.5	120.0	126.2	132.6	138.9

regular gasoline prices, and the right represents unleaded premium gasoline prices. The area between these lines represents the cost-competitive area for methanol during the transition period. If crude oil prices rose to $25 per barrel, the premium gasoline price will be about $1.30 per gallon (as indicated by the column on the right side of the graph). At this price, methanol could compete if the landed costs do not exceed $0.35 per gallon.

Figure 15.3 indicates that landed methanol prices must remain below an upper limit to remain competitive with gasoline. Above $0.50 per gallon, crude oil prices would have to exceed $50 per barrel. In fact methanol's competitive landed price falls within a relatively narrow range of $0.35 to $0.43 per gallon given the crude oil price forecast by the CEC.

The estimated retail price of methanol is calculated in Table 15.2 using the 1:1.76 energy equivalency ratio for methanol. The cost components related to the fuel station are based on the experience with the methanol fueling station network sponsored by the CEC. The "landed cost" estimates are subject to considerable uncertainty; Table 15.3 shows a range of calculated M85 equivalent prices for various landed methanol costs.

While the CEC crude oil projections tend to indicate that methanol would have difficulty competing with gasoline fuels until the year 2000, several factors may intercede to bring the competitive period sooner. First, these price forecasts are highly uncertain. Unforeseen events could change crude oil prices rapidly, as the 1970s demonstrated. Second, pending air-quality regulations will raise the cost of gasoline production. These regulations, including tightened Reid Vapor Pressure standards and reduced aromatic content in gasoline, may increase costs by $0.03 to $0.10 per gallon.[16]

Third, the feasibility of new methanol production process technology has become an important consideration for determining cost effectiveness. At least three new processes have been proposed for new plants: catalytic partial oxidation, combined reforming, and fluidized bed synthesis. These new processes could reduce capital costs of the facility by 50 percent or production costs by 30 to 40 percent.[17] While built at the pilot plant stage, these new processes still require full-scale buildup before these cost savings are realized. Based on preliminary indications, California could receive methanol at $0.35 per gallon from new plants using these processes. This would allow methanol to compete with premium unleaded gasoline when crude oil prices reach $25 per barrel.

GOVERNMENTAL ROLE IN ENCOURAGING A TRANSITION TO ALTERNATIVE FUELS

Government will have to play an important role in bringing alternative transportation fuels to California. While the market by itself may allow some penetration by alternative fuels, a market-driven transition strategy will probably have an inconsequential effect on air quality. At the same time aggressive governmental policies, regulations, and implementation actions could prove costly and politically undesirable.

California has taken a relatively active approach to alternative fuels. Relying on research and laboratory testing by others, the state program first tested and evaluated the technical feasibility and in-service acceptability of a limited number of alternatively fueled vehicles. This laid the foundation for introducing factory-produced vehicles, establishing a rudimentary fueling station network, and involving a number of different fleets. These activities are critical for building commercially viable alternative fuels. Vehicles were tested for emissions, increased wear, and driver acceptance. So far, California's program has involved only government fleets and focused on methanol.

The California demonstration fleets have not gone beyond demonstrating technical feasibility. In 1987, proposed legislation requiring all new underground fuel storage tanks to be methanol-compatible was vetoed by the governor. In addition, legislation that would have required automobile manufacturers to sell at least one line of methanol-capable vehicles by 1993 was amended to mandate further study of alternative fuels. Yet, $5 million was approved for further FFV demonstrations and a heavy-duty methanol truck demonstration. Thus, the state's role is to provide ample opportunities for showing the capability of these new technologies but stops short of mandating alternative fuels for the transportation market.

Despite the absence of explicit mandates, California benefits from related regulations. The one important set of regulations are those promulgated by the Environmental Protection Agency (EPA) concerning particulate emissions from buses in the 1991 model year and from trucks in 1994. While particulate traps may allow new diesel buses to meet this standard, alternative fuels provide another option.[18] The EPA regulation does not specify the means of achieving the standard, leaving the problem to industry. As noted earlier, methanol buses have already been placed in service in California and elsewhere.

An important incentive is legislation that allows automobile manufacturers to obtain a credit in meeting corporate average fuel economy (CAFE) standards by selling methanol-capable vehicles (S. 1518, Rockefeller). This concept of providing incentives is attractive to automobile manufacturers who will have difficulty meeting the CAFE standard, while at the same time stimulating the introduction of alternative fuels without government subsidies.

THE ROLE FOR ALTERNATIVE FUELS OTHER THAN METHANOL

Methanol holds considerable promise for both improving air quality and aiding energy diversity goals. Of course, proponents for other alternative fuels have made similar claims. Our view does not necessarily conflict with these other claims, although we tend to emphasize methanol.

Early in the CEC's program for alternative fuels, the CEC performed a comparative evaluation to determine which alternative showed the most potential benefits.[19] This evaluation concluded that methanol could meet the requirements of a transportation fuel with fewer changes in the vehicle or fuel distribution infrastructure. In addition, methanol was recognized in early 1980 for its potential to improve air quality. Studies by the Port Authority of Allengheny County and the U.S. Department of Energy concur in their preference for methanol fuel.[20]

Meanwhile, improved technology for other alternative fuels has increased their attractiveness. Batteries for electric vehicles have become more efficient and have greater energy capacity. Similarly, compressed natural gas vehicles have improved primarily through lighter storage tanks and the use of fuel injection and microprocessor controls. As a result of these improvements, the CEC intends to conduct CNG and EV demonstration programs in the early 1990s.

While these fuels have some disavantages, they also are attractive in specific applications. EVs are attractive for short distance, stop-and-go delivery jobs; vehicles powered by internal combustion engines in these uses spend considerable time idling, wasting fuel, and spewing emissions. Finding these attractive applications will help the use of these other alternative fuels.

CONCLUSION

The most unlikely approach for introducing alternative fuels is widespread government subsidies or mandates. However, with a strong commitment to improved air quality, government is likely to offer incentives such as seed loans or grants for demonstrations or specially designed CAFE credits. In addition, stricter emission standards may add another incentive for alternative fuels.

California has been a stronger supporter of alternative transportation fuels than any other state. Not only has it sponsored demonstration programs and formed private and public partnerships with major corporations, it has actively solicited air quality regulations to encourage alternative fuels. But is it enough?

The transition to alternative fuels requires the commercialization of vehicles capable of operating on alternative fuels. Firm governmental resolve for improving air quality needs to be maintained so as to encourage manufacturers to increase their research and development of alternative-fuel vehicles. The first stage of commercialization could lie with new transit buses which must meet clean particulate standards by the 1991 model year.

Since the infrastructure problem will act as a major barrier for dedicated vehicles, an FFV strategy will provide an important mechanism to increased commercialization for alternative fuels. However, the FFV strategy entails risks as well as benefits. The biggest risk is that the consumer may perceive more costs than benefits in using methanol. The price of converting to alternative fuels must be low enough to compensate for added inconveniences, risks, and uncertainty.

Clearly, consumer acceptance needs to be better understood and needs to be studied further. We need to understand the role of price and other variables in the consumer decision process. This understanding will play a key role in developing strategies to encourage the introduction of alternative fuels.

In conclusion, while we have used methanol as the basis of our discussion of transitions to alternative fuels, we believe the issues identified here apply to other alternatives as well. The success or failure of the alternatives demonstration will be determined by the technology, evaluation of its performance in service, analysis of costs and benefits, development of infrastructure, the role of government, and consumer acceptance. By adding CNG and electric vehicle demonstrations to the CEC program, we can look forward to aiding the prospects for energy diversity.

NOTES

The authors would like to thank David Solomon, Lisa Darling, and Lilly Ghaffari of the CEC, and Doug Lowell and Stefen Unnasch of Acurex for providing data and comments on previous drafts of this chapter. In addition we appreciate the management support of Kenneth D. Smith and Mike Jackson for allowing us to complete this project on staff time.

1. California Energy Commission (1987), *Fuels Report,* CEC P300–87–016, Sacramento, Calif.

2. Sierra Research, Inc. (1988), *The Feasibility and Costs of More Stringent Mobile Source Emission Controls,* prepared for U.S. Congress, Office of Technology Assessment, Washington, D.C.

3. Acurex Corp. (1988), "Making the Transition to Methanol," Draft Report, prepared for the California Energy Commission, unpublished.

4. Acurex Corp., *Evaluation Report.*

5. See Chapter 8 in this volume.

6. K. Koyama et al. (1986), "California's Methanol Fleet," in *Proceedings of the 7th International Alcohol Fuels Technology Symposium* vol. 3, Paris, France.

7. C. B. Moyer, M. D. Jackson and S. Unnasch (1987), "Issues in a Market-Oriented Methanol Transition Strategy Based on Fuel Flexible Vehicles," presented to *1987 World Methanol Conference Proceedings,* December 1–3, 1987, San Francisco, Calif. pp. IV-I–IV-14.

8. Ibid.

9. B. McNutt, and E. Ecklund, (1986), "Is There a Government Role in Methanol Market Development?" Society of Automotive Engineers, Paper No. 861571.

10. U.S. Department of Energy (1988), *Assessment of Costs and Benefits of Flexible and Alternative Fuel Use in the U.S. Transportation Sector, Progress Report One: Context and Analytical Framework,* Office of Policy, Planning, and Analysis, PE–0080, Washington, D.C.

11. Michael Brandman Associates and R. P. O'Toole (1987), *Methanol: An Overview of an Alternative Vehicle Fuel,* prepared for Western Oil and Gas Association, Los Angeles, Calif.; California Council for Environmental and Economic Balance (1987), *Alternative Fuels as an Air Quality Improvement Strategy,* Sacramento, Calif.; D. Sperling and K. S. Kurani (1988), "The Rise and Fall of Diesel Cars: A Consumer Choice Analysis," in *Transportation Research Record* 1175; C. Defligio and M. F. Lawrence (1987), "Economic and Security Issues of Methanol Supply," Society of Automotive Engineers, Paper No. 872062.

12. See Chapter 4 in this volume. Jackson, M. D.; Webb R., and Moyer, C. B.

13. Ibid.

14. U.S. General Accounting Office (1983), *Removing Barriers to the Market Penetration of Methanol Fuels,* GAO/RCED–84–36, presented to the Chairman, Subcommittee on Fossil and Synthetic Fuels, Committee on Energy and Commerce, House of Representatives. R. J. Nichols, (1986), "Update on Ford's Methanol Vehicle Experience" presented at the European Fuel Oxygenates Association, Brussels, Belgium. Acurex Corp., *Evaluation Report.*

15. CEC *1987 Fuels Report.*

16. T. Cackette, (1987), California Air Resources Board, Testimony to the California Energy Commission, Fuels Committee, June 17, 1987.

17. J. D. Korchnak and M. Dunster (1987), "Reduced Methanol Production Cost" presented to *1987 World Methanol Conference Proceedings,* December 1–3, 1987, San Francisco, Calif; Y. Saito, M. Kuwa, and O. Hashimoto (1987), "Development of a Fluidized-Bed Methanol Synthesis Process," presented to the 1987 Annual Meeting of the American Institute of Chemical Engineers, New York, NY.

18. S. Unnasch, et al., "Emission Control Options for Heavy Duty Engines," Society of Automotive Engineers, Paper No. 861111; C. S. Weaver et al. (1986), "Feasibility of Retrofit Technologies for Diesel Emissions Control," Society of Automotive Engineers, Paper No. 860296.

19. Acurex Corp. (1981), *Clean Coal Fuels: Alternate Fuel Strategies for Stationary and Mobile Engines,* prepared for the CEC, Sacramento, Calif.

20. Booz, Allen and Hamilton, Inc. (1983), *Evaluation of Alternative Fuels for Urban Mass Transit Buses,* prepared for the Port Authority of Allegheny County, Pittsburgh, Penn.; U.S. DOE, *Assessments of Costs and Benefits for Alternative Fuels.*

16

BARRY McNUTT

Alternative Fuels Market Development: Elements of a Transition Strategy

Introducing alternative fuels to reduce transportation's oil dependency has been widely discussed and analyzed from national security and environmental perspectives.[1] Technical possibilities for the transition away from oil have been outlined, the rationale for a government role has been developed, and the costs of a widespread distribution and utilization system have been quantified.[2] However, a strategy to achieve a national transition, one that spells out specific policies and roles, has yet to be defined, although California has made some progress on a limited urban-based program to achieve air quality goals.

The alternative fuels problem is difficult. We are discussing a system that does not exist, has potentially high costs, is sensitive to unpredictable oil prices, and if implemented, has diffuse benefits but concentrated costs. It is not surprising that individuals and institutions are reluctant to articulate a policy and strategy that acknowledges these high costs and risks of failure.

This chapter presents the elements of a transition strategy designed to achieve the energy policy goal of fuel flexibility, which is not necessarily the same as the environmental policy goal of clean-burning alternative fuels. The federal government plays a crucial role in this transition strategy: it must mandate that vehicle manufacturers and fuel suppliers establish an alternative fuels infrastructure of multifueled (includes fuel-flexible, variable-fueled, and dual-fueled) vehicles and widespread retail fuel outlets. This strategy does not mandate fuel use, but it considers some fuel price adjustments that may increase or stabilize demand for the alternative fuel. Several approaches are proposed to increase market flexibility and spread costs. While this strategy is likely to be effective in bringing the infrastructure into place, its benefits are dependent on uncertain oil prices.

We make some critical assumptions. The first is that a political consensus must form on the necessity of an alternative fuels transition to meet energy policy objectives. Obviously, if this is not the case, alternative fuel proponents will have to seek some other avenue, such as an environmental policy mandate. In any case, environmental needs may motivate development of the infrastructure in the fuel flexibility strategy, provided there are no unreasonable conflicts between environmental and energy policy goals.

The second assumption is that the alternative "fuel of choice" is methanol, which requires less infrastructure investment and less disruption to vehicle manufacturers, fuel suppliers, and consumers than a gaseous fuel or electricity. If another fuel, such as compressed natural gas (CNG), is chosen, this strategy would have to be modified substantially.

DEFINING THE PROBLEM

The Goal

Our goal is an in-place alternative fuel distribution and utilization (alternative fuel compatible vehicles) infrastructure, capable of displacing a significant amount of oil (one to two million barrels per day) by the time oil prices are expected to rise in the late 1990s or early 2000s.[3] This goal can be expanded to include some use of alternative fuels in other markets to encourage development of fuel production facilities.

The Barriers

The principal barriers to realizing our goal are economic, institutional, and behavioral, but not technological. Further technology development may reduce vehicle cost, increase performance, and improve the control of unregulated pollutants, but the general technology is well known and available. While some development work remains, dual-fueled and multifueled vehicles can be built at costs similar to those of gasoline vehicles.

First, alternative fuels are, for the most part, uneconomic in the face of currently low oil prices. Even if oil prices rise, uncertainty about future prices and price volatility will discourage investment in infrastructure.

Second, a distribution infrastructure does not exist, and a full-scale system will be difficult to put in place because of the high cost (on the order of $20 billion) and lack of demand (i.e., no vehicles). Limited systems, for instance to serve fleet users, would be less costly but would have proportionally less benefit to energy security.

Third, consumers have relatively short vehicle-investment time horizons; expected payback is one to two years.[4] They are concerned about fuel costs only when fuel prices are rising rapidly, and some are suspicious of new technology and new fuels.[5] Therefore, consumer demand for alternative fuels will be limited unless petroleum prices rise, consumers perceive clear tangible performance benefits, or environmental concerns or mandates become very strong.

A fourth barrier is the limited incentive of existing transportation fuel suppliers to invest in alternative fuels infrastructure, since a successful alternative fuels transition will depress oil prices—an unattractive prospect for companies with large oil reserve assets.

Risks, Costs, and Benefits

A fundamental problem is that the potential benefits of an alternative fuel initiative are diffuse and the risks and costs tend to be concentrated. While this diffusion of benefits can create the conditions for political consensus, the economic benefits resulting from lower oil prices worldwide are so small for any one user that no activist political constituency arises.

The risks associated with alternative fuel initiatives, on the other hand, are large and concentrated in a few corporations. If the opportunity for large increases in market share or revenue existed, then these businesses would actively pursue opportunities, but that is not the case. The transportation fuels market is stagnant, and the market penetration rate is likely to be slow because of slow vehicle turnover. Thus, the lure of profits is small while the risk of failure is high. In a policy- (not market-) driven transition, the government policy makers and legislators also share in the risk with little expectation of enjoying the political benefits that may occur years later.

The costs of a mandated system, where all current market participants are required to participate in the transition infrastructure development, will fall on new-vehicle buyers and all retail fuel purchasers. While this is a wide base over which to spread cost, these vehicle and fuel purchasers account for less than one-half of all direct oil use in the United States. The costs are further concentrated if other countries with large oil-using transportation sectors do not participate in the transition investments.

POLICY OPTIONS

Once the political decision is made to undertake an alternative fuels transition, the government has a wide range of policy levers or tools available to effect the required changes. Those policy tools range from tax subsidies to regulatory mandates. Which of those tools are used will depend on the strength of the political convictions behind the decision, the degree to which government is willing to intervene in the market for energy policy purposes, and the speed and certainty with which the transition is to be made. In making a choice among the options, one needs to consider how they would apply specifically to an alternative fuels transition and affected players.

Economic Subsidies

This policy tool can include price supports or tax benefits; until the recent increase in concern over the federal budget, this was one of the favorite tools of policy makers for a wide range of initiatives. While subsidies generally represent a modest level of intervention, the costs can be very high.

Price subsidies for alternative fuels could be used to control the price relative to gasoline or diesel fuel. Subsidies could take the form of reduced fuel taxes, direct price subsidies to fuel suppliers, or a government role as a fuel wholesaler, as in the current California program. Such subsidies would mitigate high prices and future oil price volatility, and if they were perceived as certain and long term, they could be the key to an alternative fuels transition. Although their costs are open ended and could be very high, the benefit of subsidies is that they incur no cost unless alternative fuels are used.

On the vehicle side, a price subsidy in the form of a tax/rebate program, similar to the gas-guzzler tax as it was originally proposed in the 1970s, could be used. The current tax could be increased, conventionally fueled vehicles could be taxed, or fuel tax revenues could be used to provide the funds for a rebate to high mpg or alternative fuel vehicles. This could be critical if alternative-fuel vehicles with high purchase costs, such as dual-fuel CNG vehicles, were introduced.

Tax subsidies for the alternative fuels infrastructure are another possibility. Vehicle manufacturers, vehicle purchasers, fuel suppliers, and possibly fuel producers could be given one or more tax subsidies (credits, faster depreciation, etc.) to cover the cost of an alternative fuels infrastructure. While this would spread the costs to all taxpayers and be predictable in its size, such benefits may not be large enough to overcome the market uncertainty of getting enough fuel used to give an adequate return on investment. These subsidies would also do little to directly encourage fuel use, although the selling price of the fuel could be lowered.

In general, price and tax subsidies spread the costs and shift risk from the market participants to the government and taxpayers.

Regulations

A regulatory approach to an alternative fuels transition has precedent in the energy and environmental areas. The Clean Air Act was used to bring unleaded gasoline into the market. The Fuel Use Act and other state and local regulations have controlled types of fuel used in certain applications.

Regulatory mandates could require vehicle manufacturers to produce multifuel or dedicated alternative fuel vehicles and require fuel suppliers to have facilities capable of handling the respective fuels. Such mandates could further require that alternative fuel be available at some or all fuel sales outlets. This approach has been considered in an environmental context in California and might work well from an energy-policy perspective in conjunction with environmental strategies. At the national level, these mandates would have to apply to all cars but could be limited, initially, to fuel suppliers in major urban areas with air pollution problems.

Regulatory incentives are another possibility. The existing regulatory requirements would be relaxed or modified to encourage alternative fuels infrastructure or use. The CAFE credits for alternative-fuel vehicles are an obvious example.[6] Others might include emission tradeoffs (e.g., lower HC for higher NO_x) or reduced annual vehicle emissions inspection requirements (an incentive to vehicle purchasers).

Research and Development

Government funded R&D remains one of the principal mechanisms to assist the private sector and serve the national interest. While there are some clear R&D needs in the alternative fuels area, most are longer term (e.g., fuel cells and combustion research) and would do little to effect a transition in the next few decades. There are some exceptions, including work on C_1 chemistry, lower cost CNG storage, fuel-flexible vehicle emissions performance, and engine durability on methanol. Most are being addressed in the private sector, but government efforts could be helpful, particularly if carried to the demonstration phase for more in-use experience.

Government "Leadership"

This area of policy is often discussed but poorly defined. Obviously, the first step is to set a clear national policy direction. We assume that will have occurred before we consider other government roles. Frequently, the government role is seen as government use of alternative fuels at the federal, state, or local level. While this can be useful in a technology development/demonstration phase, the government fleet is too small and diffuse to be a major market development force.

There are other more appropriate "leadership" roles the government can play. Perhaps the most important of those (after setting clear national policy direction) is to encourage international participation in an alternative fuel transition. It may be possible to work bilaterally with countries such as Canada and Japan where we have strong automotive ties. In addition, the U.S. government could use forums such as the International Energy Agency and OECD to encourage other oil-using countries to share some of the costs and risks by initiating and coordinating their own transitions.

This could increase the demand for vehicles and fuel, encouraging manufacturers and suppliers to enter the market. While this might increase alternative fuel prices in the short run if demand for the fuel outstrips initial supplies, it will encourage more fuel producers to enter the market and stabilize the supply/

price situation in the mid- to longer-term. It would also increase the market size for alternative fuel vehicles, further spreading development costs.

The government can assist in marketing once the infrastructure is in place; but, unfortunately, government has not been very successful at influencing marketing, particularly in the energy area. Information on the unit cost and fuel economy of alternative fuels must be standardized so that consumers are not confused or deceived in their comparisons. Clearly a single specification for methanol (or CNG) fuel will be needed before fuel suppliers or vehicle manufacturers can successfully enter a consumer market. More important from the consumer perspective will be a fair and easy-to-understand system for labeling fuel volumes and prices. Government can take the lead here, working through existing organizations such as ASTM or SAE to establish an acceptable approach before the transition starts. Any other marketing efforts must supplement and not conflict with private sector marketing. California's experience will be instructive.

Providing analysis and technical information to market participants also is an important function for government, particularly when participants are diverse. A number of federal and state agencies are now providing this support.

These government promotion and marketing activities are likely to be more important in a market-driven transition than in a policy-driven transition, unless one wants to encourage actual alternative fuel use. The combination of policy tools chosen depends on a number of factors, the most important of which is the balance between need for certainty and the risks and costs associated with market intervention. Fuel choice is a key to this dilemma and must be addressed before laying out a strategy.

ONE FUEL OR MANY?

An earlier paper argued that methanol is the most logical *transition* fuel.[7] It also seems to be the "fuel of choice" for several of the key players in transportation. With the development of relatively inexpensive fuel-flexible vehicle technology, the auto manufacturers have committed to methanol, at least for light-duty vehicles. Environmental agencies are pursuing methanol as a clean fuel to achieve air quality goals. While these actions do not foreclose other fuels (e.g., compressed natural gas [CNG]), it is unlikely that two alternative fuels will simultaneously enter widespread use.

A central question is whether energy policies should be fuel specific. If the policy is fuel specific, how is that choice made, institutionally or politically? Or, in the case of methanol, is there a *de facto* government ratification of the auto manufacturers' choice as occurred in the early 1970s for unleaded gasoline?

While the strongest arguments can be made for methanol, it seems that CNG and electricity may have small but important niche markets where they can dominate economically. Since fuel flexibility and diversity serve national energy-policy interest, there is little reason to exclude other fuels except where it will substantially weaken the strategy or raise its costs without commensurate benefits. This means that the overall policy should define the major fuel for the transition, and that a few key elements of the policy may need to be fuel specific. However, other elements can be fuel-choice neutral, allowing the market place to sort out the "fuel of choice" for certain market segments.

It is not clear that the institutional mechanism or political will exists to make a fuel choice. Fuel-specific interests are strong and vocal. If the legislative or administrative policy-making mechanisms cannot choose beforehand, we will be left with an after-the-fact ratification of the decision by the auto and fuel industries. While this may not be fatal to a transition strategy, it will certainly take additional time and, potentially, money as parallel efforts are pursued. It is our view that duplicative efforts are unnecessary and may in fact stop any effective transition from occurring. Our strategy assumes that methanol is the fuel of choice for the initial transition.

THE STRATEGY

A successful strategy should have a variety of elements, targeted at various players and barriers, to make it robust (if one element fails, others may succeed) and comprehensive (all players are involved). This may also make it easier to achieve political consensus since there will be "something (positive and negative) for everyone." However, the elements of the strategy described below are designed to increase effectiveness (certainty), decrease player risk, and spread costs, not achieve something as elusive as political consensus.

Vehicle manufacturers and buyers are the main targets of this strategy. They have the most to gain if the strategy is successful: lower and more stable fuel prices to reduce the cost of vehicle use, more vehicle sales, and (potentially) reduced need for stringent emissions standards. The vehicles also represent the longest lead-time item. It has been shown elsewhere that, even with an aggressive program of fuel-flexible vehicle introduction, it would take 10 years to put enough vehicles on the road to displace 2 million barrels per day of gasoline.[8] The following policies are proposed to achieve introduction of fuel-flexible vehicles.

1. Mandated alternative fuel capability (AFC) in new light-duty vehicles, starting at 10 percent of each manufacturer's production and rising to 100 percent after 10 years, or possibly some lower level commensurate with alternative fuel production capability.[9] These could be either fuel-flexible or dedicated vehicles.

2. AFC trading rights that would allow one manufacturer to avoid the production of AFC vehicles by paying another to produce more. This should increase the efficiency of the program and decrease the burden, particularly for small manufacturers.

3. Tax breaks (accelerated depreciation on equipment and credits on other expenditures) on the research, development, and manufacturing equipment for AFC vehicles. This would spread the cost of the effort more widely, but not provide incentives, since the activity would be mandated.

4. CAFE credits for AFC vehicles to provide manufacturers some choice about where to focus their greatest efforts and some incentive to focus on alternative fuels. This would be particularly important in the context of increased CAFE standards in the mid–1990s. CAFE credits for AFCs should also be tradeable.

5. Vehicle purchase incentives. An enhanced "gas guzzler"-type tax or gasoline-fuel vehicle tax and a rebate program aimed at AFC vehicles would help spread costs to all vehicle buyers. At the state and local level, lower annual vehicle taxes for AFC vehicles could encourage their use in those areas that might benefit environmentally.

6. Optional accelerated scrappage program (ASP) with credits good only toward purchase of AFC cars. ASPs have been considered in the past and were found very expensive relative to the oil saved by prematurely scrapping old inefficient cars.[10] However, such a program might be justified if one considered the broader benefits of increased economic activity associated with increased auto sales, environmental benefits, fuel-efficiency gains, and the fuel flexibility of AFC vehicles. In this case, limiting the credits to fuel-efficient AFC vehicles could increase the benefits.

Fuel suppliers and distributors are our secondary target. The following elements are proposed.

1. Mandatory alternative fuel capability in the fuel distribution system. This would be a two-phase program with large retail outlets to be compatible after five years and most smaller outlets after ten years.

2. Mandatory alternative fuel availability (AFA) in the second five years at all larger retail outlets. This requirement would be phased in at 20 percent of the outlets per year.

3. AFA trading rights. A fuel retailer could pay another retailer to meet its requirements for alternative-fuel availability. Smaller retailers not under the availability mandate would have additional incentives to enter the market. Larger retailers (or their oil company owners) could trade so that the lowest-cost supplier made the fuel available. As with the vehicles, at the end of the period all larger retailers would be required to have AFA, and trading of rights would stop or be limited to smaller retailers.

4. Alternative fuel tax reduction. A substantial fuel tax exemption (e.g., all of the federal tax and some or all of the state tax where possible) in the second five years of the program would encourage alternative fuel use. The exemption could be variable and keyed to oil prices to stabilize relative gasoline and alternative fuel process. This will be very important if retailers are required to have the fuel available. Combined with mandatory AFA, it will send strong signals to alternative fuel producers to be prepared to enter the market in the second five years of the program. While this is an obvious price subsidy, it has a predictable cost that is incurred only if the alternative fuel is used.

Government has two key leadership functions in our transition strategy beyond implementing the policies described above: establishing standardized fuel specification (labeling and pricing information, which will be critical to consumer acceptance of alternative fuel and important to vehicle and fuel suppliers), and encouraging other countries to join in the overall program.

SUMMARY AND CONCLUSIONS

The alternative fuels transition strategy we have proposed considers vehicle manufacturers and buyers, fuel distributors, and fuel producers, but focuses on the vehicle and fuel distribution infrastructure. It is intended to serve energy policy goals, which may be different from environmental needs. It is policy driven, not market driven, and is based on conventional policy tools. It is designed to spread but not reduce the costs and risks of an alternative-fuels transition. It assumes a clear decision to proceed with the transition prior to an oil price increase and, therefore, requires a high level of government involvement to effect the necessary changes. Our major conclusions are the following:

1. The greatest barriers to an alternative fuels transition are economic market and institutional. Further technology development will be helpful but is not critical.

2. Clear government energy policy direction must precede any specific element of the transition strategy. Environmental policy goals may serve as a secondary motivation.

3. The transition strategy must address all players in the transportation fuels market.

4. There is no low-risk way to introduce alternative fuel capability.

5. The success of an alternative-fuels transition strategy is directly related to the commitment to and aggressiveness of the strategy.

6. The benefits of the transition strategy will be determined by uncertain oil prices.

7. Specific attention should be given to policy elements that can help spread the high cost of an alternative fuel strategy to a greater fraction of the oil users who will benefit from lower oil prices.

8. A high degree of flexibility and market choice is possible, even in a policy-driven, government-mandated transition, and can be achieved without weakening the strategy.

NOTES

1. U.S. Department of Energy, *Assessment of Costs and Benefits of Flexible and Alternative Fuel Use in the U.S. Transportation Sector, Progress Report One; Context and Analytical Framework,* Office of Policy, Planning and Analysis, PE–0080 (1988); U.S. Environmental Protection Agency, "Air Quality Benefits of Alternative Fuels," a report prepared for the Vice-President's Task Force on Alternative Fuels (1987).

2. B. McNutt and E. Ecklund, "Is There a Government Role in Methanol Market Development?" SAE Technical Paper Series 861571, Society of Automotive Engineers (1986); B. McNutt, J. Dowd, and J. Holmes, "The Cost of Making Methanol Available to a National Market," *Methanol—Promise and Problems,* Society of Automotive Engineers SP–726 (1987).

3. U.S. Department of Energy, *Energy Security, A Report to the President of the United States* DOE/S–0057 (1987).

4. D. Greene, "A Note on Implicit Consumer Discounting of Automobile Fuel Economy: Reviewing the Available Evidence," *Transportation Research* 17B, no. 6 (1983).

5. A. Rothman, "Gasoline Price Expectations Update," Analysis Memorandum, Office of Policy, Planning and Analysis, U.S. Department of Energy (1986).

6. U.S. Congress, *Alternative Motor Fuels Act of 1988*, 100th Congress, 2d sess.

7. McNutt, "Is There a Government Role?"

8. McNutt, "The Cost of Making Methanol Available."

9. C. Difiglio, "Timing of Methanol Supply and Demand: Implications for Alternative Transportation Fuel Policies." Office of Policy, Planning and Analysis, U.S. Department of Energy (1988).

10. Energy and Environmental Analysis, Inc., "Energy Impacts of Accelerated Retirement of Less Efficient Passenger Cars," prepared for the Office of Policy and Evaluation, U.S. Department of Energy (1980).

17

DANIEL SPERLING
and MARK A. DeLUCHI

Is Methanol the Transportation Fuel of the Future?

Because the transportation sector, unlike other energy-consuming sectors, has remained almost completely dependent on petroleum fuels, transportation has gradually gained an increased share of the petroleum market. In the United States, transportation increased its share from 53 percent of petroleum consumption in 1977 to 63 percent in 1987.[1] Already, the U.S. transportation sector by itself consumes more petroleum than is produced in the entire country. Eventually the transportation sector will have to be shifted to other energy sources. But to what fuel or fuels will it be switched?

According to the president of the United States, the U.S. Environmental Protection Agency, Ford Motor Company, General Motors, Toyota, the California Energy Commission, and other influential organizations and individuals, the transportation fuel of the future in the United States will be methanol.[2] This belief that methanol will replace petroleum as the dominant transportation fuel has several explanations: methanol can be made from a large number of materials, many of them available in abundance in the United States; it can be made less expensively than almost all other options; it burns more cleanly than petroleum fuels; and, because it is similar to gasoline and diesel fuel, it does not require costly changes in motor vehicles and the fuel distribution system.

We examine this growing support for methanol in a historical context with the objective of analyzing whether methanol will be the primary transportation fuel of the future. In the analysis we will compare methanol with compressed natural gas (CNG). We conclude that methanol is not a clearly superior option for replacing petroleum fuels and that until compelling new evidence is provided, public policy should promote a diversity of fuel alternatives.

We address highway applications of alternative fuels and focus on the near- and medium-term future, roughly the next 30 years. Barring unforeseen changes, our findings also hold for a longer time-frame since methanol and CNG come from the same resources and their production and delivery costs are fairly well known and unlikely to shift relative to each other. Of course, other energy options may become more attractive relative to methanol and CNG.

THE ASCENDANCE OF METHANOL POPULARITY

The growing enthusiasm for methanol is partly explained by historical circumstance. In the mid–1970s, just after the 1973 Arab oil embargo, nations began searching for ways to attain energy independence. The major nonpetroleum domestic energy resources in the United States were coal, oil shale, and biomass. Natural gas (NG) was virtually ignored since it was considered to be even scarcer than petroleum. Curtailments of NG deliveries to customers in accordance with the U.S. government's allocation scheme during the winter of 1976–77 served to reinforce the notion that NG was a scarce resource that should be reserved for winter heating needs.[3]

For the transportation sector, the most attractive options seemed to be petroleum-like fuels produced from coal and oil shale, methanol produced from coal, and ethanol made from corn and other biomass. Ethanol was quickly discarded as a major option by most energy analysts because it is far too expensive, although this assessment is not endorsed by the agricultural community, who sees ethanol as an answer to excess production and low prices of farm goods.

At a fall 1973 conference on Project Independence sponsored by the U.S. Department of Interior, oil and automotive industry representatives voiced sharp opposition to an initial proposal that national energy policy emphasize methanol over synthetic gasoline fuels.[4] In a major 1974 energy study, methanol was rated below oil shale and other coal-liquid options because it would have required major changes in motor vehicles and pipeline and fuel distribution systems and would not have supported existing investments in oil refineries.[5] SRI International prepared a 1976 report for the predecessor agency of DOE that rated synthetic gasoline a far more promising alternative than methanol. They argued that oil companies would be extremely unlikely to adopt methanol because synthetic crude could simply be added to the natural crudes still available to refineries, serving the needs of oil companies wishing to maintain the usefulness of present investments and insulating the consumer from change.[6]

The authors of virtually all of the major energy studies in the 1970s and early 1980s, as well as government energy policy, favored petroleum-like fuels from coal and oil shale.[7] Public and private R&D was heavily weighted toward direct liquefaction of coal.[8] Indeed, as late as 1981, only 5 of the 31 most-advanced synthetic fuels projects in the United States were intended to produce methanol as a primary product, and of these several were intended to coproduce high-Btu, pipeline-quality substitute NG.[9] Two additional projects were designed to manufacture methanol and to convert the methanol into synthetic gasoline in order to make the fuel compatible with the existing motor-vehicle and fuel-distribution systems, thus essentially downgrading the methanol into a lower-octane, higher-polluting fuel at additional cost.

In the early 1980s, perceptions began to shift, motivated by two insights: first, the cost of manufacturing petroleum-like fuels was greater than had been anticipated and, second, petroleum-like synthetic fuels did not help reduce persistent urban air pollution. The cost problem became salient as world petroleum prices stabilized and then dropped and as feasibility studies performed by project sponsors for the U.S. Synthetic Fuels Corporation began to indicate that the cost of producing refined shale oil and petroleum-like liquids from coal would be $60 to $100 per oil-equivalent barrel in first-generation plants.[10] Later-generation plants were projected to have much lower costs.

The air-pollution benefits derived from methanol first gained attention, although as a secondary issue, in the early 1980s. A study prepared for the California Energy Commission (CEC) played a key role, not because it gained wide circulation but because it laid the basis for the commission's organizational commitment to methanol fuel.[11] The authors of this landmark study concluded that, given California's high priority for reducing air pollution, the most attractive use of coal, then thought to be the most promising future source of portable fuel, was to convert it to methanol for the transportation and electric-utility sectors. This study was important because the CEC has proven to be the most influential advocate of methanol through the 1980s, its major justification being the air-quality argument.[12]

As the expensive synfuels projects floundered, attention began to shift toward methanol, at first because of the relatively advanced state of coal-to-methanol conversion technology and shortly thereafter because of a growing realization that much more NG existed than had been recognized. Although estimates of domestic and worldwide natural gas reserves began to be revised sharply upward in 1979, this was not widely acknowledged until several years later. The changed perception of natural gas availability was crucial because methanol can be manufactured much more cheaply and cleanly from NG than from coal.

Interest in methanol began to surge around 1985 as methanol proponents shifted their arguments away from energy security, a diminishing concern, to urban air quality, a stubborn problem for which most of the easy solutions had already been exhausted. Proponents, especially in California, argued that the transition to methanol transportation fuels represented the most significant opportunity for improving urban air quality.[13] At that time ozone air quality standards were being violated in virtually all major metropolitan areas, affecting over 80 million people.

From this historical review of informed opinion, an important question emerges: if NG is the preferred feedstock for making methanol, then shouldn't we reconsider the option of using NG directly in compressed or liquefied form? Analysts and decision makers remark that gaseous fuels are too different from liquid fuels, requiring too many costly changes in motor vehicles and the fuel distribution system to be widely used—exactly the same argument used against

Table 17.1
Natural Gas Reserves in 1985 in TCF

Country	Proven Conventional Reserves[b]	Undiscovered Recoverable Conventional Reserves[b]	Potentially Recoverable Unconventional Reserves[c]	Total Recoverable Reserves	Total Recoverable Reserves/Annual Production[d]
USA	200	775	130-1145	1100-2126	62-121
Canada	100	300	NA[e]	---	36
Mexico	77	200	NA[e]	---	56
World	3400	7600	NA[e]	---	120

[a]One TCF = 10^{12} SCF = 1 quadrillion Btu; one TCF per year = 0.5 million oil-equivalent barrels per day.

[b]Conventional gas comes from onshore and offshore and offshore proven and inferred reserves and is recoverable with current or near-term prices and technology.

[c]Unconventional reserves include low-flow or tight gas-bearing sands, coal seams, shales, geopressurized brines, and methane hydrates. Gas recovery depends on the state of technology and gas prices; lower values reflect current exploration and development technology and low gas prices (below about $4.50 per million Btu); higher values reflect advanced technology and higher gas prices.

[d]Based on mid-1980s production.

[e]NA = not available; treated as zero in calculations of the R/P ratios in the last column.

See nn. 16, 46.

methanol ten years earlier.[14] Experience indicates that these assumptions should be carefully scrutinized. Indeed, other countries, especially Canada and New Zealand, have deliberately chosen CNG over methanol.[15]

We suspect that policy inertia may be a major factor favoring methanol. This suggests the need for a careful reconsideration of methanol's perceived superiority. The salient criteria to consider in an evaluation of new transportation fuels, and that will be used in the following comparative analysis of methanol and CNG, are market costs, air-quality impacts, national security impacts, start-up barriers, and vehicle performance attributes.

NATURAL GAS RESERVES

First, as background, we note that worldwide proven NG reserves are increasing. They will surpass proven petroleum reserves in energy content by 1990 and are expected to widen the gap in the foreseeable future. A significant proportion of these NG reserves are in the United States and its two neighboring countries (see Table 17.1).

Proven gas reserves in the United States were over 30 percent greater (in energy content) than proven oil reserves in the 1980s. At present rates of consumption, the United States has enough economically recoverable conventional reserves of NG to last almost 60 years and, possibly, as much as 120 years if economically recoverable unconventional reserves are included. If the transportation sector were to switch 100 percent to NG, then the gas would be used up about twice as fast; somewhat faster if the gas is used for methanol, somewhat slower if used directly as CNG or LNG. The reason it would be exhausted faster as methanol is because methanol production is more resource-intensive than CNG: only about 57 percent of the original NG energy is available to the motorist, compared to 84 percent for CNG (taking into account losses in extraction, transport, and compression).[16] Thus it is clear the United States could sustain an aggressive CNG initiative for a prolonged period; for how long is still uncertain.

The worldwide gas supply situation is even more promising. As a result there will be little economic incentive for the United States to use coal or oil shale as a transportation energy feedstock for many years, perhaps 70 or more. If at that time coal were to become an important source of transportation fuel, the coal could be used to manufacture a substitute NG just as easily and for the same or lower cost as methanol from coal.

COSTS

According to most scenarios, the full cost of owning and operating a CNG automobile or truck will be slightly less than for a comparable methanol vehicle. A comparative cost analysis is summarized below. The analysis is conducted from the perspective of the owner of the motor vehicle. The following assumptions are made: the automobiles are optimized for neat (100 percent) methanol and CNG, respectively; the fuels are produced and used on a large scale; refueling station costs are fully incorporated; and costs are calculated on a per-kilometer basis to take into account differences in total life-cycle vehicle costs,[17] including differences in thermal efficiency, maintenance, and engine life.

The assumptions are based on an exhaustive review of the literature, including experiences in Europe, Canada, New Zealand, and the United States, and extensive discussions with vehicle and equipment manufacturers. The analysis is based on a near-term scenario (see Appendix) for single-fuel vehicles optimized to run on their respective fuels. The costs associated with CNG vehicles are somewhat more uncertain than those for methanol since the development of CNG-vehicle technology has lagged; relatively little effort has gone into designing and testing an optimized CNG vehicle, including the development of advanced storage tanks, and there is little reliable evidence from which to estimate the operating costs and life of such an optimized vehicle.

The baseline gasoline vehicle, against which single-fuel CNG and methanol vehicles are compared, has the following attributes: 35 mpg, 1,150 kg, 420 km range, and a vehicle life of 213,000 km at 16,000 km per year. It is assumed that a methanol car costs the same as a gasoline car and that a dedicated single-fuel CNG car costs $700 to $800 more (for fuel storage cylinders). The retail price of gasoline, including taxes, is assumed to be $1.15 per gallon, compared to an estimated $0.74 to $1.13 per gallon for methanol and $8.90 to $14.10 per million Btu for CNG. The cost parameters and vehicle attributes are listed in the Appendix and fully documented in note 16.

The methanol and CNG cars are comparable to the baseline gasoline vehicle; they have the same size, range, and weight (excluding the extra weight for CNG tanks and methanol fuel) and, because we are assuming optimized single-fuel vehicles, similar power. They are assumed to be 10 to 20 percent and 10 to 25 percent, respectively, more fuel efficient than the baseline gasoline car.

The costs of owning and operating these methanol and CNG cars, relative to gasoline, based on the foregoing assumptions, are as follows: methanol car, $+0.06$ to $+1.42$ cents per km; and CNG car, -1.86 to $+2.60$ cents per km. The analysis showed that the life-cycle cost of a CNG auto tends to be less than for a methanol vehicle, although not for all assumed values. The ranges in values correspond to uncertainties in cost estimates and vehicle attributes, as presented in the Appendix. The lower life-cycle costs for CNG are attributable to the lower fuel and maintenance costs and potentially longer engine lives, which more than offset the extra cost of CNG containers.

The next criteria are nonmarket costs such as national security and air quality. These costs are not included in the private or consumer costs calculated previously. These nonmarket costs are important because they are the primary justification for government intervention to support new fuels.

NATIONAL ENERGY SECURITY

The security risk associated with dependence on foreign gas for methanol or CNG is that once the very low-cost foreign gas is used up, which could occur fairly quickly (depending on the rents sought by foreign governments), the gas remaining is mostly controlled by OPEC countries and the USSR.[18] This OPEC-controlled gas may be subject to the same price and supply disruptions as petroleum.

Neither methanol nor CNG will provide significant energy security benefits. However, because methanol and CNG are made from the same feedstocks, with CNG requiring similar or less feedstock to provide the same amount of usable energy, CNG would use similar or less feedstock from any particular geographical source. It follows that CNG is preferred to methanol from an energy-security perspective.

AIR QUALITY

Perhaps the most important externality of vehicular fuel use is air pollution. Motor vehicles are the principal cause of urban air pollution, accounting for 57 percent of nitrogen oxide (NO$_x$) emissions, 44 percent of reactive hydrocarbons, and 75 percent of carbon monoxide (CO) in California.[19] As indicated earlier, the continuing failure of most metropolitan areas in the United States to meet ambient ozone standards (ozone is formed from reactions involving NO$_x$ and hydrocarbons) has been offered as a justification for introducing methanol. While it seems certain that some air quality benefits would occur with either methanol or CNG, data and modeling results are not in agreement on how large those benefits would be, especially for ozone.[20]

For a host of reasons it is difficult to specify accurately the differences in emissions and air quality impacts between different fuels, especially concerning ozone. First, emission rates are not simply predetermined by combustion technology but vary for a given technology (and fuel) according to tradeoffs between emissions, costs, performance, and driveability. A particular fuel may be potentially less polluting than gasoline, but the constraints of automotive design may result in these potential benefits not being realized. Under typical circumstances, engines are configured to emit the maximum allowed by statute in order to enhance other attributes, such as cost (by reducing the cost of pollution-control equipment) or engine power. Second, pollutant production is sensitive to the air/fuel ratio of engines. If future engines are designed to run "lean" (high air/fuel ratio) to gain higher fuel efficiency, then NO$_x$ levels would be relatively higher (because reduction catalysts do not work well with excess air), and CO and HC emissions and engine power would be lower than for an engine operating at stoichiometric ratios, as are most of today's gasoline engines. Third, a distinction must be made between optimized single-fuel engines and retrofitted or bifuel engines; we focus on optimized single-fuel engines as the desirable ultimate technology because they are superior in emissions, costs, and performance.

Fourth, exact fuel composition must be specified since some methanol emission and modeling data are based on a fuel consisting of 100 percent methanol, while others mix 10 percent or 15 percent gasoline into the methanol (which results in much more pollution); this becomes even more complicated for multifuel methanol/gasoline engines since they will operate on varying blends of methanol and gasoline. Fifth, the ozone-formation process is highly complex and is sensitive to meteorological and topographic conditions; even the most sophisticated photochemical air quality models have error margins of 30 percent or more.[21]

Sixth, no reliable estimates have been made of life-cycle formaldehyde emissions, a critical consideration because formaldehyde is a product of methanol combustion and a highly reactive hydrocarbon. Seventh, only in the Los Angeles area has sufficient meteorological and spatial pollutant concentration data been collected to operate multiday photochemical airshed models; results from Los Angeles are not generalizable to other regions. Eighth, emission data for dedicated single-fuel CNG engines are much more sparse and less accurate than for methanol engines; moreover, no dispersion or photochemical modeling of CNG emissions has ever been conducted.

This list could continue. The point is that emission and air-quality data for

CNG and methanol are highly uncertain and should be interpreted with care. Nevertheless, current knowledge suggests that the use of both CNG and methanol would probably lead to lower ozone levels than gasoline use (although the authors of one study argue that methanol may increase ozone).[22] CNG may be slightly better than methanol because methane, the principal organic pollutant from CNG vehicles, is 100 times less reactive than unburned methanol, the principal organic pollutant from methanol vehicles. In addition, the secondary organic emissions may be less reactive than formaldehyde, the secondary organic emissions of methanol vehicles.

Compared to gasoline, CNG will emit much less carbon monoxide, which is a major wintertime problem in most cities, and similar or possibly higher levels of NO_x, while methanol will emit less carbon monoxide (but more than would CNG) and, depending on the air/fuel ratio, possibly less NO_x.[23]

Again, because of the lack of credible ozone modeling, the effects of methanol and CNG vis-a-vis gasoline and vis-a-vis each other are uncertain. Some of the pitfalls of this type of air quality analysis are apparent from recent efforts to study the effects of methanol use. In the mid–1980s several studies concluded that the use of methanol would reduce peak ozone concentrations in urban areas by 10 to 30 percent.[24] These conclusions depended strongly, however, on several assumptions. For example, each assumed a different volume of gasoline blended with the methanol (from 0 to 15 percent), and all assumed that NO_x emission levels would remain unchanged relative to gasoline and that the reactivity of methanol pollutants would be the same in multiday smog episodes as in single-day episodes.

A careful assessment based on more recent evidence suggests that the substitution of methanol (M85) for gasoline in all motor vehicles may result in a *maximum* reduction in peak ozone levels of 0 to 15 percent.[25] The most recent and sophisticated modeling effort, conducted at Carnegie-Mellon University, found that in the Los Angeles area, the use of M85 methanol in all mobile sources except motorcycles and planes would result in only a 6 percent reduction in peak ozone levels.[26] If M100 were used, the reduction would be 13 percent; if M100 were used, but assuming higher formaldehyde levels in the exhaust emissions (55 mg per mile instead of 15 mg per mile), the reduction would be 7 percent; and if M100 were used in advanced-technology engines, a 15 percent reduction results (or 9 percent if compared to an advanced-technology gasoline engine). The 9 percent reduction with advanced technology M100 represents 43 percent of the maximum ozone reduction attainable from motor vehicles; that is, if all vehicle emissions were eliminated, Harris and others found that ozone would be reduced only 21 percent.[27] In practice, even these reductions would not occur for several decades because of the slow turnover rate of vehicles, the initial use of multifuel cars, and the presumably low initial market penetration rate.

In summary, while methanol provides the potential for achieving a large part of the ozone reduction achievable through changes in the transport sector, the magnitude of these potential improvements is modest; moreover, these potential

Table 17.2
Start-up Barriers for Methanol and CNG Relative to Gasoline

	METHANOL	CNG
Refueling station	$40,000	$300,000 or more
Multifuel vehicle		
Initial cost	+$0 to 200	+$1600 (retrofit)
		+ $750 (factory)
Operating cost	the same or more	less
Lifecycle cost	the same or more	the same or less
Performance	the same or better	worse
Luggage space	the same	less
Cold start	worse	the same

See nn. 14 ("California Energy Commission, Fuels Report"), 16, 29, 30.

reductions with methanol require the use of M100 and very low formaldehyde emissions, two conditions that may not be attainable.

In contrast to methanol, very little research has been conducted on natural gas vehicles and none on the ozone impacts of CNG emissions.[28] Published assessments of emissions from CNG vehicles have often been overstated because they were based on retrofitted dual-fuel cars, not optimized single-fuel vehicles.[29] Such assessments offer little help in evaluating the relative attractiveness of different energy paths. What little data do exist, as summarized above, suggest that there is no scientific basis for claiming that either fuel is superior to the other from an air quality perspective.

A related concern is emissions of greenhouse gases. CNG is slightly superior to methanol. A full systems analysis of carbon dioxide and trace greenhouse gases emitted during production, transport, and combustion of both fuels indicates the following.[30] If the feedstock is natural gas, methanol generates about the same quantity of CO_2-equivalent greenhouse gases as gasoline, while CNG generates about 20 percent less. If the feedstock is coal, both fuels produce about 60 percent more greenhouse gases; if the feedstock is biomass, the net production is close to zero for both fuels (i.e., 100% less).

START-UP BARRIERS

Until now, this chapter assessed the relative attractiveness of vehicles optimized for a particular fuel because such an evaluation is important for selecting which fuel will ultimately be superior. But the ultimately superior option may never be reached because of start-up barriers; they are therefore considered here as one more factor to consider in the evaluation. The principal start-up barriers, listed in Table 17.2, are the cost of establishing new fuel stations and the higher cost and inferior attributes of vehicles that are required to operate on both gasoline and the new fuel.

Multifuel vehicles are addressed in Table 17.2 because they are the vehicles generally considered in analyzing the initial period of a fuel transition. The qualitative and quantitative judgments in the table are based on the assumption that large numbers of vehicles are produced and large volumes of fuel are sold. In general, methanol faces smaller start-up barriers than CNG. For limited vehicle and fuel sales, methanol would tend to have an even larger advantage.

One major barrier is the absence of retail outlets for each fuel. With minor modifications, a retail gasoline outlet can accommodate methanol; the cost is somewhere between $5,000 and $60,000 per station, depending upon whether a new underground tank is needed. A new tank is needed if the existing tank is corroded (in which case it should be replaced anyway), if new government rules require the tank to have double walls, or if the tank is made of fiberglass that is incompatible with methanol (as are about 10 percent of the tanks in the United States). The cost for a CNG station is much greater—up to $250,000 just for a compressor and over $300,000 per station.[31] In some cases the CNG refueling facilities could be established on the site of gasoline stations as long as they are located near a high-pressure natural gas pipeline.

The start-up constraints related to the vehicle are also substantial for CNG. The problems are that CNG, a gas, is much more different from gasoline than is methanol, a liquid; and, during the initial stages of a transition, it will be of critical importance that multifuel vehicles be used to reduce the disadvantage of limited availability of fuel at retail outlets.

Multifuel methanol/gasoline vehicles have the advantage of involving the use of a single fuel system and, if manufactured in large quantities, of costing just a little more than a gasoline vehicle. Also, from the vehicle operator's perspective, a methanol multifuel vehicle would be indistinguishable from a gasoline vehicle. The fuel, whether methanol or gasoline, would be put in the same tank, and the driver would not need to do anything different. The only difference to the driver would be slightly greater power (maximum of +10 percent) if mostly methanol was being burned. This multifuel vehicle would be only somewhat compromised from an optimized-for-methanol vehicle.

A dual-fuel CNG/gasoline vehicle, on the other hand, would be far inferior to a dedicated CNG, gasoline, or methanol vehicle. The dual-fuel vehicle would have less power, redundant fuel tanks and fuel delivery systems (one for natural gas and one for gasoline), and would therefore cost considerably more than a gasoline vehicle; the cost would be $1,600 or more extra for an aftermarket retrofit, or about $750 extra if made in the factory, according to industry estimates. It would also have much less trunk space because of the extra fuel tank. A transition CNG vehicle would therefore be acceptable only to those consumers who accumulated very high mileage (allowing them to pay off the higher initial capital cost with the lower fuel cost), did not require much trunk space, and did not demand high performance.

This analysis of start-up costs clearly indicates methanol's superiority. Several caveats are in order, however. If transition vehicles operate on both gasoline and the new fuel, which indeed would be the case unless government were to mandate that all fuel stations or a percentage of them supply the new fuel, as was the case with unleaded gasoline in 1975, then two drawbacks appear, both of which work against methanol.[32] First, methanol's air quality benefits will be negligible, even less than for optimized single-fuel vehicles, because chemical reactions take place when methanol and gasoline are mixed, resulting in higher levels of evaporative hydrocarbon emissions, and because catalyst performance, even with redesigned catalysts, would probably be compromised when obligated to handle both fuels. There is no evaporative emission problem with a bifuel CNG/gasoline vehicle because NG is mostly methane (CH_4), which is essentially nonreactive.

Second, it is unclear how often methanol/gasoline vehicles would be fueled with methanol since methanol is expected to be more costly (per vehicle mile) than gasoline into the foreseeable future. CNG, on the other hand, will be less expensive than gasoline—the price ratio in Canada and New Zealand of CNG to gasoline is roughly 0.5 to 0.8, and is likely to be similar in the United States— and, thus, consumers will have a strong incentive to use CNG on a regular basis once they make the initial vehicle purchase or retrofit.[33] Surveys in New Zealand and Canada indicate that bifuel CNG vehicles operate on CNG about 75 percent to 90 percent of the time, and that the rate would be even higher, especially in Canada, if fuel were more readily available.[34] Thus, a small number of bifuel CNG vehicles would use as much CNG fuel as would a much larger number of bifuel methanol vehicles use methanol, and may therefore generate proportionately greater air quality benefits per vehicle.

Third, as will be discussed later, the methanol start-up advantages, while apparently significant, tend to be transitory when scrutinized in the context of actual transition conditions.

WHICH VEHICLE TECHNOLOGY IS MORE DESIRABLE?

A fair comparison of the relative attractiveness of CNG and methanol vehicles is difficult because the evaluation is sensitive to whether the vehicle is optimized for the new fuel, whether it operates on multiple fuels, and a determination of comparability. Unfortunately, as we have documented elsewhere, most evaluations have been sloppy in making these distinctions in a way that is systematically biased against CNG.[35]

The bias comes from the fact that CNG technology is less advanced than methanol technology, and that multifuel CNG technology, as discussed in the previous section, is more inferior to single-fuel CNG technology than is multifuel methanol technology relative to single-fuel methanol technology. Current CNG vehicles (about 400,000 worldwide in 1988) are retrofitted gasoline vehicles. They were designed and built for gasoline, are burdened by redundant fuel systems, and use carbureted fuel control and heavy steel tanks for fuel storage; they are far inferior to single-fuel vehicles designed for CNG and equipped with modern lightweight composite-material storage cylinders and electronic fuel control. CNG vehicle and storage technology has languished because the auto manufacturing industry has not taken much interest in CNG.

Methanol vehicle technology is also primitive, but it has received more attention from the auto industry and is further advanced than CNG vehicle technology; it benefits from about 15 years of intermittent research on alcohol vehicles by Ford and Volkswagen and lesser efforts by other manufacturers, and from the experience of producing over 3 million ethanol cars in Brazil.

But even if both technologies had received equal attention, gaseous fuel technology would be still be less advanced than liquid fuel technology for the simple fact that virtually all motor vehicles have been designed to operate on liquid fuels for over a century.

While a bifuel methanol/gasoline vehicle has important advantages over a bifuel gas/gasoline vehicle, an optimized single-fuel CNG vehicle would compete well against an optimized methanol vehicle. An optimized CNG vehicle would have the disadvantage of about 60 percent shorter driving range for a comparable fuel-tank volume and, depending on differences in such parameters as the compression and air/fuel ratio, about 0 to 10 percent less power than a comparable methanol vehicle. The range can be extended by using more or larger fuel tanks or by increasing the pressure in the tanks, but more tanks means less interior space and more weight, while higher pressure incurs greater compression and fuel tank costs.

On the other hand, optimized CNG vehicles would have similar emissions, similar or lower lifecycle costs, and no problems with cold starts. Methanol's cold-start problem, its inherent difficulty in starting in temperatures less than about 5° C, can be mitigated by various techniques, including automatic heating of fuel lines, small gasoline tanks to be used only for starting, and fuel dissociation, but it is uncertain whether methanol vehicles (using either M85 or M100) will ever be fully satisfactory in cold climates, where about half the United States and all the Canadian population lives.[36]

In summary, past comparisons of CNG and methanol have often been biased against CNG because they used retrofitted bifuel CNG autos as the basis of comparison. These primitive technologies are not representative of what is likely to be commercialized in the future. Future CNG vehicles will be far superior to the retrofitted vehicles now operating in New Zealand, North America, and Italy. The major disadvantage of single-fuel CNG vehicles is their limited range, although there are no consumer choice studies that specify the importance of this disadvantage, while the disadvantages of methanol vehicles are cold-start difficulties and, relative to CNG, perhaps slightly higher life-cycle costs.[37]

DIESEL ENGINES

This chapter has addressed only spark-ignition engines to this point, but CNG and methanol may also be used in compression-ignition (diesel) engines. CNG has roughly the same advantages and disadvantages relative to methanol in both types of engines: similar improvements in emissions, lower cost, somewhat less power, and redundant fuel systems. The important difference is that both methanol and CNG provide major emission improvements, dramatically reducing particulate and sulfur oxide emissions. Methanol will also significantly reduce nitrogen oxide emissions.[38]

From an energy-transition perspective, diesel vehicles are much less important than spark-ignition vehicles. Diesel urban transit buses will probably be the first market penetrated by methanol and/or CNG because of recently promulgated emission regulations that take effect in 1991, but this market is dispersed and tiny, a total of only about 30,000 barrels per day in the United States.[39] Further penetration is likely to lag behind penetration of the gasoline market because diesel engines do not turn over as quickly, CNG and methanol are more economically attractive relative to gasoline than to diesel fuel, and diesel trucks tend to travel over larger areas and therefore are more sensitive to limited fuel availability. In the United States the entire diesel fuel market is about one-fifth the size of the gasoline market, although it is expected to increase in both absolute and relative size.

SO WHY NO INTEREST IN CNG?

If CNG is as attractive as methanol, why has it received so little attention in the United States? The probable explanation is a negative perception of CNG based in part on incomplete knowledge. This has a direct parallel in the professional community's attitude toward methanol during much of the 1970s. In that era, analysts and researchers rejected methanol as being too difficult to implement, requiring new fuel stations, pipelines, and new or modified vehicles. So it is with CNG in the 1980s. In the early 1980s, resistance to methanol by the auto manufacturers and others weakened when it became clear that methanol was the cheapest nonpetroleum liquid fuel option readily available. Key actors began to recognize that under some conditions the barriers would not be that significant. Indeed, in Brazil, with the full involvement of Ford, General Motors, Volkswagen and others, over 90 percent of new cars have been operating exclusively on alcohol since 1983.

It is clear that analysts and researchers were deceived by the apparent start-up costs of methanol into making a negative overall assessment, despite methanol's longer-term benefits. CNG is now subjected to the same criticisms. Though the start-up barriers may be more substantial, the long-term potential may be greater. History suggests that we should give CNG a more thorough reexamination.

A second explanation for methanol's prominence is an understandable resistance to CNG by oil marketers. Gasoline and diesel-fuel distributors would lose control of fuel marketing if natural gas, currently distributed by a network of pipeline transmission companies, were to replace methanol, a liquid that eventually could be fully integrated into the petroleum distribution system. Also, in the short term, methanol (unlike CNG) can be blended in small quantities into gasoline, further enhancing methanol's relative attractiveness to the oil industry.

Third, despite forecasts that real natural-gas prices will rise and therefore stimulate more exploration that will result in discoveries keeping pace with production and demand, there is still some skepticism about the price elasticity of gas supply. This skepticism is rooted in the long history of restrictive regulation which began to be phased out in 1978. Some industry and government officials are not convinced that sufficient gas will be domestically available at competitive prices for enough time to warrant developing the transportation fuels market.

Fourth, NG companies have not promoted CNG. This is a root cause of general lack of interest in CNG. If gas utilities, who are the local suppliers of NG in the United States and would be the principal or sole marketers of gas to motorists, are not interested, then others understandably will not be willing to invest the time and resources necessary to initiate this new fuel.

The chief economist of the American Gas Association argues that state regulatory bodies have built a web of rules that effectively removed the incentive for gas utilities to market CNG to the transportation market.[40] These regulations were created many years ago to protect the captive users from monopolistic pricing and supply cutoffs. In the 1970s these regulations resulted in moratoria on new hookups and bans on most types of advertising and marketing. These rules are being phased out in the 1980s, but significant impediments to gas marketing remain. In a 1982 American Gas Association survey of gas utilities, 78 percent said that their state public utility commissions still imposed restrictions on natural gas advertising; although the survey has not been repeated, it is believed that similar restrictions are still common.[41] Hay and McArdle give the example that in many states only advertising that includes an energy conservation message can be included in the utility's cost of service.[42] They point out that promotional rates for development of new markets and applications remain a regulated matter for utilities, though not for their competition.

In practice, as a byproduct of continuing natural gas deregulation in the 1980s, impediments to the entry of gas utilities into the transportation fuels market have been reduced. Regulatory impediments may be more perceived than real. Nonetheless, many years of restrictive regulation have left gas utilities with little expertise in marketing and strategic planning and an inertia that is hard to overcome. The fact is that natural gas utilities have not played a leadership role in exploring and developing the CNG option.

Overall, then, CNG has had few proponents, and therefore no constituency, to help it overcome the lack of imagination that thwarts its introduction.[43] This lack of a constituency may be reason to discard CNG, since any new fuel will need forceful support if it is to move forward, as illustrated by the farm lobby's success in introducing ethanol fuel, an economically inferior option. We argue, however, that because the merits of CNG are potentially large, concerned organizations, including state and federal governments, should (i) support further analysis and development of CNG technology so that decisions about its merit and desirability can be made on a more informed basis, and (ii) impose changes in the regulation of natural gas utilities and encourage those utilities to be more aggressive in promoting CNG.

WHEN IS METHANOL PREFERRED?

Relative to CNG, methanol is not superior economically or environmentally, does not offer more energy security, and, in an optimized vehicle, will not necessarily be more attractive to consumers. The only clear advantage of methanol relative to CNG is in terms of initial start-up barriers.

This advantage will not be an important factor in most situations, however. Consider the scenario of a rapid transition away from petroleum, presumably under crisis conditions. In this case, the network of fuel outlets for the new fuel would be expanded rapidly enough that most buyers would be able to opt for the single-fuel vehicle fairly soon after the transition was initiated. As a result, multifuel vehicles would be a short-lived phenomenon, and the advantage of methanol in a multifuel configuration would be fairly minimal. The advantage of lower costs for establishing methanol fuel outlets would also be minor in this scenario. The reason is that initiating a CNG energy system would cost the same as or less than a methanol system in the sense that the higher fuel station costs of CNG are incorporated in the retail fuel price, which is lower for CNG than methanol.

Thus higher station costs are only an important issue if fuel consumption is low, resulting in fuel station owners not earning a return on their capital investment. If fuel demand is high, as it would be in a rapid transition, then investments in new CNG stations would be readily forthcoming. The high initial station cost would not slow the rate at which fuel stations were established because even though CNG stations are more costly than methanol stations, it is difficult to imagine that capital availability (about $300,000 per station) would be a problem if a reasonable return on investment was expected. The important issue is return on investment, not capital cost.

If a slow transition is being pursued, the start-up advantages of methanol again are not important. In this scenario, there would be time to exploit those market niches where particular alternative fuels are attractive; indeed, barring major governmental intervention, the market would dictate this incremental approach. In this scenario, optimized CNG vehicles would be attractive as high-usage fleet vehicles because the low cost of fuel would offset the high vehicle cost, electric vehicles would be attractive in areas with severe air pollution problems because of their very low emissions, even including emissions from the powerplant, and CNG and methanol would be attractive in urban diesel trucks and buses in polluted areas because of their pollution-reduction effectiveness in diesel engines.[44] Note that methanol is attractive in only one niche in this slow-transition scenario (urban-based diesel vehicles in polluted areas).

The lower start-up costs of methanol therefore do not appear to be instrumental either in a rapid transition to alternative fuels or during the initial stages of a slow transition. Methanol may prove to be superior in the latter stages of a slow transition when the limited availability of fuel makes multifuel vehicles preferable to most consumers, or in some fuzzy intermediate scenario, but the inability to forecast accurately the future and cognitively to work through sets of uncertain conditions suggests that any such determination is speculative at best. In the face of this uncertainty, it seems clear that lower start-up costs of methanol are not substantial enough to render it an obviously superior choice under any set of conditions.

RECOMMENDATIONS

On the basis of the preceeding discussion, we make the following recommendations. First, new fuels should be introduced in a gradual fashion by targeting them to market niches and regions where they have comparative advantages. For instance, CNG should first be introduced in cold-weather areas where methanol would have cold-start problems, and in diesel engines in areas with air quality problems. Electric vehicles, which we address elsewhere,[45] should be introduced in urban areas with air quality problems in situations where range and power are not major concerns, because in most locations, even using a full-system perspective, electric vehicles will be greatly superior to methanol and CNG vehicles in terms of air-quality impact.

Second, government support of R&D for other cleaner fuels (e.g. hydrogen or clean electricity) should be increased. Neither natural gas nor methanol will provide large air-quality (or greenhouse effect) benefits, except perhaps for limited niches in the diesel market.

In concluding, we wish to emphasize that our purpose is not to conduct a vendetta against methanol or its proponents or to suggest that only one alternative fuel be selected. We strongly believe that a concerted effort should be made to remove barriers and to provide incentives for methanol use. But the same should be done for CNG, liquefied natural gas (LNG), electric vehicles and, perhaps further in the future, hydrogen vehicles, because all new fuels different from petroleum face considerable start-up barriers.

While it appears that the methanol and natural gas options are not going to yield huge environmental benefits, we still believe that both should receive support from government for the environmental benefits they do offer, for their reduction of petroleum imports, for their restraining influence on world oil prices, and because there is a possibility that the overall economic and environmental benefits of these natural gas-based fuels may be greater than we realize. At this point, in the face of a global greenhouse threat, the potential for political instability in key resource supply areas, the U.S. trade deficit, and health-threatening air pollution, it would be irresponsible to proceed with "business as usual" in transportation fuels. But let us proceed in an informed fashion.

NOTES

This research was supported, in part, by grants from the University of California Transportation Center and the University of California Energy Research Group. We thank R. Dunstan, N. Hay, P. McArdle, B. McNutt, R. Johnston, G. Harvey, and others who prefer to remain anonymous for their comments on earlier drafts.

1. Energy Information Administration, Annual Energy Review 1987, Washington, D.C. (1988).

2. For documentation, see D. Sperling, *New Transportation Fuels: A Strategic Approach to Technological Change* (Berkeley, Calif.: University of California Press, 1988); B. McNutt and E. E. Ecklund, "Is there a government role in methanol market development?" *SAE* 861571 (1986); C. Gray and J. Alson, *Moving America to Methanol* (Ann Arbor. Mich.: Univ. of Michigan Press, 1985).

3. U.S. Department of Energy, "Energy Security," DOE/S–0057, Washington, D.C. (1987).

4. R. L. Bechtold, "Compendium of Significant Events in the Recent Development of Alcohol Fuels in the United States," ORNL/Sub/85–22007/1, NTIS, Springfield, Va. (1987).

5. F. Kant, A. Cohn, A. Cunningham, M. Farmer, and W. Herbst, "Feasibility Study of Alternative Fuels for Automotive Transport," prepared by Exxon Research and Engineering Co. for U.S. EPA, PB–23 5580, 3 vols., NTIS, Springfield, Va. (1974).

6. SRI, "Synthetic Liquid Fuels Development: Assessment of Critical Factors," ERDA 76–129/1, GPO, Washington, D.C. (1976).

7. Kant, "Feasibility Studies"; Purdue University, "Transportation Energy Futures: Paths of Transition (1981–82)," West Lafayette, Indiana; partial results were published in J. K. Binkley, W. Tyner, and M. Mathews, *The Energy Journal* 4, no. 91 (1983).

8. H. Perry and H. Landsberg, "Factors in the development of a major U.S. synthetic fuels industry," *Ann. Rev. Energy* 6, no. 248 (1981).

9. Pace Co. Consultants and Engineers, "Comparative Analysis of Coal Gasification and Liquefaction," prepared for Acurex Corp. and California Energy Commission, 650 S. Cherry St., Denver, CO 80222 (1981).

10. Synthetic Fuel Corporation, "Comprehensive Strategy Report," Appendices, Washington, D.C. (1985).

11. Acurex, "Clean Coal Fuels: Alternate Fuel Strategies for Stationary and Mobil Engines, Executive Summary," P500–82–020, California Energy Commission, Sacramento, Calif. (1982).

12. K. D. Smith, D. W. Fong, D. S. Kondoleon, and L. S. Sullivan, "The California State methanol program," *Prcs, Int Alc Fuel Symp on Alc Fuel Tech,* Ottawa, Canada (1984) pp. 2–373 to 2–383; Three-Agency Methanol Task Force, "Report," prepared jointly by California Air Resources Board, California Energy Commission, and South Coast Air Quality Management District, Sacramento, Calif. (1986).

13. L. Berg, Testimony to U.S. Congress, H. of Rep., Comm. on Energy and Commerce, Subcomm. on Fossil and Synthetic Fuels, *Methanol as Transportation Fuel,* 98th Cong., 2d sess., GPO, Washington, D.C. 98–145 (4 and 25 April 1984), p 26.

14. U.S. Department of Energy, "Assessment of Costs and Benefits of Flexible and Alternative Fuel Use in the U.S. Transportation Sector," Washington, D.C. (1988); California Energy Commission, Energy Development Report, Sacramento, Calif. (1986);

California Energy Commission, Biennial Fuels Report, Sacramento, Calif. (1986); California Energy Commission, Fuels Report, Sacramento, Calif. (1987).

15. D. Sperling, *New Transportation Fuels,* chap. 5.

16. M. A. DeLuchi, R. A. Johnston, and D. Sperling, "Methanol vs. natural gas vehicles: a comparison of resource supply, performance, emissions, fuel storage, safety, costs, and transitions," *SAE* 881656 (1988).

17. Ibid.

18. Faucett Associates, "Future Natural Gas Supply from Selected World Regions: Projected Quantities and Costs," Bethesda, Md. (1988).

19. California Air Resources Board, "Emissions Inventory 1983," Technical Support Division, CARB, Sacramento, Calif. (1986).

20. J. D. Murrell and G. K. Piotrowski, "Fuel economy and emissions of a Toyota T-LCS-M methanol prototype vehicle," *SAE* 871090 (1987); W. P. L. Carter et al., "Effects of Methanol Fuel Substitution on Multi-Day Air Pollution Episodes," prepared for California Air Resources Board, ARB–86, Statewide Air Pollution Research Center, University of California, Riverside, Calif. (1986); J. N. Harris, A. G. Russell, and J. B. Milford, "Air quality implications of methanol fuel utilization," *SAE* 881198 (1988).

21. T. W. Tesche, "Photochemical dispersion modeling: review of model concepts and applications studies," *Environment International* 9:465 (1984).

22. Sierra Research, Inc., "Potential Emissions and Air Quality Effects of Alternative Fuels," SR88–11–02, Sacramento, Calif. (1988).

23. J. A. Alson, J. M. Adler, and T. M. Baines, in chapter 8 of this volume.

24. Systems Application, Inc., "Impact of Methanol on Smog: A Preliminary Estimate," SAE Publication no. 83044 (1983). Also see SAE, "The Impact of Alcohol Fuels on Urban Air Pollution: Methanol Photochemistry Study," DOE/CE/50036–1, NTIS, Springfield, Va. (1984); Jet Propulsion Laboratory, "California Methanol Assessment," #83–14, Pasadena, Calif. (1983); R. J. Nichols and J. M. Norbeck, "Assessment of Emissions from Methanol-Fueled Vehicles: Implications for Ozone Air Quality," presented at 78th Annual Meeting of the Air Pollution Control Association, Detroit, Mich. (1985).

25. DeLuchi, Johnston, and Sperling, "Methanol vs. natural gas"; Harris, Russell, and Milford, "Air Quality Implications"; Sierra Research, "Potential Emissions."

26. Carter et al., "Effects of Methanol Fuel."

27. Harris et al., "Air Quality Implications."

28. See chapter 8.

29. See, for example, Environmental Protection Agency, "Air Quality Benefits of Alternative Fuels," report prepared for the vice-president's Task Force on Alternative Fuels, Ann Arbor, Mich. (1987).

30. M. A. DeLuchi, R. A. Johnston, and D. Sperling, "Transportation Fuels and the Greenhouse Effect," *Transportation Research Record,* 1175 (1988):33–44. Initial studies performed by the World Resources Institute and by Acurex for the California Energy Commission have similar results.

31. DeLuchi, Johnston, and Sperling, "Methanol vs. natural gas."

32. D. Sperling and J. Dill, "Unleaded Gasoline in the U.S.: A Successful Model of System Innovation," *Transportation Research Record,* 1175 (1988): 45–52.

33. See chapter 13 of this volume.

34. G. Harris, P. Phillips, L. Richards, and L. Arnoux, "CNG Market Development Study," Pub. P86, New Zealand Energy Research and Development Committee, Univ.

of Auckland, Auckland, New Zealand (1984); Canadian Facts, "Management Summary: Natural Gas for Vehicles—Conversion Motivation Study," British Columbia Hydro Gas Operations, Burnaby, B.C. (1986).

35. DeLuchi, Johnston, and Sperling, "Methanol vs. natural gas."

36. For recent progress see R. Pefley, R. Caldeira, P. Russell, and K. Suresh, "Methanol-Fueled Vehicles for the Federal Government," presented at the 10th Annual Energy Sources Technology Conference, Dallas, Texas (1987).

37. See D. Sperling and K. Kurani, "Refueling and the vehicle purchase decision: the diesel car case," *SAE* 870644 (1987); D. Greene, "Estimating daily vehicle usage distributions and the implications for limited-range vehicles," *Transportation Research* 19B, 347 (1985).

38. See chapter 8 of this volume.

39. Oak Ridge National Laboratories, "Transp. Energy Data Book: Edition 9," ORNL–6325, NTIS, Springfield, Va (1987).

40. See chapter 13.

41. J. P. Sprowl, "Gas utility advertising: an overview of recent commission decisions," *The Natural Gas Lawyer's Journal* 1, no. 137 (1985); see chapter 13 of this volume.

42. See chapter 13.

43. An indication that the natural gas industry may be taking a more aggressive role in penetrating the transportation fuels market is indicated by the establishment in August 1988 of the Natural Gas Vehicle Coalition, with the intent of soliciting participation not only from the gas industry but from the vehicle manufacturing industry, state and local governments, the environmental community, and the oil industry.

44. M. DeLuchi, Q. Wang, and D. Sperling, "Electric Vehicles," *Transportation Research A*, 23A, no. 3, 1989, in press; D. Sperling, *New Transportation Fuels*.

45. M. DeLuchi, Q. Wang, and D. Sperling, "Electric Vehicles."

46. Institute of Gas Technology, "IGT World Reserves Study," Chicago, Illinois (1986); American Gas Association, "The Gas Energy Supply Outlook Through 2010," Arlington, Va. (1985); Energy Information Administration, "Annual Energy Outlook 1984," U.S. DOE, Washington, D.C. (1985); L. D'Andrea, "National Petroleum Council study on unconventional gas," *Procs., 1981 International Gas Research Conference,* Government Institutes, Rockville, Md. (1982) 621–28; U.S. Department of Energy, "An Assessment of the Natural Gas Resource Base of the United States," Washington, D.C. (1988).

APPENDIX: Cost Parameters and Vehicle Attributes Used in Cost Analysis

Gasoline	Methanol	CNG	
0.95	—	—	Retail price of gasoline, $/gallon, excluding taxes
—	0.50–0.80	—	Methanol price, $/gallon, plantgate or at the port if imported
—	0.14–0.23	—	Domestic transportation cost and retail mark-up
—	—	5–8	Cost of gas to station, $/mmBtu
—	—	2.3–4.5	Station mark-up, $/mmBtu[a]
0.20	0.10	1.60	Fuel taxes, $/gallon for liquids, $/mmBtu for natural gas
35	—	—	Lifetime vehicle fuel efficiency, mpg
—	+10–20	+10–25	Thermal efficiency relative to a gasoline-powered car, percent
9.5	9.5	10.2–10.3	Vehicle price, $10³ (1985)
213	213	213–262	Life of vehicle, 10³ km
1150	1150	1197	Weight of vehicle, kg
9	9	9	Real interest rate for a car loan, percent
400	400	300–400	Maintenance costs, $/year

[a] Station costs for CNG were calculated independently, taking into account 15 different cost and operations factors. For details, see n. 15.

18

DEBORAH L. BLEVISS

The Role of Energy Efficiency in Making the Transition to Nonpetroleum Transportation Fuels

In these days of relatively low oil prices and renewed demand for performance and power in vehicles, interest in fuel economy often appears to be a characteristic of the past. This phenomenon, in turn, has fueled a growing perception that we have reached the optimum fuel economy that can be achieved and that seeking to go any higher would prove both costly and risky. Instead, it is argued, we should turn national attention to converting our wholly petroleum-dependent transportation fleet to alternative fuels so as to reduce vulnerability to a future oil crisis.

Stated enough times, as it has, this perception that we have approached the fuel economy limits is now increasingly accepted as truth. Yet the facts indicate otherwise. It is indeed ironic that an industry that boasts it will one day be able to have "high technology" radars warning drivers of impending collisions and holographic images projected onto vehicle windshields simultaneously cries technology limitations when discussion turns to improved fuel economy.

The fact is, though, that the technology revolution spawned by the space age and the race to build "SDI" (the Reagan Administration's Strategic Defense Initiative) has barely begun to be applied to the light-transportation industry. Indeed, the enormous strides seen in the past decade in improved light-vehicle fuel economy have resulted primarily from optimization of existing technologies rather than application of new ones.

However, the technology revolution is now making its way into the light-transportation industry, and as it does so, new opportunities for fuel economy improvements are materializing, opportunities that will be both cost effective and could well yield other benefits for consumers. The pursuit of these new technologies will be critical to making the transition to alternative fuels, for they will provide the bridge to that transition.

THE TECHNICAL POTENTIAL FOR FUEL ECONOMY

Before summarizing these technologies, it is appropriate to review where energy losses occur in today's modern light vehicle so as to be able to assess where the opportunities for efficiency improvements lie. Vehicle fuel is converted to useful work in the engine where its combustion is used to turn the crankshaft. About 70 to 80 percent of the fuel's energy content is lost directly at this point to coolants, exhaust gases, convection, and radiation. Much of this loss can be reduced, but thermodynamic laws dictate that generally over 50 percent of the energy content of a fuel cannot be recovered.[1]

Crankshaft energy is then transmitted to the vehicle's wheels through the transmission, with further losses occurring along the way. Today's manual transmissions, when driven optimally, have efficiencies of 90 to 95 percent, while the increasingly popular automatic transmissions enjoy lesser efficiencies of 80 to 85 percent. Some crankshaft energy—about 10 percent over a typical driving cycle—is also used to operate the vehicle's accessories, including its air conditioner and alternator.

Once power reaches a car's wheels, it must be used to overcome the aerodynamic drag of the vehicle (as determined by its drag coefficient, frontal area, and speed) and its rolling resistance (as determined by the friction of its tires, the road surface, and the vehicle's weight). In addition, energy is lost in braking. In urban driving, the fraction of energy dissipated by each of these functions is about equal—one-third each for overcoming aerodynamic drag, overcoming rolling resistance, and braking. On the highway, however, the fraction of energy used to overcome aerodynamic drag rises to more than 60 percent, while less than 10 percent is dissipated in braking.

Hence, the opportunities for reducing the fuel consumption of a vehicle lie in improving the efficiency of its engine and transmission; reducing its weight, aerodynamic drag, and rolling resistance; improving the efficiency of its accessories; and lessening the energy dissipation of braking.

New opportunities are now arising for improving the efficiency of engines. The advent of electronics and better sensor technology, for example, has enabled the development of ultra-lean-burn gasoline engines in which very lean air/fuel mixtures can be burned in low-load conditions; Toyota has already developed such an engine for commercial use, finding a 20 percent improvement in fuel economy.[2] Similarly, the development of ceramics has facilitated the evolution of "adiabatic" or "low-heat-rejection" diesel engines that may eventually obviate the need for any engine coolants. If the waste energy of the exhaust gases from such engines can be effectively recovered, fuel efficiency improvements of 30 to 60 percent are possible.[3] Technology strides are even bringing back the stratified engine for reevaluation. While such an engine is multifuel capable and enjoys an efficiency comparable to a diesel, its widespread use has been hampered in the past by high emissions rates. Improved combustion techniques, however, in combination with better electronic controls promise to reinvigorate interest in this engine, possibly as a two-stroke rather than conventional four-stroke machine. In fact, Ford Motor Company recently signed a licensing agreement with the inventor of a high technology two-stroke stratified charge engine.[4]

Similar opportunities for efficiency improvements are arising with transmissions. Electronics advances have enabled better coordinative control of the transmission with the engine so as to maintain the latter in its more efficient full-load state as much as possible. Even more advanced has been the development of better conventional automatic transmissions. These are characterized by having more gears—hence maintaining the engine in full-load for a larger fraction of operations—or by lock-up clutches on all gears—hence reducing frictional losses of the torque converters used in such transmissions. Most advanced, however, are continuously variable transmissions (CVTs) which have the potential to maintain the engine in full-load continually. CVTs are not new, but they have been dogged in the past by poor control methods, low power densities, and endurance problems. New electronics and materials advances have the promise to solve these difficulties.

Perhaps in no other arena will the technology revolution have a greater impact than with weight reduction, for a host of new materials have come forward with the desirable structural characteristics for use in a light vehicle. These range from ceramics to composite and polymer plastics to aluminum and magnesium alloys. The impact of these materials on weight reduction is substantial. One hundred pounds of steel replaced by the appropriate amount (to yield the same degree of strength) of aluminum yields a 55-pound reduction; replaced by a composite plastic a 65-pound reduction; and replaced by magnesium a 75-pound reduction.[5]

Of all these new materials, plastics has perhaps been the most widely publicized. Three automotive manufacturers, General Motors, Honda, and Renault, have already produced vehicles with body panels almost entirely composed of composite plastics, and work is proceeding on chassis composed of plastics. Work is also progressing in applying other materials to light vehicles. Nissan has already commercialized turbocharger blades composed of ceramics; not only is the weight of the resulting turbocharger substantially reduced, but its performance characteristics are noticeably improved. Audi is working cooperatively with Alcoa Aluminum to produce an aluminum body for its 5000 series sedan that will achieve at least a 10 percent overall weight reduction. Volvo has developed a prototype vehicle in which magnesium accounts for an unheard-of 7 percent of the vehicle's weight as compared with less than 1 percent for most of today's commercial vehicles.

Aerodynamics is another arena that has seen tremendous improvements in recent years. In the late 1970s, the average drag coefficient of American cars was about .48, with European vehicles achieving a slightly better .44.[6] In contrast, today the best drag efficient achieved by a production vehicle is .28.[7] And experimental vehicles are doing better still; Ford has developed an experimental vehicle, the Probe V, with a drag coefficient of .137, besting that of the F–15 military jet aircraft.[8]

Improvements in tires are also being pursued vigorously. New tire-making techniques developed by Goodyear and Bridgestone are delivering less rolling resistance while also improving steering precision, riding comfort, and braking performance. In addition, new materials for tires are being investigated, exemplified best by a small tire company in Kittsee, Austria, Polyair Maschinenbau GmbH, which has developed a tire made of liquid-injection-molded polyurethane. Tests conducted by the Venezuelan government have determined that cars with such tires use almost 10 percent less fuel.[9]

Light-vehicle accessories are not being ignored by the technology revolution. Energy demand by power-guzzling air conditioners is being reduced with better electronic control, the use of variable displacement compressors, and window glazings that decrease the amount of solar radiation into the vehicle. In addition, high efficiency alternators, electric power steering to replace energy-guzzling hydraulic versions, and low-wattage "electroluminescent" interior lighting are being energetically sought. Finally, efforts are being launched to substitute solar energy for the engine as the source of power for accessories.

Perhaps the greatest challenge in improving the energy efficiency of light vehicles lies in reducing the energy loss associated with braking. Volkswagen was one of the first companies to address this issue by seeking to tap the unused engine energy associated with deceleration and idling. Its prototype Glider Automatic uses the engine to charge a flywheel storage device; in decelerating or idling conditions, when sufficient energy is stored within the flywheel, the engine is then turned off and the flywheel used to power the necessary accessories and restart the engine when power is next required.

A more advanced concept than this approach has been under development at the University of Wisconsin, with support from a Toyota-based company. In this case, energy is stored whenever demand is less than what can be delivered at full-load conditions. Not only is this stored energy used instead of the engine in deceleration or idling, but it supplements the engine's power when additional boost is required. Finally, during braking, a large fraction of the kinetic energy of the wheels is recaptured and stored. Under some circumstances, this approach is believed to be able to improve the fuel economy of some light vehicles as much as 100 percent.[10]

The examples just listed are only a fraction of the research activity going on now. These are being pursued for reasons that extend well beyond their fuel economy benefits for they offer a multiplicity of benefits. Plastics, for example, are attractive not only because they are lightweight but because they can reduce assembly costs and have a longer lifetime than their corrosion-prone steel counterparts. Similarly, CVTs are attractive to many not only because they are more fuel efficient but because they offer a "jerk-free" ride compared to conventional automatic transmissions.

Inevitably, the question arises as to when these advanced technologies could reach commercialization. As already noted, some, such as plastic body panels, are already on production vehicles, albeit in limited numbers. Others, however, will require several more years of development. In truth, the rate at which these technologies will be introduced will depend heavily upon the aggressiveness of research and development efforts. Should an aggressive stance be taken, most of these technologies could probably be introduced into commercialization by the end of this century, with a few entering production in the early years of the 21st century (see Table 18.1).

But will these technologies really work in on-the-road vehicles? And what types of efficiency improvements could they produce in real vehicles? In the past these questions have been difficult to answer until the technologies were commercially available. However, answers have become noticeably easier to come by lately because of the plethora of fuel-efficient "prototype" vehicles that have been brought forward in recent years. As shown in Table 18.2, these prototypes have been introduced principally by European light-vehicle manufacturers, often at the instigation of their individual governments.

Generally speaking, these prototypes combine low aerodynamic drag with improved engine and transmission designs and innovative substitution of light-weight materials. As a result, with few exceptions, the vehicles enjoy city fuel economies in excess of 60 mpg and highway economies in excess of 75 mpg.

Moreover, even these innovative prototypes do not incorporate the most advanced fuel-economy technologies now under development. For example, not one makes use of a ceramic engine. Similarly, no such vehicle utilizes advanced energy storage and regenerative braking systems. Hence, the fuel-economy potential for light vehicles by the end of the century is probably even higher than reflected by these prototypes.

Not only has the development of fuel-efficient vehicles been useful in demonstrating what efficiency technologies can produce, but it has also illustrated that other consumer amenities need not be sacrificed to achieve high efficiency levels. Perhaps the best example of this can be seen in the prototype developed by Volvo, the LCP 2000.[11]

Volvo's engineers began their prototype project by setting minimum criteria for the LCP 2000 to ensure that it was safe, nonpolluting, comfortable, affordable, performed well, and could be manufactured easily. As a result, the vehicle, which has a fuel economy of 63 mpg city/81 mpg highway, can accelerate from 0 to 60 miles per hour in 11 seconds (the average for the 1986 American fleet was 13.1 seconds), and comfortably seat four to five persons. In addition, it was designed to have a greater crashworthiness than is required today in the United States of production automobiles—American safety requirements being the toughest in the world—and to have emissions levels close to American requirements—these again being the most stringent worldwide. Finally, the car was engineered such that the increased costs of its materials and associated technologies could be offset by a streamlined assembly process, a side benefit of the

Table 18.1

Estimated Dates of Production Readiness for Advanced Light-Vehicle Fuel-Efficient Technologies

1990	1995	2000
ENGINES		
Variable Turbocharger	Torch Ignition	Solid Lubricants
Pressure Wave Supercharger	1st Generation Ceramic Diesel	Advanced Ceramic Diesel
Variable Valve Engine	Advanced Variable Engine Displacement	Oxygen Separating Engine
	Dual Circuit Cooling	
Electronically Controlled Cooling		
Peak Pressure Ignition Sensor		
	"Clean" Stratified Charge	
Electronically Controlled IDI Diesel	"Clean" DI Diesel	
	"Clean" Ultra Lean Burn	
TRANSMISSIONS		
1st Generation ALL-Gear Lock-Up Clutch	2nd Generation ALL-Gear Lock-Up Clutch	
Electronically Controlled Automatic Transmission With Torque Converter		
	Electronically Controlled Automatic Transmission With Synchromesh Clutches	
Lubricated Belt CVT With Hydraulic Control	Electronically Controlled "Dry" Belt CVT	Beltless CVT

Table 18.1 (continued)

1990	1995	2000

MATERIALS

1990	1995	2000
Composite Body Panels	Polymer Body Panels	
Composite Load-Bearing Parts	Composite Chassis	
	Fiber-Reinforced Plastic Engine Parts	
Increased HSLA Steel Use		
Aluminum Load-Bearing Parts	Aluminum Body	
Fiber-Reinforced Aluminum Engine Parts		
Magnesium "Interior" Parts	Magnesium Engine, Load-Bearing Parts, And Wheels	
Ceramic Liners and Turbocharger Blades		

AERODYNAMICS

1990	1995	2000
Drag Coefficients In The Middle ".20s"	Drag Coefficients In The Low ".20s"	

TIRES

1990	1995	2000
New "Dynamic" Tire-Making Processes	"Plastic" Tires	

ACCESSORIES

1990	1995	2000
Variable Displacement Air Conditioner		Solar Power for Accessories
Advanced Electronically Controlled Air Conditioner		
Laminated Reflecting Film Auto Glass		
"Smart" Alternator High Efficiency Alternator		
Variable Power Steering Electric Power Steering		
Electroluminescent Lighting		

Source: Deborah L. Bleviss, The New Oil Crisis and Fuel Economy Technologies: Preparing the Light Transportation Industry for the 1990s (Westport, Conn.: Greenwood Press, Quorum Books, 1988).

Table 18.2
High Fuel Economy Prototype Vehicles

Company	General Motors	British Leyland	Volkswagen	Volkswagen	Volvo
Model	TPC (gasoline)	ECV-3 (gasoline)	Auto 2000 (diesel)	VW-E80 (diesel)	LCP 2000 (diesel)
Number of passengers	2	4-5	4-5	4	2-4
Aerodynamic Drag Coefficient	.31	.24-.25	.25	.35	.25-.28
Curb Weight (lb)	1040	1460	1716	1540	1555
Maximum Power (hp)	38	72	53	51	52, 88
Fuel Economy (mpg)	61 city 74 hwy.	41 city 52 hwy	63 city 71 hwy	74 city 99 hwy	63 city 81 hwy
Innovative Features	Aluinum body and engine.	High use of aluminum & plastics.	DI with plastic and aluminum parts, flywheel stop-start.	Modified DI 3-cyl. Polo, flywheel stop-start supercharger.	Hi magnesium use; 2 DI engines developed, 1 heat insulated.
Development Status	Prototype complete, no production plans.	Prototype complete.	Prototype complete.	Ongoing research, possibility of production.	Prototype complete, adaptable to production.

Source: Deborah L. Bleviss, *The New Oil Crisis and Fuel Economy Technologies.*

Table 18.2 (continued)

Renault	Renault	Peugeot	Peugeot	Ford	Toyota
EVE+	VESTA2	VERA+	ECO 2000	---	AXY
(diesel)	(gasoline)	(diesel)	(gasoline)	(diesel)	(diesel)
4-5	2-4	4-5	4	4-5	4-5
.225	.186	.22	.21	.40	.26
1880	1047	1740	990	1875	1430-target
50	27	50	28	40	56
63 city 81 hwy	78 city 107 hwy	55 city 87 hwy	70 city 77 hwy	57 city 92 hwy	89 city 110 hwy
Super- charged DI stop-start	High use of light material.	DI engine, high use of light materials.	2-cylinder engine, high use light material.	DI engine	Weight is 15% plastic 6% aluminum, has CVT & DI engine.
Prototype complete.	Program completed.	Ongoing development.	Ongoing develop- ment.	Research.	Ongoing development.

use of these materials. Hence, the LCP 2000's engineers estimate this vehicle could be manufactured at a cost comparable to today's cars of equivalent size.

LIMITATIONS OF THE MARKET IN EXPEDITING THE EFFICIENCY TRANSITION

The news on the technical front, therefore, is quite hopeful. Not only are new technologies coming forward that expand the technological frontier for fuel economy, but they are also increasingly enabling fuel economy to be achieved while meeting, and in some cases improving upon, other consumer amenities. However, the rate at which these technologies will be brought into commercialization is subject to considerable uncertainty, for the "market" acting alone is not providing much incentive for commercialization at the present time.

First, low oil prices are discouraging both consumer and manufacturer interest in fuel economy. While consumers have truly lost their love affair with the large car (contrary to recent press and manufacturer reports) their tastes are still switching to more energy-intensive options. Demand for subcompacts has dropped in favor of the larger compacts. Within each size class there is an increasing demand for engine performance.[12] As a result, six-cylinder engines, and, to some extent, eight-cylinder engines are making a comeback. In addition, demand has grown for energy-guzzling options such as air conditioning, automatic transmissions, and power windows.

Exacerbating the problem of low oil prices is the fact that as the fuel economy of the national fleet increases the economic impact made on a car purchaser of buying an even more efficient car diminishes. Trading in the 10 mpg car of a decade ago for a model that achieved 10 more mpg effectively meant doubling the owner's driving range, a noticeable change. However, trading in the 30 mpg car of today for one that achieves 10 more mpg only increases the driving range for the new car a third. Hence, regardless of the price of fuel, today's drivers are not receiving the same message on the virtues of fuel economy when they refuel.

Of course, oil prices need not be the only market force in operation. Competitive pressures can spur introduction of new technologies—many of which could well have fuel economy benefits—in order to gain market share. And indeed it does look as if in the highly competitive Japanese and European markets some new fuel-efficient technologies may move forward for this reason. But manufacturers do not bring forward new products unless they are fairly certain the demand for them will be there. Hence, to the degree they introduce fuel-efficient technologies, these innovations are likely to be first introduced in smaller car lines, where polls have repeatedly shown that buyers are more interested in fuel economy.[13] Perhaps the only technology to escape this stricture may be aerodynamic improvements in Europe. The high road-speeds on that continent mean that consumers are interested in aerodynamics, perhaps not so much for its impact on fuel economy as for the benefit it offers in reducing wind-related noise.

While competitive pressures may have some role to play in bringing forward some fuel-efficient technologies in Japan and Europe, similar activity is unlikely to occur in the United States. As already noted, to the degree these technologies are introduced, they will probably concentrate in smaller cars. Yet, the decreased market share anticipated for domestic companies in the years ahead is expected to lead to a significant reduction of small-car capacity by these companies in favor of large-car lines, where the competition is anticipated to be less intense. Hence, domestic companies will not have the very market segment where introduction of fuel-efficient technologies would be most attractive.

MARKET LIMITATIONS EVEN GREATER FOR ALTERNATIVE FUELS

While the market, under present conditions, is anticipated to have a limited role to play in encouraging fuel economy, its role in encouraging alternative fuels is likely to be even less. This is a critical point to consider, for devising a strategy that pushes alternative fuels even though market signals are leaning in the opposite direction is difficult enough. But pushing alternative fuels, when the market signals, weak as they may be, are showing a preference for fuel economy is an even more difficult task.

To begin with, even though low oil prices, in and of themselves, discourage the introduction of new technologies that improve fuel economy, the fact that these technologies often offer other benefits still maintains the interest of manufacturers. Even in the midst of the oil glut, for example, Subaru introduced a CVT on its mini Justy. The reason was not so much the benefit to fuel economy of this technology as the fact that it provides a more comfortable ride without the jerks that characterize other automatic transmissions. Similarly, plastics are increasingly being introduced in light vehicles in the United States. General Motors plans a plastic-paneled van in the next few years, and rumors abound once again that the new Saturn may also have plastic panels. The reasons that manufacturers are turning to plastics are principally that they are corrosion-resistant, thus longer-lived than their predecessors, and they can reduce the number of parts in a vehicle and thus assembly costs. The fact that plastics are also lighter weight, and therefore can improve fuel economy, has been almost incidental in the decision to use these materials.

In contrast, alternative-fuel vehicles rarely offer these multiple benefits. One benefit they do offer is better short-term air quality (in the case of compressed natural gas [CNG] or alcohol fuels). But questions abound as to whether consumers will show sufficient interest in this one benefit for manufacturers to justify the production of these vehicles. In truth, some type of nonmarket, governmental interest will probably be required before automakers can be persuaded to put alternative-fuel vehicles into production.

Another reason why the present market favors fuel efficiency over alternative fuels lies in the impact on vehicle operating costs. With few exceptions, even if the buying price of a vehicle is higher with a fuel-efficient technology, manufacturers still have a strong selling point in the fact that the operating costs of the vehicle, due to its fuel economy, are likely to go down. Yet, as a result of uncertainties in the price of alternative fuels, manufacturers of vehicles propelled by such fuels cannot make similar claims with any assurance. In fact, in some cases, in particular with alcohol fuels, operating costs could well rise.

Perhaps the greatest market limitation to the introduction of alternative fuels is consumer acceptance. The multiplicity of benefits that accompany many fuel-efficient technologies means that it is easier to persuade consumers of the value of the change being made in the vehicles they buy. Moreover, often they do not even know there is a new technology in their vehicle, just that its fuel economy is better than the one they had before.

But, with alternative fuels, every time a consumer pulls up to a filling station, he or she is reminded there is something different about the vehicle, that it is in some sense experimental. Hence, the consumer is often waiting for something to go wrong, just to blame it on the fuel. Early experience in the California methanol test fleet illustrates this characteristic; complaints about the frequent shifting required in the manual transmissions of these cars at commonly-used speed ranges were unjustifiably blamed on the fuel.[14]

Unfortunately, once consumers blame the fuel for whatever wrong, they are likely to shift their loyalties dramatically. Such has been the experience in numerous cases with both gasohol and diesel fuel. In the case of diesel, in particular, bad publicity about the General Motors diesel cars is credited with switching consumer loyalties away from diesels.[15]

INTEGRATING FUEL ECONOMY AND ALTERNATIVE FUELS

Clearly, regardless of the market limitations to alternative fuels, these fuels will definitely be required in the future as the world moves to a postpetroleum economy. And governments must begin planning now for that transition. But, as important as governmental involvement will be to alternative fuels planning, it will be just as important for fuel economy planning. Indeed, to effect the transition, governments will need to capitalize on market pressures for change as much as possible, pressures that will inevitably push fuel economy more than alternative fuels. For this and the following reasons, then, the continued pursuit of fuel economy will play a critical role in preparing for and indeed easing the transition to alternative fuels.

First, fuel economy will furnish the bridge for that transition. The passage to a postpetroleum era will not occur overnight. New infrastructures need to be built, market limitations overcome, consumer prejudices turned around. Indeed, it is highly unlikely that even with an aggressive alternative-fuels drive starting tomorrow that half of the petroleum used for transportation will be able to be displaced by the turn of the century. Yet, within precisely this time frame, a new oil crisis is increasingly expected to occur. The pursuit of improvements in the fuel efficiency of light vehicle fleets, which account for more than half of all transportation oil use, will play a critical role in maintaining a stable transportation system while the transition is being made. These improvements can help to maintain the "softness" of the oil market and hence reduce the risk of an oil shock.

Second, for the most likely near-term alternative fuels, namely methanol or CNG, improved fuel economy can increase vehicle range and thus make these alternatives more attractive to skeptical consumers. Of course, the range of these alternative-fueled vehicles can also be increased by increasing the size of the tanks, but such action would mean foregoing valuable trunk space, an option consumers are not likely to embrace.

Third, fuel economy can extend the lifetime of the resource base from which the alternative fuel is produced. Such a characteristic is particularly important for the alternative fuel most discussed today, methanol. At this time, methanol looks quite attractive because it can be produced from inexpensive natural gas at remote production sites around the world. However, since the disproportionate share of the world's reserves of natural gas can be found in the Middle East and the Soviet Union, eventually a dependence on natural gas-produced methanol will return the United States to dependence on strategically-insecure regions. In such an eventuality, the production of methanol from the United States' vast reserves of coal will look increasingly attractive, even though both the costs and environmental degradation will be considerably higher. Therefore, it is important to extend the lifetime of the secure natural gas reserves as long as possible. Improving the efficiency of the light vehicle fleet is critical to such an effort.

Fourth, energy efficiency can reduce the greenhouse effect resulting from the production and use of alternative fuels. This is an especially important consideration for some of the most likely near-term fuels, methanol and natural gas. Rising levels of carbon dioxide (CO_2) gas are believed to be causing unprecedented increases in global temperatures. As a result, a consensus is slowly building that action should be taken to reduce the overall emissions rate of this gas. However, the use of CNG reduces the rate of CO_2 production only marginally relative to gasoline, and the use of methanol from natural gas produces about the same amount of this gas.[16] Taking into consideration the expected increase in both the number of light vehicles in this country as well as the number of miles driven annually per vehicle, this means that the use of either of these two fuels is unlikely to yield even a stabilization of CO_2 emissions from the transportation sector, much less a reduction. Only continued and aggressive pursuit of fuel economy will render that desired result.

Finally, considering the strategic and environmental difficulties associated with a transition to near-term fuels, fuel economy offers the option of "buying time" until more benign alternative fuels options can be brought on-line. As already noted, the use of CNG and methanol from natural gas may ultimately return the

United States to a dependence on strategically undesirable suppliers. Moreover, these fuels pose concerns about their impact on global warming. Alternatively, methanol from coal, while obviating the problem of reliance on undesirable suppliers, is much more expensive to produce and releases twice as much CO_2 as methanol from natural gas, not to mention other assorted environmental problems.[16] And ethanol, another possibility, is both expensive to produce and requires diversion of food production.

But there are more benign alternatives, although none will be ready in the short term. These include hydrogen, methanol from biomass, and electric vehicles. An aggressive program of fuel economy, which is certainly feasible technically, could buy the time needed to bring these alternatives on-line without the disruption of another oil crisis, rather than suffering through the problems associated with today's near-term fuels. For illustrative purposes, if the United States exercised its leadership such that the world light-vehicle fleet approached 45 mpg for cars and 35 mpg for light trucks by the end of the century, the expected savings would more than compensate for the projected drop in industrialized-country oil production capability by that time. Thus, the impact of increased dependence on Middle East oil would be diluted. If the world light vehicle fleet approached 60 mpg for cars and 45 mpg for light trucks by the year 2000, the percentage of production capacity being pumped by OPEC, which otherwise could exceed 90 percent by that time, could drop below 80 percent, the threshold that has historically characterized a stable oil market (see Figure 18.1).

CONCLUSION

As this country approaches the last decade of this century with rather bleak forecasts for the oil future, it is important to study policy options carefully and rationally. The travesty of the experience with the Synthetic Fuels Corporation illustrates the perils of acting otherwise.

Clearly, this country must start preparations for the transition to a nonpetroleum-based era. This means developing alternative fuels, particularly for the almost wholly dependent transportation sector. But these fuels must be selected prudently so as to minimize unwanted and unintended future problems for the environment and national security.

The continued pursuit of fuel economy for the transportation fleet, particularly for light vehicles, will have to play a major role in the transition. While the subject of fuel economy has unfortunately been highly politicized in recent years, the fact is that improving the fuel efficiency of light vehicles is technologically feasible, has fewer market constraints than alternative fuels, and can mitigate some of the problems associated with the use of alternative fuels. To ignore this option would be a major disservice to the nation and the world as a whole.

Figure 18.1
Non–Eastern Bloc Light-Vehicle Fuel Use for Various Levels of Fuel Economy

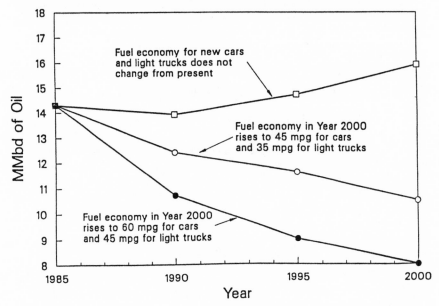

NOTES

1. D. E. Cole and J. H. Johnson, *Spark Ignition and Diesel Engines as Applied to Passenger Cars and Light Trucks,* working paper prepared for the U.S. Congressional Office of Technology Assessment, December 4, 1979.

2. Takashi Kamo et al., "Lean Mixture Sensor," Society of Automotive Engineers paper #850380, 1985.

3. Isuzu is working on a first-generation ceramic engine, which it claims will improve fuel economy 30 percent (*Automotive News,* October 20, 1985). Recent analysis by Energy and Environmental Analysis, Inc. for the Office of Technology Assessment places the potential for fuel economy improvements from a truly adiabatic engine at 60 percent (*Developments in the Fuel Economy of Light-Duty Highway Vehicles,* June 1988).

4. Roger Rowand, "The 2-Stroke," *Automotive News,* July 4, 1988, p. 1.

5. Merton C. Flemings et al., *Materials Substitution and Development for the Light Weight, Energy Efficient Automobile,* prepared for the U.S. Congressional Office of Technology Assessment, February 8, 1980.

6. R. Buchheim et al., "Contribution of Aerodynamics to Fuel Economy Improvements in Future Passenger Cars," *Proceedings of the First International Automotive Fuel Economy Research Conference, 1979,* U.S. Department of Transportation, 1979.

7. This drag coefficient characterizes the Renault 25.

8. *Probe V News Conference,* October 28, 1985, Tokyo, Japan.

9. Stuart Marshall, "First Test Report on the Revolutionary Plastic Tire," *Popular Science,* April 1983.

10. Steve M. Rohde and Neil A. Schilke, "The Fuel Economy Potential of Heat

Engine/Flywheel Hybrid Automobiles," *1980 Flywheel Technology Symposium Supplement*, October 1980.

11. Deborah Bleviss, *The New Oil Crises and Fuel Economy Technologies: Preparing the Light Transportation Industry for the 1990s*, Westport, Connecticut: Greenwood Press, 1988).

12. Phil Patterson, U.S. Department of Energy, unpublished data.

13. J. D. Powers poll, reported in U.S. Department of Energy Monthly Data Report, no. 27, February 4, 1986.

14. Conversation with Ken Smith and Dan Fong, California Energy Commission, April 1985.

15. Kenneth Kurani and Daniel Sperling, "The Rise and Fall of Diesel Cars: A Consumer Choice Analysis," *Transportation Research Record*, 1175 (1988): 23–32.

16. M. A. DeLuchi, R. A. Johnston, and Daniel Sperling, "Transportation Fuels and the Greenhouse Effect," *Transportation Research Record*, 1175 (1988).

19

Conclusion

Fuels different from gasoline and diesel fuel will not enter the marketplace on the basis of lower costs in the foreseeable future. Even if they were less expensive to produce, distribute, and use than petroleum fuels, they still would be hindered by start-up barriers. These barriers must be reduced if new fuels are to overcome resistance and enter the marketplace; that is the challenge facing government. These barriers result from the conservative and disjoint nature of transportation energy systems. These barriers can be categorized in two groups: those confronting vehicle and fuel users and those confronting manufacturers and suppliers of fuels and vehicles.

An underlying factor contributing to the resistance to new fuels in the United States magnifying the importance of the barriers is the stagnant gasoline market. Gasoline sales are expected to remain constant or decrease slightly at least for the next 10 years or so and probably much longer, depending on the rate of improvement in vehicular fuel efficiency. The stagnant gasoline market discourages energy companies from taking the risk of investing in fuels dissimilar to petroleum. It results in a "not-me-first" phenomenon.

Unlike investors in growing markets such as computers in the 1980s or petroleum through the 1970s, investors in energy companies have few opportunities to create new markets and new demand. The risk for the first company that enters the market is huge. The risk for later companies is lower, since they can wait to see if a market develops. Due to the homogeneity of the product, it is unlikely that the first company will be able to gain and retain a large share of the alternative-fuels market solely because it was first into the market. Thus the risk of being first is large, while the potential benefits are modest at best.

Reluctance to be first into the market also extends to vehicle manufacturers. The absence of large rewards to vehicle manufacturers is due to two factors: it is widely known how to build alcohol and CNG vehicles, and initial nonpetroleum vehicle sales will undoubtedly be small because of limited fuel availability.

Slow market-penetration is due to the appropriate conservativeness of vehicle owners. Most vehicle owners, including fleet operators, are reluctant to purchase a new type of vehicle unless the risk is minimal. In the case of an alcohol or gaseous fuel vehicle, this risk may involve the possibility of fuel shortages and a high fuel price in the future, low resale value of the vehicle, and technical failings that result in high repair bills, safety problems, inferior performance, poor fuel efficiency, starting problems, and high noise levels. In addition, the shorter driving range of alcohol, gaseous fuel, and battery-powered vehicles also detracts from their attractiveness.

The initial absence of a network of retail outlets serving the respective fuels and the slowness in expanding the network play a particularly important role in discouraging the purchase of nonpetroleum-fueled vehicles. Based on an analysis of the U.S. diesel car market, one study concluded that at least 10 percent or so of all fuel stations, representing over 10,000 nationally, would need to provide a new fuel before a significant number of people would purchase a vehicle that operates on that fuel.[1] The sale of a multifuel vehicle would mitigate but not overcome this disincentive to purchase a nonpetroleum vehicle.

Virtually all evidence indicates that significant numbers of individuals and fleets will *not* purchase a nonpetroleum vehicle unless there is a strong economic incentive. They need that incentive to offset the various uncertainties and negative attributes associated with those vehicles. For instance, as indicated in earlier chapters, even though compressed natural gas was available at half to three-quarters the price of gasoline in New Zealand and Canada, market penetration was slow. In the United States, diesel cars increased their share of new car sales from practically zero in 1976 to only 6 percent in 1981, even with a fuel cost advantage of about 25 percent.[2] The life-cycle cost advantage was less than 25 percent, but still substantially greater than zero because diesel cars cost somewhat more than gasoline cars, but what is surprising is that market penetration was not greater.[3] These CNG and diesel experiences illustrate the conservative nature of consumers.

JUSTIFICATIONS FOR GOVERNMENT INTERVENTION

Active government involvement is absolutely critical, not only because of the large start-up barriers but also because energy markets fail to incorporate the four important societal concerns (''imperatives'') identified in Chapter 1: energy security, indirect economic costs, global warming, and air pollution. All of these are important, but their role in introducing alternative fuels is best explained in political and institutional terms.

As indicated in Chapter 2, the energy supply issue is not one of availability but of dependence on foreign suppliers. To what extent this energy security issue is a problem is difficult to determine. It depends on a multitude of factors. If, for instance, foreign suppliers are highly reliable, it is less of a problem, while if domestic alternatives are expensive, it is more of a problem. One factor exacerbating the problem from the point of view of importers is that much of the exported oil, especially from the Persian Gulf, costs only a few dollars per barrel to produce, giving exporting nations considerable discretion in pricing. As a result, oil-importing nations must have a firm resolve in any attempt to introduce alternative fuels, because oil exporters in the Persian Gulf can easily price oil below the cost of any alternative fuel and still make a profit.

A resolve to reduce oil imports did form in the United States in 1973–74 and 1979–81, but quickly dissolved each time because it was based on energy-security concerns for which there was no powerful constituency or interest group. The oil industry, the only natural constituent, is composed of integrated multinational corporations that now have large holdings not only in petroleum but also in many other energy resources. The U.S. oil industry is not tied to domestic production and the domestic market. The oil industry is not a strong advocate of energy independence, and there is no other advocate. Energy security, while still an important concern, is therefore not likely to play a major role in forming the political resolve to support new fuels. The fact that methanol and perhaps CNG will be imported from foreign sources further undermines the role of energy security concerns.

Indirect economic concerns have an even less influential constituency because the costs and benefits are diffuse and so difficult to measure. The measurement problem is one of determining the impact on world oil prices of replacing, say, 1 million barrels per day of oil with alternative fuels. How much would that lower world oil prices? What effect would that have on the volatility of oil prices? Even if these benefits of lower and less volatile oil prices could be reliably and convincingly measured, the benefits would be dispersed among many individuals and organizations. As McNutt points out in Chapter 16, no single group or set of organizations would benefit enough from these indirect oil-price changes for them to take a strong position in support of alternative fuels.

A more broad-based concern with a potentially much larger and more influential constituency is air pollution and global warming. The many people living in polluted areas and the powerful environmental lobbying groups who articulate their concerns are the air pollution constituency. The constituency for reduced global warming is even broader, cutting across the entire population and including not only environmentalists but urban dwellers near the seacoast, farmers who would be affected by changing climatic patterns, residents of the western U.S. who depend on precarious annual rainfall and snowfall for water supply, and perhaps many others. In the near term the driving force for introducing new fuels will be air quality and perhaps the greenhouse effect, not energy security or economic concerns.

NEAR-TERM PROGNOSIS (TO 2005)

While the future for alternative fuels remains highly uncertain, certain trends and forces, as noted above, are clear. During this near-term time period, some pressure to introduce alternative fuels will be felt as oil imports increase in the United States and elsewhere, especially in less-developed countries. Increasing imports will create unease because of growing dependency on foreign suppliers but will also have the eventual effect of causing the oil market to tighten, creating greater awareness of the economic costs of inaction. This market tightening may occur by the mid- to latter–1990s.

Global warming caused by the greenhouse effect will also create political pressure during this time period to replace petroleum with alternative fuels. The effect of alternative fuels on global warming is more difficult to discern, however, principally because the most attractive near-term fuel alternatives, methanol and CNG, produce almost as much greenhouse gases as petroleum fuels.

The most important force in the near-term future, however, as indicated above, is likely to be air quality rules and regulations—particulate emission limits for buses and trucks and incentives and mandates for reducing carbon monoxide and ozone in areas not attaining ambient standards. Even if concerns for import dependency, economic inefficiency, and global warming prove to be more prominent than air quality concerns, the existence of a well-established regulatory framework and institutional structure for air quality control allows air quality control initiatives to bypass the difficult consensus-building and institution-building process faced by those other concerns.

Driven by diesel emission standards and general concern for air quality in several regions, small amounts of methanol and CNG will be used in transit buses (whose total U.S. market is only 30,000 barrels per day) and centrally fueled urban truck fleets. Larger amounts of methanol and CNG may be used in certain regional markets with severe CO and ozone problems including Denver and New York, but especially in California. Severe ozone problems in Los Angeles, Sacramento, and other metropolitan areas in California, the strong commitment to environmental quality in that state, and the existence of powerful air quality regulators statewide and in Los Angeles, suggest that methanol and to a lesser extent CNG and electric vehicles are likely to gain some share of the California market in the near future.

Meanwhile, in this near-term future, corn-based ethanol, an alternative briefly addressed in this book, will probably continue to be used in the United States as a blending agent with gasoline, slowly increasing its market penetration beyond the late–1980s level of 55,000 barrels per day. Ethanol is much more expensive than most other options, however, and will not play an important role in supplying transportation energy.[4] Its continued use will be due strictly to generous subsidies resulting from continued lobbying by politically powerful agricultural interests.

Electric vehicles will also have a presence. Promotional activities by the Electric Vehicle Development Corporation (see Chapter 14) will assure some market penetration, initially in vans operated by large urban-based fleets, especially in more polluted cities where government incentives are made available. Market penetration will be modest, however, because the performance of electric vehicles is greatly inferior to internal combustion engine vehicles.

By 2005, alternative fuels, including methanol, CNG, electricity, and ethanol may account for a substantial share of the California highway fuels market and have a sizable presence elsewhere. CNG, methanol, and electric vehicles will be used in areas where ozone and carbon monoxide standards are being violated; CNG would be preferred in inland areas served by natural gas pipeline and in cold climates, while methanol will be more attractive near coastal and inland ports in mild climates. A lower penetration level would occur if legislators and regulators are passive; a higher level if strong incentives and/or mandates are instituted.

Strong government initiatives to support plausible alternative fuels during this period include the following: modified vehicle fuel-efficiency standards that count only the gasoline consumed and thereby encourage manufacturers to market alternative-fueled vehicles; requirements that urban-based fleet operators convert their vehicles to clean-burning alternative fuels; and much lower vehicle emission standards that may be more easily met by nonpetroleum vehicles.

During this transition period through 2005, the difficulty and cost of establishing a network of retail outlets serving alternative fuels will result in virtually all methanol and CNG use being in multifuel vehicles that also run on gasoline. These vehicles will not be optimized for CNG nor methanol. To assure that drivers actually use methanol and CNG fuel in their multifuel vehicles, it will be necessary to keep the price of the alternative fuel below the price of gasoline on a mileage-equivalent basis.

In conclusion, it is likely that by the turn of the century, alternative fuels may be used in diesel engines and in a few regional markets where air quality is a problem, principally the Los Angeles region but also several others.

LONG-TERM OUTLOOK (2020 AND BEYOND)

By 2020 a full-scale transition to alternative fuels will likely be in progress. Methanol and natural gas fuels will be the primary alternatives, but electric vehicles will be common, and initial investments in oil shale and perhaps coal liquids will likely be proceeding. In southern California, where ozone standards will still be violated, it is plausible that virtually all new vehicles, representing sales of over one million vehicles per year, would be designed to operate on methanol, CNG, or electricity. Unlike in the past, a multiplicity of transportation fuels will likely prevail, sometimes several in the same local market.

The less-efficient multifuel vehicles will by then have been superseded by low-polluting, efficient, single-fuel vehicles optimized to either methanol or CNG (or LNG).

By 2020, air quality will likely be supplanted as the primary force driving the introduction of alternative fuels except in areas still in nonattainment. High oil prices and concerns over growing oil imports and vulnerability to erratic market behavior will be the more powerful forces.

If global warming continues, as expected, there will also be strong pressure to switch to biomass and nonfossil fuels, including methanol made from wood, hydrogen from water using cleanly generated electricity, and perhaps electrified roadways. Increased attention to the greenhouse effect may be the catalyst that creates a resolve in government to support alternative fuels.

TOWARD AN ALTERNATIVE-FUEL POLICY

The debate surrounding alternative fuels, as reflected in this book, is not *if* alternative fuels are needed, but rather when and how they should be introduced. There is a widespread commitment to two fundamental goals that support alternative fuels: reduced dependency on oil imports and reduced urban air pollution. Growing commitment to a third goal, reduced production of greenhouse gases, will lead to further support for alternative fuels.

The central question, though, for which there is not a simple answer is: How pressing is it that new clean-burning fuels be introduced? The interested parties in this debate will never fully agree on what and when. The reason is that accurate measurements of the cost of oil-import dependency, the cost-effectiveness of air quality control strategies, and the relationship between greenhouse gas emissions and global warming do not and cannot exist; sufficient data do not exist, relationships are not well understood, the future cannot be accurately predicted, and individuals and organizations have different values and beliefs.

There are many groups with differing interests in alternative fuels. The oil industry wants to remain a profitable marketer of transportation fuels but does not want to make large premature investments in a fuel that consumers will not buy; likewise, the automobile industry is fearful of marketing a vehicle that remains unsold. Electricity and natural gas suppliers would like to expand their sales. The U.S. Department of Energy is concerned about national energy security and the State of California about economical supplies of cleaner-burning fuels. Environmental public interest groups emphasize the importance of dramatically improving urban air quality and averting a greenhouse effect.

I would characterize the major participants and interest groups in the United States as moving toward the following common beliefs. Most believe that methanol is the most acceptable alternative and that compressed natural gas may play an important role. Electric vehicles are seen as a minor option that may be attractive in small niche markets where air pollution is especially severe. There is also widespread agreement among policy analysts and representatives from both the public and private sectors that the existence of various barriers and large start-up costs inhibits the private sector from taking the initiative—that government must take the lead role initially. This belief is seen as appropriate given the large market externalities involved: air pollution, energy security, and the greenhouse effect.

In summary, there remains disagreement over how fast we should be moving away from petroleum and how strong a role government should play, but there is a commitment to the overall goals of environmental quality and national energy security, and a widespread belief that government must play a prominent role. While premature or inefficient government intervention may incur large costs for a region or nation, tardiness also carries a large cost, economically and environmentally.

NOTES

1. D. Sperling and K. S. Kurani, "Refueling and the vehicle purchase decision: the diesel car case," *SAE* 87064 (1987).

2. K. S. Kurani and D. Sperling, "The rise and fall of diesel cars: a consumer choice analysis," *Transportation Research Record*, 1175 (1988): 23–32.

3. D. Greene, "The market share of diesel cars in the USA, 1979–83," *Energy Economics* 8:1 (1986): 13–21.

4. D. Sperling, *New Transportation Fuels: A Strategic Approach to Technological Change* (Berkeley, Calif.: University of California Press, 1988).

Index

Acid rain, 106

Air pollution, 247, 249, 251, 259, 278–81, 311–14; benzene, 104; carbon monoxide, 83, 106, 110, 112, 117–42, 279–81; formaldehyde, 104, 110, 112, 117–42, 279–81; hydrocarbon, 110, 111, 114–16, 125, 127–28, 136, 138–42, 279–81; lead, 110; nitrogen oxides, 89, 90, 97, 104–5, 110, 112, 117–42, 279–81; ozone, 92–93, 101, 110–12, 114, 121, 127–28, 138–42, 275, 279–81, 312; particulates, 104, 110, 112, 128–32, 136; sulfur oxides, 90. *See also* Ambient air quality standards; Emission standards; Greenhouse effect

Alcoa Aluminum, 295

Ambient air quality standards, 6, 110–12, 250, 275, 279, 312

American Gas Association, 122, 123, 125, 228, 285

Atlantic Richfield (ARCO), 104, 249

Audi, 295

Automobile industry: Brazilian, 194; Canadian, 151; United States, 217. *See also under specific automobile manufacturers*

Benzene. *See* Air pollution, benzene

Biomass, 106, 314

Bitumens, 18

Blending. *See* Motor vehicles, multifuel *and specific blends*

Brazil, 163–85, 192–97, 210, 284, 285

Bridgestone, 296

Brooklyn Union Gas (BUG), 134, 135, 137

Buses: compressed natural gas, 197, 227; diesel, 102, 112, 128–137; methanol, 132–34, 136, 248

CAFE. *See* Corporate average fleet economy rules

California, 260–74; Air Resources Board, 122–23; Energy Commission (CEC), 8, 79, 104, 247–60, 273, 275; Legislature SB 620 (Mills), 248; Los Angeles, 101–17, 247, 279; Riverside County, 101, 104

Canada, 145–62, 197–210, 225, 275, 283

Canadian Gas Association, 159, 161

Carbon dioxide. *See* Greenhouse effect

Carbon monoxide. *See* Emissions, carbon monoxide

Carcinogens, 102, 104, 121

Cars. *See* Motor vehicles

Catalysts, 133–34

Catalytic converters, 118, 119, 132–34

CEC. *See* California Energy Commission

Chevron, 51–63, 104, 249

"Chicken-and-egg" problem, 61

Chloride EV Systems, 236

Chrysler, 236, 240

Clean Air Act, and amendments, 132, 247, 266

CNG. *See* Compressed natural gas

Coal: conversion to methanol, 62, 306; as a feedstock, 2, 83, 277, 313; liquefaction, 212–13, 274–75

Colorado, 104

Compressed gaseous hydrogen. *See* Hydrogen

Compressed natural gas (CNG), 273–88, 312–14; government intervention, 150–62; start-up barriers, 136–37, 221–22, 281–83, 287

Compressed natural gas (CNG) fuel, 65–69, 80–81; distribution and retail, 44–47, 152–53, 156, 190, 200, 210, 230–31; production, 40–46; transportation, 40–46, 210. *See also* Natural gas, as feedstock

Compressed natural gas (CNG) vehicles: air pollution impact, 114–16, 122–28, 138–42, 278–81; consumer cost/benefit, 155, 189, 201, 277–78; demonstration projects, 227–28, 248, 259, 260; driving range, 125, 127, 210; efficiency/performance, 122, 126–27, 211, 305; replacing diesel, 69, 134–37, 159, 197; research and development, 226–27, 231; single versus multi-fuel, 122–28, 277–88

Compression ignition engine. *See* Diesel engines

Consumer perspectives and behavior, 149, 249–53, 260, 304

Conversion of automobiles. *See* Retrofitting motor vehicles

Corn, 104, 312

Corporate average fleet economy (CAFE) rules, 214, 231, 258–59, 267, 270

Costs, capital. *See* Return on investment; Costs, production; *and specific fuel production costs, transportation costs and distribution costs*

Costs, marginal supply curves, 21–22, 29–48

Costs, opportunity, 30

Costs, private, 97. *See also* Costs, production

Costs, production, 51–60, 62–63, 65–67, 80, 92–95, 97. *See also particular kinds of fuel or vehicle*

Costs, societal, 3, 97, 271, 310. *See also* Air pollution; Greenhouse effect; Import dependency

Crude oil. *See* Oil

Cummins, 134

Denver, Colorado, 104

Detroit Diesel Corporation (DDC), 131–34, 136–37

Detroit Edison, 239

Developing countries: capital needs, 208, 215; food versus fuel, 212; fuel supply, 24–25, 207–8; fuel use, 206; market restrictions, 217

Diesel: air pollution, 102, 112, 128–37, 284–85, 287; blending, 212; cetane rating, 132, 134; General Motors, 304; research and development, 294; Stirling engine, 213

DOE. *See* United States Department of Energy

Driving range: compressed natural gas, 125, 127, 210; electric, 103, 242; methanol, 121, 136, 210, 248, 249, 252

Dual fuel system. *See* Motor vehicles, multifuel

Electricity generation, 46

Electric Power Research Institute (EPRI), 235, 236, 240, 244

Electric utilities, 235, 239, 240

Electric Vehicle Development Corporation (EVDC), 235–45

Electric vehicles: air pollution impact, 244, 287, 288, 312, 314; commercialization strategy, 236–37; demonstration projects, 103, 105, 237–39, 248, 259, 260; driving range, 103, 242; fleets, 236, 242; government intervention,

244–45; Griffon, 237–39; G-Van, 239–42; research and development, 103, 235, 244; service and support, 238–39, 240–41; TEVan, 240
Electrified roadways, 314
Electrolysis, 84. *See also* Hydrogen
Emissions. *See* Air pollution
Emission standards: 1981 passenger car, 113–14; 1991 bus, 69, 129; 1994 truck, 129
Energy independence. *See* Import dependency
Energy security. *See* Import dependency
Enhanced Oil Recovery (EOR), 18
Environmental regulation. *See* Ambient air quality standards; Emission standards
Ethanol, 163–65; government intervention, 104, 167–69, 193–97
Ethanol fuel: distribution and retail, 167; food versus fuel, 196, 212; production, 52, 62, 167, 169–72, 306; transportation, 172
Ethanol vehicles, 284; air pollution impact, 104; consumer cost/benefit, 177, 194, 274; efficiency, 194–95; replacing diesel, 165; single versus multifuel, 193–94
Exploration and development. *See particular energy resources*
Externalities. *See* Costs, societal

Federal Energy Regulatory Commission (FERC), 226
Fleets. *See* Motor vehicles, fleets
Florida Power Company, 239
Food versus fuel issue, 196, 212
Ford, 104, 248, 251, 273, 284, 285, 295; compressed natural gas vehicles, 125–27; electric vehicles, 236; methanol vehicles, 104, 248, 251, 284, 285; preference for methanol, 273; stratified charge engine, 295
Formaldehyde. *See* Air pollution, formaldehyde
Fuel availability. *See* Retail fuel outlets
Fuel cells, 102

Fuel dispensing pumps. *See* Retail fuel outlets
Fuel distribution systems and fuel distributors, 66–81
Fuel efficiency and economy of vehicles, 61–62
Fuel-flexible vehicle. *See* Motor vehicles, multifuel

Gas bubble, 224
Gas companies. *See* Natural gas companies
Gas flaring, 61
Gas guzzler tax, 266, 270
Gas Research Institute, 226
Gasoline: air pollution effect, 112–14; consumer cost/benefit, 277–78; efficiency/performance, 17, 293–306; marginal cost supply curve, 147–48; market, 309; prices, 17; supply, 21; synthetic, 274
Gasoline additives, 104
General Motors: diesel engine, 304; electric vehicles, 236–39; ethanol vehicles, 285; plastic panel vehicle body, 295, 303; preference for methanol, 273; prototypes, 104
Global warming. *See* Greenhouse effect
Goodyear, 296
Government intervention, 6, 103, 258–60, 265–71, 286–88, 303, 304, 310–15; compressed natural gas vehicles, 150–62; electric vehicles, 244–45; ethanol vehicles, 104, 167–69, 193–97; liquefied petroleum gas vehicles, 150–62; methanol vehicles, 247–60; natural gas industry, 222–33, 286. *See also* Ambient air quality standards; Emission standards; Level playing field
Greenhouse effect, 5–7, 83, 90–93, 104, 106, 107, 281, 305, 306, 311–14

Heavy oil, 18
Honda, 295
Hotelling Principle, 31–32
Hydrocarbon emissions. *See* Air pollution, hydrocarbon
Hydrogen, 104, 106

Hydrogen fuel: compressed gaseous, 87; distribution and retail, 85–87, 96; from fossil fuels, 83–84, 90, 92–94; gasoline blends, 83, 87; production/costs, 92–95; transportation, 85, 96; from water, 84, 92–94

Hydrogen vehicles: air pollution effect, 89–93, 97; consumer cost/benefit, 92–94; efficiency/performance, 87–88, 96; research and development, 96–97; safety, 86, 88–89, 96; storage/fueling, 86, 96; storage in hydrides, 85–87, 89, 94

Import dependency, 3–4, 164, 247, 263–71, 274, 278, 305, 306, 310–11, 313–14

Incentives. See Government intervention

Italy, 210

Kentucky Agricultural Energy Corporation, 62

Lents, James, 101–2, 104, 105

Level playing field, 52, 228. See also Costs, societal; Government intervention

Liquefied natural gas, 40–46

Liquefied petroleum gas, 150–62

LNG. See Liquefied natural gas

LPG. See Liquefied petroleum gas

M.A.N., 132, 137

Market externalities. See Costs, societal

Market research, 242. See also Consumer perspectives and behavior

Mass transit. See Buses

Metal hydrides. See Hydrogen vehicles, storage in hydrides

Methane. See Compressed natural gas

Methanol, 273–88, 312–14; government intervention, 247–60; start-up barriers, 136–37, 281–83

Methanol fuel: from coal, 2, 62, 306; fuel distribution and retail, 44–47, 78–81, 136, 248, 250–52, 257, 258; from natural gas, 2, 40–48, 71; production/

production costs, 58–59, 62; transportation, 40–46, 69–77, 251

Methanol vehicles: air pollution effect, 102–4, 114–22, 132–34, 138–42, 247, 249, 251, 258, 259, 278–81; buses, 132–34, 136, 248; cold start problems, 121–22, 136, 206, 211, 217; consumer cost/benefit, 27–28; conversions or retrofits, 117–19, 250; demonstration projects, 248, 258; driving range, 121, 136, 210, 248, 249, 252; efficiency/performance, 46, 206, 214, 252, 305; fleets, 251; fuel, flexible (or variable), 248, 250–52; gasoline blends, 78–79, 117–19, 121; prices, 253–58; single fuel versus multi-fuel, 127–28, 277–88, 313; trucks, 258

Methyl-t-butyl-ether (MTBE), 104

Motor vehicles. See specific fuel

Motor vehicles, fleets, 68, 78–79, 236, 242, 251, 310

Motor vehicles, multifuel, 122–28, 277–88, 313

MTBE. See Methyl-t-butyl-ether

Natural gas, as fuel. See Compressed natural gas fuel

Natural gas, as feedstock: curtailments, 274; economic rent, 30; excess production capability, 224; fields in Brazil, 197; marginal cost supply curves, 21–22, 29–48, 223; market price, 29–30; price elasticity of supply, 286; production costs, 32–40, 223–25; production forecasting, 27–40, 223–25; remote, 2, 53–58; reserves, proven, 23–26, 146, 209, 275–77; resources, unconventional, 224

Natural gas companies, 150, 159, 286

Natural Gas Policy Act of 1978, 222, 226

Natural Gas Vehicle Coalition, 221

New Zealand, 188–92, 212, 275, 283

Nissan, 295

Nitrogen oxide emissions. See Air pollution, nitrogen oxides

Nonattainment regions, 111, 112

Ocean transport, 41–46, 71–77, 251
Oil: forecasting, supply, 11–13; price controls, 13–14; prices, 11–17, 19, 208–9; production, 15–16, 18; reserves, 15; supply disruption, 17–18, 208
Oil industry, 311. *See also specific oil companies*
Oil sands, 18
Oil shale and shale oil, 2, 52, 62, 277, 313
OPEC (Organization of Petroleum Exporting Countries), 3, 12, 18, 19, 30, 206, 209, 216, 278, 306
Ozone. *See* Air pollution, ozone

Pacific Energy Resources, 106
Particulates. *See* Air pollution, particulates
Pennsylvania, Port Authority of Allegheny County, 259
Petroleum, 1–2, 273. *See also* Gasoline; Oil
Photochemical oxidants. *See* Air pollution, ozone
Photolysis, 84
Photovoltaics: for electric vehicle power, 103, 106; for hydrogen fuel production, 84, 97; to power accessories, 296
Pipeline systems: hydrogen, 85, 96; methanol, 70–71, 76; natural gas, 66–67, 210
Polyair Maschinenbau GmbH, 296
Powerplex Technologies, Inc., 236
Price volatility, 3
Propane, 150–62

Rail transport, 71, 74–77, 85
Reactive organic emissions. *See* Air pollution, hydrocarbons
Regulation. *See* Ambient air quality; Corporate average fuel economy; Nonattainment regions
RNG. *See* Natural gas, remote
Research and development (R&D), 259–60; coal liquefaction, 274, 288; compressed natural gas, 161, 226–27, 231; electric vehicles, 235, 244; hydrogen,

96–97; government funding, 231, 267; liquefied petroleum gas, 157
Retail fuel outlets, 281–83, 287, 310, 313; compressed natural gas, 44–47, 152–53, 156, 190, 200, 210, 230–31, 287; ethanol, 167; gasoline, 102; hydrogen, 85–87; methanol, 44–47, 78–81, 136, 248–53, 257, 258, 287
Retrofitting motor vehicles, 61–62
Return on investment, 210, 217, 242, 287

Self-ignitability. *See* Diesel engines
Service stations. *See* Retail fuel outlets
SFC. *See* United States Synthetic Fuel Corporation
Shale oil. *See* Oil shale and shale oil
Smog. *See* Air Pollution, ozone
Smoke. *See* Emissions, particulates
SNG. *See* Substitute natural gas
Social costs. *See* Costs, societal
Solar energy. *See* Photovoltaics
South Coast Air Basin (SCAB), 247, 250. *See also* California, Los Angeles; South Coast Air Quality Management District
South Coast Air Quality Management District (AQMD), 101–6. *See also* California, Los Angeles
Southern California Edison, 240
SPR. *See* Strategic Petroleum Reserve
SRI International, 274
Start-up barriers, 136–37, 221–22, 281–83, 287–88, 309–10
Storage tanks, bulk, 78, 85. *See also* Retail fuel outlets
Strategic Petroleum Reserve, 3, 18–19
Subsidies. *See* Government intervention
Sulfur oxides, 90
Sunk costs, 63. *See also* Costs, capital
Synthetic Fuel Corporation. *See* United States Synthetic Fuel Corporation
Synthetic gasoline, 211, 274

Tar sands, 18
Texas Railroad Commission, 14
Tire companies, 296
Toxic materials, 121

Toyota, 120–21, 273, 296
Transit buses. *See* Buses
Transportation of fuel: ocean, 41–46, 71–
   77, 251; pipeline, 66–67, 70–71, 76,
   85, 96, 210; rail, 71, 74–77, 85; truck,
   78, 85
Trucks, 102; air pollution effect, 112,
   129; methanol, 258; to transport fuel,
   78, 85

United States Department of Energy: alternative fuel vehicle testing, 103, 236;
   compressed natural gas vehicles, 231;
   fuel-flexible vehicles, 251; methanol
   preference, 259; natural gas studies, 8,
   30, 223; research and development
   funding, 231, 244
United States Department of the Interior,
   274

United States Environmental Protection
   Agency, 101, 105, 109–42, 258, 273
United States Synthetic Fuel Corporation,
   14, 306
University of Wisconsin, 296

Vapor recovery system, 102
Vehicle conversion. *See* Retrofitting mo-
   tor vehicles
Vehicles. *See makes and types*
Venezuela, 296
Volkswagen, 248, 284, 285, 296
Volvo, 295, 297, 302

Waxman, Henry, 103, 105
Windfall profit tax, 14
World Bank, 194, 195, 209

About the Contributors

JEFFREY A. ALSON has worked at the Environmental Protection Agency's Motor Vehicle Emission Laboratory since 1978, and is currently assistant to the director of the Emission Control Technology Division. He received the EPA Bronze Medal in 1980 for establishing the initial diesel passenger car particulate emission standards. He is coauthor of *Moving America to Methanol: A Plan to Replace Oil Imports, Reduce Acid Rain, and Revitalize Our Domestic Economy*.

JONATHAN M. ADLER has worked at the EPA's Motor Vehicle Emission Laboratory since 1984. His work at EPA centers on alternative fuels, particularly gasoline/oxygenate blends and their effects on automotive emissions.

BARBARA ATKINSON is a research assistant at Lawrence Berkeley Laboratory specializing in energy and environmental issues in developing countries. She is currently a graduate student in the Energy & Resources Program at the University of California, Berkeley.

THOMAS M. BAINES has worked at the Environmental Protection Agency's Motor Vehicle Emission Laboratory since 1974, and is currently the manager of all heavy-duty engine testing. He has published 28 technical papers, primarily in the area of heavy-duty truck and bus emissions.

ORESTE M. BEVILACQUA is president of Bevilacqua-Knight Associates in Oakland, California. He has directed studies of the development and commercialization of electric vehicles for the Electric Power Research Institute and Electric Vehicle Development Corporation. Previously he was assistant professor of Civil Engineering at Purdue University.

DEBORAH L. BLEVISS is executive director of the International Institute for Energy Conservation in Washington, D.C. Previously she was president of the Energy Conservation Coalition, associate director for Energy and the Environment for the Federation of American Scientists. She has testified on numerous occasions before Congress on various aspects of energy efficiency. She is the author of *The New Oil Crisis and Fuel Economy Technologies: Preparing the Light Transportation Industry for the 1990s* (Quorum Books, 1988).

THOMAS G. BURNS is manager of the economics department in Chevron's Corporate Planning and Analysis Department. He worked for Caltex and Chevron in West Germany as petrochemical sales coordinator, assistant to the refinery manager, and general operations superintendent, before returning to Chevron's economics department in San Francisco. He was elected chairman of the Economic Affairs Committee of the Western Oil and Gas Association in 1986.

MARK A. DeLUCHI is a graduate student in Environmental Policy Analysis in the Graduate Group in Ecology, University of California, Davis. He performs research on alternative transportation fuels and technologies, and evaluates transit and transportation planning. He has authored 10 technical papers on evaluation of alternative transportation fuels.

NELSON E. HAY is chief economist and director of policy analysis at the American Gas Association. In 1975 he served as a policy analyst and speechwriter at the U.S. Congressional Research Service in Washington and in 1977 served as vice-president of Trans-Energy International, Inc. He is the editor of three books: *Guide to New Natural Gas Utilization Technologies, Natural Gas Applications for Pollution Control,* and *Guide to Natural Gas Cogeneration.*

M. D. JACKSON is manager of Alternative Fuels in the Environmental Systems Division of Acurex Corporation. He manages projects for the California Energy Commission, California Air Resources Board, South Coast Air Quality Management District, and the U.S. Department of the Navy.

JANIS K. KAPLER is a senior economist at Jack Faucett Associates. She is a doctoral candidate in economics at American University.

KENNETH KOYAMA has worked for the California Energy Commission for 11 years, the past 6 in the alternative transportation fuels program. He has authored several papers on methanol fuel economics and methanol refueling.

MICHAEL F. LAWRENCE is vice-president of Jack Faucett Associates. He has authored major reports on corporate average fuel economy standards for automobiles and various economic, energy supply, employment, and institutional aspects of alcohol and gaseous fuels.

PAUL F. McARDLE is a policy analyst at the American Gas Association, an industry trade group. He has authored seven articles in the past two years for gas industry trade publications.

BARRY McNUTT is a senior policy analyst with the U.S. Department of Energy. Previously he worked in automotive engineering at the U.S. Environmental Protection Agency. He is chairman of the conservation subgroup of the International Energy Agency and has published over 15 technical articles on transportation energy and environmental topics.

GERALD H. MADER is president and founder of the Electric Vehicle Development Corporation, a non-profit corporation organized and funded by electric utilities and motor vehicle supplier companies. He was previously manager of the electric transportation research program at the Electric Power Research Institute.

STEPHEN MEYERS is a staff scientist at Lawrence Berkeley Laboratory. He has published widely in major energy journals in the areas of residential energy use and conservation in the U.S. and Europe, and energy use, conservation, and renewable energy in developing countries.

ROBERT J. MOTAL is supervisor of the Energy and Conversion Technology Group in Chevron's Engineering Technology Department.

CARL B. MOYER is chief scientist of the Environmental Systems Division of Acurex Corporation. He is responsible for combustion-related air pollution projects, addressing cost and availability of fuels, emission impacts, and air quality plans and projects. He is the author of over 80 technical reports.

DICK RUSSELL is a free-lance writer, based in Los Angeles and Boston, specializing in environmental concerns. His articles have appeared in numerous national publications, including *The Nation, In These Times, Yankee,* and *Amicus Journal.* In 1988 he received a Chevron Conservation Award.

JAYANT SATHAYE is a staff scientist at the Lawrence Berkeley Laboratory. He has published widely in major energy journals and authored almost 20 publications on the demand for energy in the developing countries. He has lectured and consulted in Asia and Africa with international organizations.

ROBERT SAUVÉ is chief of Technology Assessment in the Transportation Energy Branch of the Ministry of Energy, Mines and Resources, Canada. He has held a number of executive positions in the petroleum and beer industries.

ALBERT J. SOBEY retired in 1987 from his position as senior director of energy and advanced product economics for General Motors Corporation. After retirement, he formed Gemino Services, Inc. and has continued to work with GM and state and government agencies on advanced technology programs. Previously, Mr. Sobey was manager of the control analysis department and head of rocket and space propulsion research in GM's Allison Engine Division, then president of Transportation Technology Inc., a company he founded (which was later purchased by Otis Elevator), and worked for Booz Allen Hamilton on transportation and energy studies.

DANIEL SPERLING is an associate professor of civil engineering and environmental studies at the University of California, Davis. He is also director of the Transportation Research Group at UC Davis, vice-chairman of the committee on Energy and Environmental Aspects of Transportation of the American Society of Civil Engineers, chairman of the committee on Alternative Transportation Fuels of the Transportation Research Board (National Research Council), and member of a National Academy of Science Committee on Liquid Fuel Options. Professor Sperling has authored over 40 articles and reports in the past seven years, including *National Transportation Planning* (with Adib Kanafani) and *New Transportation Fuels: A Strategic Approach to Technological Change*.

SERGIO C. TRINDADE is assistant secretary-general and executive director of the United Nations Center for Science and Technology for Development. He previously was a consultant on new and renewable sources of energy and financial planning for science and technology development to many public and private organizations, including the United Nations, World Bank, Inter-American Development Bank, and the Research and Development Financing Agency of the Brazilian government. He is a citizen of Brazil. He is the author of two books and over 30 articles.

ARNALDO VIEIRA DE CARVALHO, JR. is general manager of Promon Engenharia S.A. in Rio de Janeiro, supervising a staff of 170 professionals that provides engineering services in many aspects of energy technology. He has consulted for various international organizations throughout Latin America.

R. F. WEBB is president of his own consulting firm with offices in Ottawa and Los Angeles. The firm specializes in the development and evaluation of alternative fuel programs and policies for government and industry. Before establishing his company in Ottawa in 1972, he worked for Allied Corporation and Standard Oil. Webb has a doctorate in chemistry from London University and University of Cambridge.